FLAT EARTH

CHRISTINE GARWOOD studied history as an undergraduate
and was later awarded a doctorate in History of Science.
This is her first book.

CHRISTINE GARWOOD

FLAT EARTH

The History of an Infamous Idea

PAN BOOKS

First published 2007 by Macmillan

First published in paperback 2008 by Pan Books
an imprint of Pan Macmillan Ltd
Pan Macmillan, 20 New Wharf Road, London N1 9RR
Basingstoke and Oxford
Associated companies throughout the world
www.panmacmillan.com

ISBN 978-0-330-43289-4

Typeset by SetSystems Ltd, Saffron Walden, Essex
Printed and bound in Great Britain by
Mackays of Chatham plc, Chatham, Kent

Visit **www.panmacmillan.com** to read more about all our books
and to buy them. You will also find features, author interviews and
news of any author events, and you can sign up for e-newsletters
so that you're always first to hear about our new releases.

This book is dedicated to

the founder of zetetic astronomy,

Samuel Birley Rowbotham, a.k.a. 'Parallax' (1816–84),

his one-time adversary

Alfred Russel Wallace (1823–1913)

and pioneering researcher and long-time writer in the field

Robert J. Schadewald (1943–2000)

'Tis strange, – but true; for truth is always strange;
Stranger than fiction: if it could be told,
How much would novels gain by the exchange!

BYRON,
Don Juan, Canto 14, Stanza CI, 801–3

Create a belief in the theory and the facts will create themselves.

JOSEPH JASTROW,
Wish and Wisdom: Episodes in the Vagaries of Belief (1935)

Quixotism is a folly when the energy which might have achieved conquests over misery and wrong, if rightly applied, is wasted fighting windmills.

PARALLAX [Samuel Birley Rowbotham],
Zetetic Astronomy: Earth not a Globe! (1881)

A little Learning is a dang'rous Thing;
Drink deep, or taste not the Pierian Spring:
There shallow Draughts intoxicate the Brain,
And drinking largely sobers us again.

ALEXANDER POPE,
'An Essay on Criticism' (1711)

Contents

Acknowledgements

The making of this book has been a story in itself. Along this winding road, I have amassed many debts of gratitude, too many to mention but none forgotten, and I am happy to have the opportunity to detail some of these finally in print.

Among many such debts, my first is to the numerous librarians and archivists who have answered my sometimes peculiar research enquiries with a straight face and further assistance in the shape of photocopies from some of the most unexplored corners of their collections. Among the most generous with their time and resources have been Andy Sawyer, Claire Lyng and Lucie Barnes at the University of Liverpool, Robin Rider and staff at the University of Madison-Wisconsin, Mary Flagg and Linda Baier at the University of New Brunswick, Apollonia Steele and Marlys Chevrefils at the University of Calgary, Melissa Conway, Darian Davies and Mary Jones at the University of California, Riverside, Mary Chibnall at the Royal Astronomical Society, Adam Perkins at the University of Cambridge, Jonathan Harrison at St John's College, Cambridge, Stella Brecknell at Oxford University Museum of Natural History, Carol Ruesch of Zion Historical Society, Allender Sybert of Maryland Genealogical Society, Jean Hankins of Otisfield Historical Society and Roy Bailey and members of the John Hampden Society. In addition, I am extremely grateful to Robert Gibbs and the Alden Nowlan Estate, the estates of George Orwell, George Bernard Shaw and G. K. Chesterton, the Royal Geographical Society, the University of Liverpool, the British Library, the Bodleian Library, Oxford, the University of New Brunswick, the University of Calgary, the University of California, Riverside, the University of Wisconsin-Madison,

Maine Historical Society, Oxford University Museum of Natural History, the Syndics of Cambridge University Library, and the Particle Physics and Astronomy Research Council for permission to quote from various correspondence held in their collections.

Besides this invaluable research assistance, I have also received financial support from several educational institutions and grant-giving bodies. In particular I would like to thank Tom Garver and the Friends of the Library of the University of Wisconsin-Madison, the Authors' Foundation, the Open University and the Andrew W. Mellon Foundation for funding my research and employment at various points. I am also indebted to William H. Brock, Chris Chilvers, Steve Cloutier and Ronald L. Numbers for reading and commenting on the manuscript and to my anonymous referees for their helpful advice. Special thanks are also due to James R. Moore, without whom this book would not have been written.

Meanwhile, in the publishing world, I'm grateful to Kate Harvey and Georgina Morley at Macmillan, and my agent, Peter Tallack at Conville and Walsh, for their patience, editorial skills and belief in the book. Chief among the many others who have provided encouragement, inspiration and assistance in numerous ways are Roger Balfour, Andy Barnes, Matt Barnes, Peter Bartrip, Janet Brian, Ian Cramp, Tim Coley, Paul Collins, Ian Firla, Michael Shere-Gill, James Gregory, Mary Hopper, Martin Hughes, Alan James, Tony Jasper, Dave Jones, Aaron and Lucy Kaplan, Marie King, Michael Lovett, Robin Mackie, Helen McBurnie, Jo McDermott, Francis Neary, Claire Ormandy, Ian Parrish, Leslie Price, Hazel Sampson, Chris Smith, James Sumner, John and Dee Vint, Richard and Audrey Wallace, and Nigel Warburton. Especial gratitude is due to Sir Walter Blount, Ann Coltman, Felicity Coltman, Leo Ferrari, Raymond Fraser and Wendy Schadewald for their personal insights and feedback on chapters, and to Paul Cox for all he has done. Finally I would like to pay testament to my parents, Bruce and Kathleen Garwood. My greatest debt is to them.

Picture Acknowledgements

1 – From N. M. Sarna, *Understanding Genesis*, 1966. 2 – With thanks to Paul Cox. 3 – Mary Evans Picture Library. 4 – Mary Evans Picture Library. 5 – Reproduced by kind permission of Ann and Felicity Coltman. 6 – Author's collection. 7 – Mary Evans Picture Library. 8 – Mary Evans Picture Library. 9 – From the *Mechanics' Magazine*, 29 March 1861. 10 – With thanks to Paul Cox. 11 – Mary Evans Picture Library / College of Psychic Studies. 12 – From T. W. Wallace, *Autobiography of Thomas Wilkinson Wallis, Sculptor in Wood* (1899). 13 – From Alfred Russel Wallace, *My Life*, vol. II, 1905. 14 – Author's collection. 15 – Reproduced courtesy of Liverpool University Library. 16 – Greenwich Royal Observatory Collection, Cambridge University Library. Reproduced by permission of the Syndics of Cambridge University Library and of the Particle Physics and Astronomy Research Council. 17 – Reproduced courtesy of the University of Liverpool Library. 18 – John Johnson Ephemera Collection. Reproduced by permission of the Bodleian Library, Oxford. 19 – From *Knowledge*, 23 March 1883. 20 – From *Adrian Galileo or a Songwriter's Story*, 1898. 21 – Author's collection. 22 – Reproduced with thanks to Pearl Pierce and Jean Hankins, Otisfield Historical Society. 23 – From David Wardlaw Scott, *Terra Firma: The Earth not a Planet*, 1901. 24 – From the *Earth not a Globe Review*, No. 4, July 1895. 25 – *Chicago Daily News*, 9 April 1911, *Chicago Daily News* negatives collection, Chicago Historical Society. 26 – *Modern Mechanics and Invention*, October 1931. Reproduced with thanks to Charles Shopsin. 27 – Reproduced by kind permission of Judy Lucero, Emanuelson Inn, Zion. 28 – *Modern Mechanics and Invention*, October 1931. Reproduced with thanks to Charles Shopsin for his generous assistance. 29 – Reproduced by courtesy of the University of Liverpool Library, Science Fiction Foundation Collection. 30 – *Sun*, 17 August 1967. 31 – Reproduced courtesy of the University of Liverpool Library, Science Fiction Foundation Collection. 32 – *Birmingham Evening Echo*, 17 April 1969. 33 – *Weekend Magazine*, 14 December 1974. Reproduced by permission of Frank Prazak / Library and Archives Canada. 34 – *Weekend Magazine*, 17 January 1976. Reproduced by permission of Frank Prazak / Library and Archives Canada. 35 – Reproduced by kind permission of Wendy Schadewald and the University of Wisconsin, Madison. 36 – Reproduced by kind permission of Wendy Schadewald and the University of Wisconsin, Madison. 37 – Copyright © the estate of Charles M. Schultz courtesy of Knight Features. 38 – Copyright © the estate of Dik Browne courtesy of Allsorts Media.

Prologue

THE COLUMBUS BLUNDER

In fourteen hundred and ninety-two,
Columbus sailed the ocean blue.
He took three ships with him, too,
And called aboard his faithful crew.
Mighty, strong and brave was he
As he sailed across the open sea.
Some people still thought the world was flat!
Can you even imagine that?

Traditional children's poem

IN 1919, SCHOOLCHILDREN in classrooms across America opened a new book. Entitled *The Boys' and Girls' Reader*, the illustrated volume by prolific children's author, Emma Miller Bolenius, was aimed at schoolchildren aged nine to twelve.[1] In true fairytale fashion, it began with a vision. 'Books are the windows through which the soul looks out,' the children were told, and browsing the *Reader* would be like climbing a winding staircase in a great tower, peering out of a different arched window with every step that they took. From this otherworldly turret, Bolenius asserted, the children would be treated to many different views of the world; images of strange countries and great heroes beyond everyday experience but real none the less.

From 'Home and Neighbourhood', to the 'Great Outdoors and the Workaday World', the 'window book' was a reading programme with special recommendations for days of the year. Among them, the guideline for 12 October, Columbus Day in the United States, was especially relevant to the presentation of sweeping world-views. On this date, Bolenius suggested, classes should turn to an account of America's discovery by Christopher Columbus in 1492. A popular topic for nineteenth-century authors eager to tell tales of great men of exploration and science, the version chosen by Bolenius was taken from Alphonse de Lamartine's swashbuckling hero-history, *Life of Columbus*, published in 1853.[2] To emphasize Columbus's achievements, she added a preface to the account for the benefit of her young readers:

> When Columbus lived, people thought that the earth was flat. They believed the Atlantic Ocean to be filled with monsters large enough to devour their ships, and with fearful waterfalls over which their frail vessels would plunge to destruction. Columbus had to fight these foolish beliefs in order to get men to sail with him. He felt sure that the earth was round. He believed that by sailing westward he would find a 'short cut' to India. The French people have always admired courage. Here a French writer, Lamartine, pays tribute to courage of spirit which is even greater than physical courage. As you read, look for the various ways in which Columbus showed his greatness . . .[3]

As children across America chewed their pencils and stared out of real windows, two conjoined 'facts' were absorbed: medieval people believed the earth to be flat, Columbus was the first to prove it was a globe. Doubtless, for many, this topped the list of ways that Columbus showed his greatness to the world.

Such storybook accounts would be quaint, were they not untrue and, more importantly, widely believed. Back in reality, educated medieval people did not believe the earth to be flat, and it was neither Columbus's intention nor the outcome of his voyage to demonstrate to doubters that it was a globe.[4] It was only in 1522, when the expedition led by Ferdinand Magellan returned from circumnavigating the earth, that rotundity was empirically proved. Just as there was no fearful waterfall in the Atlantic Ocean, so too was there no mutiny of flat-earth sailors on the *Santa María* and no globular heroism at stake. Although there were debates in the court of Queen Isabella and King Ferdinand over the feasibility of a voyage to Asia, the central question for committee members investigating the matter was the size of the earth rather than its shape. While Columbus's contemporaries assumed that it was spherical – indeed, the point was far beyond any sort of dispute – many believed that the stretch of water between Europe and Asia was uncrossable and sailors risked becoming stranded or running out of food.[5] Under these circumstances, what is widely assumed to be his greatest achievement is a chimera: no educated person in fifteenth-century Europe would have imagined that Columbus was bound to sail off the edge of the world. Yet it is one of those myths, like 'Newton and the Apple', so embedded in popular wisdom that many of us do not know where we heard it first. The imagery is golden, of treasure, of daring and globe-making adventure, and it is still paraded in various guises in newspapers and books worldwide.

Among many versions of the flat-earth myth, as the historian Jeffrey Burton Russell has noted, is the fallacy that nobody knew it was round before Columbus discovered America in 1492. Less far-fetched is the contention that the ancient Greeks realized the rotundity of the earth but this knowledge was lost to humankind through the backward and superstitious 'Dark Ages'. Then there are versions claiming that the globe idea was re-established, in Western Europe at least, at various times ranging from the first century to 1410.[6] Whatever the case, Columbus's 'discovery' that the earth is a globe is a prime slice of scientific folklore, the ultimate hero story of world exploration, a source of lessons in moral virtue or easy bylines. Nevertheless, as with many such myths, the truth behind the construction has a colourful history in its own right.

The assumption that Columbus spared humankind from foolish superstition over the shape of the earth is part of a much deeper set of assumptions about the progress of knowledge and the passage of time. In this sense, the origin and appeal of the Columbus story lie in another mythical world-view: the idea that Europe suffered a 'Dark Age' for a thousand years from *c.* 450 to 1450. According to popular mythology, this was a time of barbarism and superstition, when the earth was flat and plague, pestilence and hardship scoured the land. In intellectual pursuits, so the story goes, knowledge that survived the invasions of late antiquity was stifled by the Christian Church, which is said to have placed scriptural literalism above rational investigation, thus seriously retarding scientific progress. Possessed by 'sin, guilt and anguish', the era was supposedly a 'dark interlude' for knowledge, haunted by 'mass hysteria', 'mental disorders', 'compulsive ritualism' and the muddled confusion of a 'split mind'.[7] Although it is fair to say that the early medieval period (300–1000) experienced upheaval, and intellectual pursuits went into decline, the time did see scientific advances and the relationship between the Church and the natural sciences was not a case of a cartoon stand-off between sworn enemies.[8]

Nevertheless, the spectre of a disturbed era between the radiance of classical learning and modernity looms large in the popular imagination, and it is in this legend that the simple story of Columbus and the flat earth has its roots.

If tidiness is an occupational hazard for the historian, the neat threefold division of history into ancient, middle and modern is one example of the compulsion made real. Over time, the term 'middle ages' has also become closely associated with the idea of a 'dark age', a concept that was originally the invention of poet and scholar Francesco Petrarch. Throughout the Renaissance Italian humanists, set on glorifying the literary achievements of classical antiquity and their role as restorers of this golden age, likewise promoted the idea that the time that went before them was an era of intellectual stagnation and gloom. The idea of a 'dark age' and associated prejudices have endured to this day, and although the humanists did not accuse medieval people of believing that the earth was flat, they evoked images of ignorance, darkness and peculiar superstitions that later commentators put to ample use.[9]

During the eighteenth century, this model of a vast disparity in knowledge between two distinct periods of history was brought into even sharper relief by Enlightenment *philosophes* such as the Marquis de Condorcet and François-Marie Arouet, commonly known as Voltaire. Partisans inspired by a republican vision of post-revolutionary France, they aimed to smash the authority of the Church in social, political and cultural life and used 'Dark Age' imagery in a more explicitly anti-religious sense as part of this broader secular programme. In Enlightenment propaganda, hope and optimism pervaded the age, and man was depicted as master of his own fate: through rational empirical investigation, it was said, humankind could be improved and progress and perfectibility were achievable. Such endeavours were to be founded on critical thinking; reason and empiricism were posted as the new brooms that would sweep society clean of irrational and outmoded beliefs and institutions.

In this atmosphere, established sources of authority, Churches and governments alike, were subjected to logical examination, and even the Bible, widely accepted as the word of God, did not escape rigorous scrutiny. In certain philosophical circles, scepticism was paramount and, guided by the watchwords 'reason' and 'nature', anticlerical radicals keen to identify themselves as modern and revolutionary attempted to banish the supposed bigotry, superstition and tyranny of the 'Dark Ages' to the shadows.[10] For many Enlightenment thinkers, reason was the key to historical advance, and to promote the triumph of rationality over religion – a clash of the forces of light and dark in a sweeping vision of change over time – fitted their ideological and secular goals. Thus the scene was set for the Columbus myth through two alternative myths intertwined: through the mythical 'Dark Ages' and the triumph of reason over superstition, cultural assumptions about strange medieval beliefs were becoming established in the popular mind.

The popular version of the Columbus story has a pre-eminent

place among such historical fallacies, a point that renders its origins all the more fitting. As Jeffrey Burton Russell has noted, the tale, which has gradually infiltrated the annals of fact, was first widely disseminated by *Legend of Sleepy Hollow* and *Rip Van Winkle* author Washington Irving, as part of his *The Life and Voyages of Christopher Columbus* published in 1828.[11] From his first work, *A History of New-York from the Beginning of the World to the End of the Dutch Dynasty* (1809), Irving established a reputation as a humorist with a taste for teasing the public with clever mixtures of fiction and fact. A parody of pretentious history books, Irving's work was an overblown chronicle of the customs and institutions of the early Dutch rulers of New York, written in a mock-deadpan style and dedicated in jest to the New York Historical Society. To add to the ruse, the parody was published under the pseudonym Diedrich Knickerbocker, supposedly an elderly New Yorker of Dutch descent, who had left the musty old manuscript in his lodgings before disappearing in mysterious circumstances dressed in knee breeches and a cocked hat. 'Missing' notices appeared in the papers, designed to advertise publication of the mythical Knickerbocker's book, while his surname eventually became a household word for quaint Dutch-descended New Yorkers, and later, more specifically, their fancy trousers. In the interim, Irving's good-humoured satire was sufficient to earn him celebrity status, and he went on to write the legends and short stories commonly associated with his name.

Given the circumstances, it is unsurprising that Irving was to have a similarly quirky impact on the facts of Christopher Columbus's life. The book was his first historical biography, and the project transpired quite by chance. Irving had originally travelled to Madrid to translate some newly discovered historical documents relating to the voyages of Columbus, but found that the material was sufficiently rich to provide the basis for a full-length history of his own. Such a study would signal a new departure for Irving, who hoped to secure a more serious, hard-hitting literary reputation away from the sweet sentimentality and dark fantasy that had characterized his previous work. Inspired by the romance and ruin of the 'land of fierce contrasts', he set to work, hoping to achieve factual accuracy imbued with the colour of life. What he came up with was a rip-roaring saga, which, despite its intention to serve as reliable source-based history, fell squarely into the popular hero-myth school. One of the book's most memorable moments is Columbus's confrontation with Ferdinand and Isabella's royal commission (which Irving called the 'Council of Salamanca') where he attempts to persuade them of the feasibility of his proposed voyage to Asia. To set the scene, Irving explained,

> [The council] was composed of professors of astronomy, geography, mathematics, and other branches of science, together with various dignitaries of the church, and learned friars. Before this erudite assembly, Columbus presented himself to propound and defend his conclusions. He had been scoffed at as a visionary by the vulgar and the ignorant; but he was convinced that he only required a body of enlightened men to listen dispassionately to his reasonings.[12]

So, there Columbus stood, 'a simple mariner', according to Irving, who pleaded his cause with 'natural eloquence'. The atmosphere was tense and the debate heated, proof in itself of 'how knowledge was impeded in its progress by monastic bigotry'.[13] Even Columbus's most basic proposition, the spherical form of the earth, was shouted down with reference to scriptural passages that seemed to imply that it was flat. A devoutly religious man, Columbus allegedly began to fear that he was 'in danger of being convicted not merely of error, but of heterodoxy', but stood his ground. In this key scene in Irving's book, Columbus defended reason alongside the possibilities of his personal dream, winning through against ignorance, cowardice and Christian opposition. Of course, this line of questioning was fantasy, for educated people in fifteenth-century Europe did not believe that the earth was flat, but here Irving's talent for creating living legends from Ichabod Crane to Christopher Columbus was clearly evident. His book, supposedly historical non-fiction, displays many elements of story-book romance.

Unsurprisingly, Irving's colourful biography had immense popular appeal and the first edition of ten thousand copies sold out almost immediately on its publication in 1828. Reflecting this success, Irving went on to receive many honours, including election to the membership of Spain's Real Academia de la Historia (1828), the Gold Medal of England's Royal Society of Literature (1830) and honorary doctorates from the universities of Columbia (1829), Oxford (1831) and Harvard (1832), while his work was also exploited by writers looking for information to incorporate into their own studies of Columbus's life.[14] Although Irving had stated later in his narrative that the flat-earth opinions supposedly voiced by the Council of Salamanca were probably the views of 'but a few' and were 'outdated for their day' – thus providing himself with a get-out clause – his dramatic twist was to have a lasting charm.[15]

Like Chinese whispers, aspects of Irving's tale were regurgitated through the course of the nineteenth century in the work of several French writers, including Bolenius's source Alphonse de Lamartine, and the Voltairean academic Antoine-Jean Letronne. As anticlericalists, they had ideological reasons for using versions of the flat-earth myth to attack the Church for its alleged suppression of scientific knowledge. In terms of medieval flat-earth belief, Letronne's scholarly article 'On the Cosmological Opinions of the Church Fathers', published six years after Irving's biography, was an especially influential source for the idea that early Christians believed the earth was flat and that such views were commonplace through the 'Dark Ages' of Western civilization.[16]

While such currents flowed in Victorian literature, further myth-making was under way from a much different direction. In 1837, William Whewell, a Cambridge University vice-chancellor, Anglican clergyman and the best-known historian of science of his day published his *History of the Inductive Sciences*, which depicted the Middle Ages as an era beset by dogma and lack of scientific advance. Later editions of the book made explicit reference to two medieval flat-earth believers as proof of this progressivist world-view.[17] The first was early Christian author Firmianus Lactantius (c. 245–325). Raised in Africa as a pagan (non-Judeo-Christian) in the mid-third century, St Lactantius was a rhetorician who converted to Christianity and wrote a number of books supporting the truth of his new-found faith. An eloquent writer, if incredibly biased, he attacked pagan (Greco-Roman) philosophy on manifold points, including its teaching that the earth was a globe. In the third book of his *Divine Institutions* (c. 302–11), 'On the False Wisdom of the Philosophers', Lactantius ridiculed the notion of a sphere where people on the other side lived with their feet above their heads, where rain, snow and hail fell upwards, where trees and crops grew upside-down and the sky was lower than the ground. The ancient wonder of the hanging gardens of Babylon dwindles into nothing, he commented sarcastically, in comparison to the fields, seas, towns and mountains that the pagan philosophers believed to be hanging from the earth without support. The rotundity of the earth was a 'marvellous fiction', a lie spread by pagans for sinister motives or the 'sake of a jest', and as the Bible was somewhat unclear about the shape of the earth, he concluded that the subject was irrelevant anyway. Albeit strongly argued, Lactantius's views also lacked support in any sort of sense: he was denounced by some as heretical after his death, and his obscure views had little impact on contemporary thought about the shape of the earth.[18]

Nevertheless, in the sixth century Lactantius was joined by Byzantine merchant and Christian monk Cosmas Indicopleustes. An obscure figure, it is said that he travelled widely – possibly to Abyssinia, Ceylon, western India and more – and earned the name 'Indicopleustes', the Indian traveller, as a consequence of his experiences. Around 500 he returned to his birthplace, Alexandria, and later entered a monastery on the peninsula of Sinai where he wrote his geographical and cosmological masterpiece *Christian Topography* (c. 548). A comprehensive description of the universe based on a literal interpretation of the Bible, the twelve books denounced pagan teaching and Christians who interpreted the scriptures allegorically to hold that the earth was a globe. Cosmas was certain that the Bible taught that it was a disc or trapezium-shaped flat surface like the Tabernacle of the Old Testament, with heaven as a chest- or altar-shaped structure above.[19] However, there are only a few reasonably full manuscripts of Cosmas's work in existence from the period in which he lived, and as they were not translated from Greek into Latin until centuries later it is safe to conclude that, like Lactantius's, his radical flat-earth views had no impact in the Latin-speaking West.[20]

Although very few writers of the patristic period (*c.* first to eighth centuries) argued in favour of a flat earth, Lactantius and Cosmas were held up as typical medieval thinkers by Victorian rationalist writers who, like radical *philosophes*, were set on sidelining religious belief as damaging to the progress of scientific truth. From the late nineteenth century, Lactantius and Cosmas were assigned leading parts in accounts alleging that educated medieval people believed the earth to be flat. These two obscure figures were 'proof' that the early Church had strangled scientific progress at a critical period, a crime that was only latterly being redressed. This rationalist world-view was shaped by positivism – the idea that religion is an obstacle thwarting real knowledge – while empirical data gathered via observation, experience or experiment is the only road to absolute, unquestionable truth. Such assumptions gathered pace in the wake of Darwin's *On the Origin of Species* (1859) and subsequent evolutionary debates, for at this time authors began to use the metaphor of warfare between science and religion as the organizing principle for their books.

The battle-cry was an amplification of the 'reason versus superstition' declaration made by some Enlightenment propagandists, one that was first aired in John William Draper's *History of the Conflict between Religion and Science*, published in 1874. A chemistry professor at a New York medical school, renowned for taking the first known photograph of the moon, Draper was a polemicist with an axe to grind against the Roman Catholic Church. Above all, he was concerned about the doctrine of papal infallibility and the Church's pronouncement that public institutions teaching science were not exempt from its authority. With an agenda and a plan of attack, Draper's history attempted to show how Roman Catholicism had displayed 'a bitter, a mortal animosity towards science' since Christianity's rise to political power in the fourth century. Draper argued that from this point until the end of the fifteenth century an 'intellectual night' had settled on Europe as Christendom became obsessed with 'the merits of the saints, miracles [and] shrine-cures'.[21] When it came to scientific knowledge, theological hostility was supposedly evident in two distinct ways. Early Church fathers fixated on the ever-after had branded the natural world unimportant and had rejected out of hand any knowledge that contradicted a literal interpretation of the Bible.[22] In its day, Draper's book was something of a blockbuster, for it reshaped the history of science into a simple plot in which the evils and ignorance of religious dogma sidetracked the march of human knowledge and the natural progress of scientific truth. The case of the early Church fathers thinking the earth was flat was further ammunition in his retrospective report on the head-to-head conflict, where science had fought religious bigotry, like some David and Goliath, to come out shining in the cause of human knowledge and the final realization of glittering truth.

The image of warfare between science and religion was as powerful as it was simplistic; indeed, its simplicity was the very source of its power, for commentators now had a clear framework in which to order the development over time of deep and difficult concepts. Rather than a disparate mass of opinions and theories, cross-currents and shifting tendencies, tidy-minded historians, journalists and popular writers now had a convenient, attention-grabbing and straightforward system in which to file the events of the past. From the 'Dark Ages' of superstition to the Enlightenment era and modern science, the scheme could have been named 'progress from the past', for that was what commentators were at pains to show, and the equation of science with truth and advance, and religion with error and backwardness, served their purposes only too well.

Dramatic and attractive, Draper's work sold well, and as it did, so the book's central premise was being reshaped by Andrew Dickson White, a University of Michigan professor and soon-to-be Cornell University president. Like Draper, White had personal reasons for painting a portrait of ongoing religious opposition to scientific advance. Engaged in his own battle with theological interests to found Cornell, America's first secular higher-education establishment, White had been battered by an onslaught of innuendo and abuse from clerics defending traditions of religious control. While White crusaded for 'an asylum for science' in the face of bitter and acrimonious personal attacks, he became increasingly fixated on the idea that the relationship between science and dogmatic religion was a battlefield where science had finally triumphed over clerical adversity.[23] Despite this, unlike Draper, he did not believe that science and religion were natural enemies; for him there was at source no conflict between scientific enquiry and true religion. Rather, he branded sectarian dogmatic theology, such as the views of his clerical critics, as the arch enemy of freedom and progress, which he equated with scientific advance.[24]

The scene was set for his hefty two-volume *History of the Warfare of Science with Theology in Christendom*, finally published in 1896. The most famous expression of the warfare metaphor, this sweeping history of science from ancient civilizations to the Victorian age had all the hallmarks of a scholarly tome and was highly influential as a result. Unfortunately, White's book was a tireless exercise in backward history, where he projected the science and religion debates of his own day (such as those over Darwinian evolution) retrospectively on to the past. The result was an exhaustive compilation of hits and misses, battles and recriminations in which medieval flat-earth thinking again played a notable role as a prime example of scriptural literalism derailing 'natural' progress towards scientific truth.[25]

Specifically, White argued that the 'great majority' of early Church fathers 'took fright' at the idea that the earth was a globe, because it appeared to contradict scriptural teaching, and sought to 'crush' the concept with reference to Biblical proof. Like Draper, White also drafted in scapegoats to defend his argument: he claimed that Lactantius and Cosmas (among others) were influential thinkers on cosmology whose ideas typified early-medieval beliefs about the shape of the earth. Yet, he noted with relish, the 'sacred theory struggled long and vigorously but in vain' for ultimately the 'ancient germ of scientific truth in geography, the idea of the earth's sphericity – still lived'. By the late-medieval period, he concluded, various writers 'felt obliged' to accept the rotundity of the earth.[26] In all of this, the problem was one of emphasis: while a tiny minority advanced flat-earth arguments, White's need to create a battle between two powerful forces to fit with his thesis of all-out warfare led him to present flat and spherical as well-matched rival theories, which was certainly not the case. The conflict model therefore led him seriously to overstate the extent of flat-earth belief, both in terms of the number of believers and the timescales involved. His set-piece concludes with the ill-judged statement: it is only 'as we approach the modern period' that 'we find [the] truth [of the globular theory] acknowledged by the vast majority of thinking men', an estimate which is incorrect by twenty centuries or so.[27]

Nevertheless, the military metaphor employed by Draper and White was propaganda *par excellence*, and it seized the popular imagination at a time when Western culture was awash with the rhetoric and imagery of war. Their books turned into bestsellers, and as readers were seduced by a colourful interpretation of a dry, complex topic, historians, polemicists and popular writers likewise became inspired by this compelling idea.[28] Between 1870 and 1920, warfare became the common framework for analyses of the relationship between science and religion and, as part of this, the Columbus story invented by Washington Irving was translated into hard 'fact'. In educational terms, the damage was done. In publications from magazines to school textbooks, the idea that medieval people thought the earth was flat and Columbus discovered it was round was recycled until it became standard fare on both sides of the ocean he had crossed. Interlinked with images of warfare and 'Dark Ages', such stories are a hall of mirrors, a series of grotesque and distorted reflections of the not-so-distant past. Yet, like so many illusions, the image of dark days when superstitious people believed the earth was flat is intriguing, and this is one of the reasons why it remains a lingering assumption in the popular mind. The myth is all too convenient: the flat-earth idea has become shorthand for 'Dark Age' stupidity, a handy one-liner to capture the days of yore before the Enlightenment era of progress and science. For the purposes of history, this is a straightforward narrative, one-dimensional and over-simplified but easy to digest, and the idea has a more human appeal as a way of conceptualizing science and its role in historical advance. While the earth is flat for medieval people, we can assume a sense of superiority and security in our progress from an ignorant and deluded age. Just as we know our true position on the surface of the world, we know our true place in the history of knowledge.

From religion to reason, superstition to science, these are deep waters, yet still they run through the assumption that flat-earth belief was commonplace before Columbus successfully avoided sailing off the edge of the world.[29] Under these circumstances, it is somewhat ironic that, beneath these false images of science and history, flat-earth belief has a chronology far stranger than all the inventions.

Chapter One

SURVEYING THE EARTH

Observation of the stars ... shows not only that the
earth is spherical but that it is of no great size, since a
small change of position on our part southward or
northward visibly alters the circle of the horizon, so that
the stars above our heads change their position considerably, and we do not see the same stars as we move to the
North or South ... This proves both that the earth is
spherical and that its periphery is not large, for otherwise
such a small change of position could not have had such
an immediate effect.

ARISTOTLE, *On the Heavens* (Book II, 350 BC)

BEYOND THE FAIRYTALE LAND of heroes and villains and
momentous conflicts between truth and falsehood on an individual,
cultural and conceptual scale, flat-earth belief can be traced back to
some of the most ancient civilizations in world history. The first of
these are the Sumerians and Babylonians, who inhabited Meso-
potamia, the land between the Tigris and the Euphrates (the site
of modern Iraq) from *c.* 4500 to 500 BC. Although these peoples
left texts describing a range of cosmological theories – too many to
speak of a single, overarching Mesopotamian world-view – they
developed the idea of a tripartite universe, with the earth as a flat

surface ruled by the god Enlil, sandwiched between the sky and the underworld. For the Egyptians, the same triple-decker arrangement applied, with the sky resting on four pillars, forked poles or mountain peaks rising from the corners of the flat earth beneath. While the Egyptian system differed in detail from the Sumero-Babylonian world-view, it too personified natural phenomena, representing the earth as the earth-god, Geb, lying outstretched to create the plane. The sky, meanwhile, was the goddess Nut, mother of the sun-god Re, who was depicted as a giant standing cow or a young woman arched over the earth like a canopy. And so the situation remained; while later Egyptians made voyages to the so-called land of Punt, thought to be along the coast of East Africa, and other evidence suggests that they circumnavigated the continent, such experiences had no impact on ideas about the shape of the earth. This being the case, in the eyes of the oldest civilizations for which we have records, whether in the form of Babylonian clay tablets or Egyptian papyri, the earth continued to be a flat surface of a circular or rectangular shape.[1]

It remained the same for the ancient Hebrews, who were flanked by the Egyptians to the south-west and the Sumero-Babylonians to the north-east, and whose cosmology resembled the assumptions of their powerful neighbours about the form of the earth. In terms of cosmology and creation, the Old Testament owes much to Mesopotamian mythology, which has led to claims that the Bible is a 'flat-earth book'.[2] Although the scriptures do contain disparate passages relating to cosmology they are not a systematic study of the heavens, however, rendering any presentation of a Biblical world-view a patchwork of statements scattered through books written over several centuries. Nevertheless, it had been claimed that the Bible presents a reasonably clear and consistent view of a tiered universe based on the Sumero-Babylonian model. In this system, the cosmos consists of the vault of heaven (*shamayim*), or 'firmament', containing the sun, moon and stars (Genesis 1:14–17). The Bible teaches that these heavenly bodies move across the stationary earth (Psalms 19:1–7), while the firmament rests on pillars or mountains (Job 26:11) rooted in the flat earth below. Beneath the earth lies the underworld, *Sheol*, seen as the abode of the dead (Numbers 16:28–34; I Samuel 28:13–15; Isaiah 14:9–11; Ecclesiastes 9:10). The earth, which is generally depicted as an immovable disc or 'circle' (Job 26:10), supported on water (Psalms 24:2) or in empty space (Job 26:7), is bordered by a protective barrier, probably a mountain range. As for its shape, it is generally spoken of as a flat disc, so that if one travelled far enough one would eventually arrive at the 'ends of the earth' (Deuteronomy 13:8, 28:64; Isaiah 5:26; Psalms 135:7). Within this scheme, the four corners of the earth (Isaiah 11:12, Ezekiel 7:2) might refer to distant regions, unless taken literally, when the earth could be considered rectangular or square.[3]

Flat-earth belief was also prevalent in Ancient Greece, details of which are provided in the writings of Aristotle (384–22 BC), who had a habit of reviewing his predecessors' opinions as a precursor to demolishing them with his own. His cosmological treatise *On the Heavens* (350 BC) provides a useful, if polemical, survey of opinions dating from Thales, the Ionian geometer, astronomer and engineer (*c.* 625–*c.* 547 BC), in the sixth century BC.[4] The first speculative thinker of the Ionian school, he believed the earth was a circular disc floating like a piece of wood on the world (meaning universe) ocean, while his pupil or younger associate, Anaximander (*c.* 611–*c.* 545 BC), argued that the earth was a cylindrical column floating upright in air in the centre of the universe. According to Aristotle, Anaximander believed that this column was three times as broad as it was high with humankind inhabiting the flat uppermost surface.[5]

Meanwhile, the third philosopher of the Ionian school, Anaximenes (*c.* 585–525 BC), is reputed to have believed that the heavenly bodies were flat, and likely thought the same of the earth, although the precise details of his view remain unknown. Indeed, when dealing with pre-Socratic thinkers – each of whom seemed to design his own cosmological system but left no writings – we are reliant on fragments for clues about a number of complex views. Yet in terms of who knew what, when and how, the Ionian school has been credited with being the first to practise a broadly philosophical and naturalistic way of looking at the world; that is to say, they began to ask a new range of questions about nature in terms of its phenomena, its composition and the way that it worked. Thales and Anaximander began to think of the world as an orderly, unitary whole that was worth investigating for its own sake, beyond its role as the realm of gods, and they looked further than Zeus and his cohorts for explanations about the way things worked. Here, then, are traces of the beginnings of what we now know as 'science' and 'philosophy', although religion and astrology continued to play a central role in and around new lines of enquiry, meaning that shifts were exceedingly gradual and sporadic. The same may be said of the development of ideas concerning the shape of the earth.[6]

It was not until the late sixth century BC that the flat-earth thinking that had dominated world-views for several millennia was to take a more radical turn through the teachings of the Pythagoreans. A much-mythologized philosophical school, it was based at the other extremity of the Greek world from the Ionians, in the thriving city of Croton in what is now southern Italy. Its leader, Pythagoras (582–500 BC), the most famous mathematician in Ancient Greece, was the son of a silversmith and gem engraver and later a pupil of Thales and Anaximander at Miletus. Having travelled in Babylonia and Egypt, at some time around 530 BC, Pythagoras settled at Croton where he established the Pythagorean brotherhood, a large, close-knit religious community that held extensive political power in its hometown and the surrounding area, Magna Graecia (Greek-governed southern Italy). While Pythagoras appears to have been a visionary who, according to Athenian philosopher Plato (*c.* 427–*c.* 347 BC), taught a whole way of life, in the absence of writings it is impossible to extricate his ideas from those of his followers in a school of thought that was bound by confidentiality and which undoubtedly developed its doctrines over something in the region of two centuries. Among an abundance of ideas that have been ascribed to the Pythagoreans or Pythagoras himself are a focus on purification and the practice of silence, the belief in a mystical union with all living things and the immortality and transmigration of the soul, along with a veto on various activities, including leaving a cooking pot's imprint visible in the ashes of a fire and standing on one's own toenail clippings. Best known, however, is Pythagoras's teaching that number lies at the heart of all things; it is the realm of eternal perfection and absolutes of which all things are composed. The contemplation of geometrical forms and patterns was believed to allow the mind to surpass the earthly appearance of reality and engender a connection to the divine, although again there is no certainty about exactly what Pythagorean number-related teaching involved.[7] As a whole, Pythagorean philosophy seemed to be based on a vision of mystical unity in nature, with number at the root of all things, from the theorem governing right-angled triangles (the square of the hypotenuse is equal to the sum of the squares of the two other sides) to those governing rhythm and acoustics in music – the connection between the pitch of a note, for example, and the length of a string.

In a religious world-view based on numbers and measurement, shapes and sizes, patterns and unity, heavenly movements and the shape of the earth were obvious sources of interest as natural manifestations of the divine. It has been suggested that Pythagoras learned the basics of his astronomy and mathematics from his travels in 'the East', but whatever the case, when it comes to ideas about the shape of the earth, a feature at the heart of traditional histories of science is entirely bypassed. While Pythagoras, or the Pythagoreans, are generally credited with being the first to argue that the earth is a globe (although doubtless someone suggested the idea before), it is ironic that the name of the first individual 'discoverer' of one of the most basic scientific facts is a mystery. All that can be said is that the Pythagoreans believed that the earth was a globe floating freely in space because the sphere was the perfect shape. In addition, it is impossible to say how readily this

doctrine was accepted outside their school, in educated circles or
through society as a whole. Importantly, post-Pythagorean philosophers, most notably Anaxagoras of Clazomenae (497–428 BC)
and the atomists Leucippus of Miletus (fl. 440) and Democritus of
Abdera (c. 460 BC–c. 370 BC), were still arguing that the earth was
disc- or drum-shaped in the fifth century BC.[8] That said, Pythagorean speculations brought into play the idea of a spherical earth
and were to have a profound influence on Plato. By the time his
pupil Aristotle was writing, later in the fourth century BC, the
globe concept seems to have become widely accepted among
educated people.

In this way, the three-tiered earth, heaven and underworld
system of Near Eastern cultures gradually faded from view,
replaced by visions of a spherical earth and an all-encompassing
sky. Undoubtedly, changing perceptions were bolstered by practical, cultural and environmental factors, particularly that the
Ancient Greek world was surrounded by sea. It has been suggested
that the culture's consequent geographical knowledge and maritime
experience may have triggered the switch from the world-view
prevalent in the literal and metaphorical flatlands of Babylonia and
Egypt.[9] Certainly Aristotle invoked the proof of ships disappearing
over the horizon, hull before masts, along with the earth's circular
shadow on the moon during a lunar eclipse and the different
appearance of stars when viewed from different latitudes, to support the contention that the earth was a globe in his book *On the
Heavens*. The mainstream consensus, all of the renowned Greek
writers, from Plato to Eudoxus (c. 375 BC), Euclid (c. 300 BC),
Aristarchus (c. 310–230 BC) and Archimedes (287–212 BC),
accepted a globular earth, while Aristotle's geocentric cosmology –
centring on an immobile sphere at the centre of the universe with
the planets moving around it in perfect concentric circles – was
to dominate Western cosmological thinking until the work of
Copernicus and Galileo nineteen centuries later.[10]

With consensus reached on the shape of the earth, focus shifted
to estimating its size. By the fourth century BC, Aristotle reports,

efforts had already been made to calculate the circumference – perhaps by the different positions of the stars when viewed from different latitudes – which had resulted in the oldest existing estimate at 400,000 stadia. If a stade is taken to be *c.* 500 feet (a moot point because measurements were not standardized), this would give a figure of 39,000 or 40,000 miles at the equator.[11] Further estimates followed and by the third century BC we have details of an experiment to measure the globe's diameter. The test was undertaken by the Greek polymath, Eratosthenes (*c.* 276– *c.* 194 BC), director of the famous library in the museum of Alexandria, then the Egyptian capital and centre of Hellenistic culture and learning. Essentially, he is said to have noticed that at noon on the summer solstice the sun was directly overhead at Syene (present-day Aswan) because a vertical pointer cast no shadow and the sun's rays shone to the bottom of a deep well. At the same time in Alexandria, which Eratosthenes believed to be 5000 stadia (approximately 530 miles) due north, the sun made an angle equivalent to one-fiftieth of a circle or 7.2 degrees to the vertical. Assuming that the sun's rays are basically parallel, Eratosthenes then used geometry to calculate the earth's circumference to 250,000 stadia, possibly somewhere in the region of 29,000 miles.[12] Although there were flaws in Eratosthenes's data, and the exact length of a stade, or stadium, is unknown, his estimate for the earth's circumference is not far from the present-day value of approximately 24,860 miles.

Mathematicians continued to estimate the size of the earth, while they also focused on cracking a puzzle on a much larger scale: explaining their geocentric vision of the universe, or how the planets moved around the central, immovable and, of course, spherical earth. In the second century, the theories and findings of six centuries of astronomical research were finally drawn together by the Greek geographer, mathematician and astronomer Claudius Ptolemy (*c.* 130–75), in his encyclopedic compilation of ancient knowledge *Syntaxis*, commonly known by its Arabic name *Almagest* or 'the greatest'. Compiled from the archives of the library at

Alexandria, where Eratosthenes had earlier been based, the book's
name was fitting indeed, for the theories of *Almagest* remained the
mainstream for Arabic and Latin civilizations until set aside in
favour of the heliocentric, or sun-centred, solar system proposed by
Copernicus in the mid-sixteenth century. The Ptolemaic system
was essentially a more complex version of the Aristotelian system,
with adjustments to account for variations in the observed distances
of the planets from the earth. By late antiquity, the Aristotelian-
Ptolemaic system of circles and spheres, cycles and epicycles,
dominated perceptions of the heavens, with a static spherical
earth placed in the centre. Moreover, in common with Aristotle,
Ptolemy's *Almagest* provided a number of 'sensible' (sense-related)
proofs for why the earth must be this shape.[13]

With the earth established as spherical for a number of centuries, the question becomes the survival and dissemination of the
idea. Although factors surrounding the gradual disintegration of
the Roman Empire (180–450) caused many works of classical
antiquity to be lost to the Latin-speaking West, including Aristotle's *On the Heavens* and Ptolemy's *Almagest*, a few significant
texts were translated into Latin from the original Greek. Such
works included a partial fourth-century translation of Plato's cosmological treatise *Timaeus*, which, albeit primitive by later Greek
standards, reflected a globular view of the earth. The book was to
serve as a principal cosmological authority alongside the work of
popular Roman writers Pliny the Elder (*c.* 79), Macrobius (*c.* 400),
Martianus Capella (*c.* 420) and Boethius (*c.* 480–524), all of whom
helped the Ptolemaic view of a spherical earth to survive in the
Christian West without any input from the Islamic world.

Although the earth and its shape or size were not focal points
for education or research, Christianity had a critical role in preserving and spreading the scientific knowledge that had survived from
Greco-Roman times. Of particular importance was the study of
the quadrivium, the study of four liberal arts – arithmetic,
geometry, astronomy and music (often accompanied by medicine)
– which took place in monastic and cathedral schools between the

fifth and twelfth centuries and further disseminated knowledge of the spherical shape of the earth.[14]

Although Christian flat-earthism was a favourite theme of 'Dark Age' promoters and warfare polemicists Draper and White, early Church fathers were not Biblical literalists who believed that the Bible was the only authority on the natural world or that the earth was a plane. The majority accepted that it was a globe, took the scriptures allegorically or simply sidestepped the issue and any associated controversy. However, there are always extremists in any field, and Lactantius and Cosmas Indicopleustes were joined in the promotion of Bible-based flat-earth belief by an atypical few, most notably Severianus, Bishop of Gabala (c. 409), and his contemporary, one-time Bishop of Constantinople, St John Chrysostom (344–408). Some detractors founded their argument on scriptural passages such as those referring to the four corners of the earth, while others were also keen to denigrate pagan (pre-Christian) culture, including the teaching that the earth was a globe. Among a number of radical flat-earth arguments, the most common was the 'fable of the Antipodes' – that there could not be an undiscovered side of the earth where crops and trees grew upside-down and people walked with their feet above their heads. The term 'Antipodes' (podes being Greek for 'feet') was coined by Pythagoras and means 'feet pointed in a direction that was opposite' or 'people with their feet turned towards ours'. For flat-earth advocates, the idea of the Antipodes was absurd so they reasoned that the earth must be flat.

The problem for the majority of theologians, meanwhile, was not the existence of the Antipodes, or the associated fact that the earth was a globe, but rather the idea that people could live on the other side of the sphere. This concept of an inhabited Antipodes was opposed on a variety of grounds: it conflicted with Christian belief in the unity of the human race, descended from Adam and Eve, and the consequent universality of original sin and redemption to be resolved on Judgement Day. If there was another race living in the Antipodes, how had they got there, how could they have received the Word of God (if the Apostles were instructed to preach the Gospel to all nations) and why did the Bible not mention them? These questions were bolstered by a more practical point. Even ancient mariners had experienced an increase in temperature as they neared the equator and the fact that no one had ever voyaged into the southern hemisphere was translated into the idea that the equator was too hot to cross and the other side of the earth too sultry to be inhabited. (It was these assumptions, along with those about the distance involved, that underpinned questions about Columbus's voyage, rather than concerns that he would sail off the edge of the earth. At a time when no one was certain just what proportion of the world was land and what was sea, the revolutionary aspect of Columbus's endeavour lay in putting theoretical knowledge to a practical test.)[15]

Arguments about whether people lived in the Antipodes had no impact on mainstream acceptance of a spherical earth, and all of the most widely renowned and distributed authors of the early medieval period were in firm agreement on the point. They included St Augustine (354–430), Bishop of Hippo in Roman-controlled North Africa (now Annaba, Algeria), who confirmed his belief in a spherical earth in a number of his writings. However, like many early Christian writers, St Augustine was far more concerned with the spiritual than the natural world, and with the Church and eternal salvation at stake, questions such as the shape of the earth faded into comparative insignificance. His emphasis on an allegorical rather than literal reading of the scriptures naturally extended to the shape of the earth, and he argued that depictions of a flat earth with the sky spread over it like a tent were simply metaphors or figures of speech.[16] But where he argued most vociferously was on the Antipodes issue for, like the majority, St Augustine was convinced that people could not live on the other side of the globe because the Bible 'speaks of no such descendants of Adam', does not mention any preaching of the gospel in that region and Antipodeans would not be able to see Christ come down to earth when he returned on Judgement Day. While St

Augustine believed the earth was a globe, he followed many of his contemporaries in asserting that men were confined to the *oikoumene*, the inhabited portion of the earth, and it was impossible for anyone to cross the immense expanse of ocean that surrounded this, the known world.

In the sixth and seventh centuries, St Augustine's stance on the shape of the earth was supported, albeit vaguely, by the most popular encyclopedist of the era, St Isidore of Seville (died in 636), and more directly by the so-called father of English history, the Venerable Bede (673–735). Along with St Ambrose's and St Jerome's, their work was standard educational fare in monastic and cathedral schools and libraries, and so, too, was the lesson that the earth was a globe. As monks or those taught in monasteries constituted much of the educated class in early medieval Europe, the spherical consensus of the Greeks – represented by Plato, reflected by Pliny, St Augustine, Bede and others and disseminated through theological channels – survived, despite the objections of an extreme few.

From the twelfth century onwards, the situation became clearer still, when Latin translations of works from Greek antiquity, along with Arabic and Jewish scholarship, gradually became available in the West. The intellectual impact of such waves of translation is inestimable. Complete versions of Ptolemy's *Almagest*, Aristotle's *Physics*, *Metaphysics*, *Meteorology*, *On the Heavens* and more opened up a whole new vista of ideas, methodologies and technical knowhow, ready for integration into the curricula of the new universities that were being established around Europe. In this context, *astronomia* (astronomy) was a branch of philosophy, rather than an area for study in its own right, but with geometry and geography it was a compulsory subject taught as part of the standard university curriculum focusing on the seven liberal arts.[17] In educated circles, Ptolemy's *Almagest* and Aristotle's *On the Heavens*, complete with discussion and proofs of a spherical earth, gradually displaced Plato's *Timaeus* as the key cosmological treatises of the age. With important exceptions (not least about the creation), aspects of Aristotle's work was accepted by medieval theologians as the standard guide to the natural world.[18] Writ large, the relationship of natural philosophy and religion was one of complex interaction, assimilation, reconciliation and interchange, rather than outright suppression or unmitigated dispute.[19]

From the twelfth century onwards, the flat-earth concept is almost a non-issue, so prevalent are written and visual images of a spherical earth. From the most popular astronomical text of the Middle Ages, John of Sacrobosco's *On the Sphere* (*De sphaera*, c. 1250), to the work of scholastic philosophers such as Thomas Aquinas (1225–74) and Jean Buridan (1300–58), not to mention Dante's *Divine Comedy* and the royal 'earth' orb held by medieval kings, culture was suffused with images of *terra rotunda* to such an extent that serious promulgation of flat-earth belief would become little more than a waste of time. The only remaining cause for confusion is medieval *mappae mundi*, or maps of the world. Until the fifteenth century the most common variety of world map was the 'T and O' type, which seems to depict the earth as a flat disc consisting of a central T-shaped landmass with Asia at the top of the T, Europe in the bottom left area and Africa to the right, surrounded by an O-shaped ring of sea. To twenty-first-century eyes, these maps present strange and frequently beautiful images, decorated with mythical beings and sea-monsters in hand-painted colour embossed with gold, yet they were not intended as literal representations of the world. Unlike modern-day maps, with a more functional purpose, medieval *mappae mundi* were symbolic depictions of the known inhabited portion of the sphere, the T-shaped *oikoumene*, which was confined to the northern hemisphere because, although, like St Augustine, they were aware that the earth was a globe, monastic map-makers believed the Antipodes were uninhabitable and not worth including on their maps. With a focus set on conveying moral meaning to the illiterate masses, medieval cartographers represented the earth in its religious, political and spiritual aspects rather than in strictly geographical terms. Consequently, their maps featured images of Christ and kings, towers and turrets, Adam and Eve and the Garden of Eden, with

Jerusalem placed at the centre of the world to highlight its place as the metaphysical rather than literal heart of the world.[20]

With flat-earth belief firmly consigned to the ancient past, from the mid-sixteenth century, debate became concentrated on the supposed position and motions of the terrestrial sphere. Central to this development was the work of Polish astronomer Nicolaus Copernicus, who published *On the Revolution of the Celestial Orbs* in 1543. Shockingly, the book challenged the Aristotelian-Ptolemaic vision of a geocentric (earth-centred) universe, an idea that had dominated astronomy for approximately 1800 years. As an alternative, Copernicus posited a heliocentric (sun-centred) system, whereby the earth became just another planet in orbit round the sun, rather than the fixed centre of the universe as the subject of God's special creation. In 1616, mindful of the potential impact on Biblical interpretation, the Catholic Church banned books that argued in favour of the motion of the earth. Nevertheless, the Copernican thesis was explored and expanded by German mathematician Johannes Kepler, who discovered that the planets do not trace a circular course round the sun, as Copernicus had thought, but instead move in ellipses. Meanwhile, in Italy, Galileo Galilei was using a telescope to make a number of discoveries that contradicted the Aristotelian system with empirical proof of a Copernican sun-centred universe.

Most widely lauded in later popular imagination, however, was Cambridge-based mathematician Sir Isaac Newton, whose classic works, the *Principia* (1687) and *Opticks* (1709), underpinned the marriage of mathematics with astronomy. Fundamentally, Newton presented the world with a rational, well-regulated cosmological system; from the analysis of observed facts he seemed to impose order and regularity with general principles such as the three laws of motion and the principle of universal gravity. On the earth itself, Book III of the *Principia* provided theoretical proof that it is not a perfect sphere but an oblate spheroid that bulges at the equator due to its rotation, a result confirmed by French expeditions during the eighteenth century. Yet Newton's findings were applied far beyond geodesy and astronomy, exceeding even the realms of natural philosophy. Throughout the eighteenth century aspects of his work were adopted as key features of Enlightenment thinking: French *philosophes* like Voltaire were gripped by the possibilities it seemed to promise. Nature appeared to be reasonable, operating by laws that were waiting to be uncovered by the rational observer. Inspired by the idea, thinkers assumed that man, as part of nature, must be the product of similar principles. The ramifications of this were staggering, some *philosophes* supposed, for if overarching laws could be discovered in the external world around us, then why not in human beings and society as a whole? Spurred on by their interpretation of Newton's work, their attempt to find fundamental laws was translated from mathematics and astronomy to ethics and morality, while the path to such discovery was believed to be reason.

Henceforward, the *philosophes* proclaimed that the study of human society should be characterized by scientific methods; it was believed that careful observation and the collection of empirical data, rather than the contemplation of abstract principles, could uncover the laws governing human existence.

Armed with this new-found confidence to discover causality, radical thinkers promoted the idea that humankind was no longer at the mercy of strange forces beyond its comprehension. In some philosophical circles, scepticism was paramount and French Enlightenment thinkers, keen to identify themselves as modern and revolutionary, attempted to banish irrational beliefs of the supposed 'Dark Ages' to the shadows.[21] In this there was a great irony: Newton – Protestant, Bible scholar, astrologer and alchemist – viewed science, philosophy and theology as inseparable components of one great whole. While many chose to promote a materialist 'Newtonian' image of a mechanical, clockwork universe (in itself first promoted by French philosopher, René Descartes), Newton had not removed God from the equation: he believed that the Deity intervened in his creation from time to time.[22] In fact, the approach common to Newton and his contemporaries was much different from what we now understand as science; 'Natural Philosophy' was a diverse corpus of ideas and practices that would seem arcane, alien and peculiar to scientists practising today.[23] But Newton's personal religious views or the realities of natural philosophy were beside the point for *philosophes* with a radical, secular goal in mind. In their efforts to popularize Newton's achievement, Voltaire and his disciples glossed over the religious, astrological and alchemical aspects of his work in keeping with their programme of sweeping social reform.

Through the eighteenth and early-nineteenth centuries, authors, savants and travelling lecturers shouldered the task of disseminating accessible versions of Newton's work to educated audiences in England and abroad. It was a weighty undertaking: among many rival cosmologies, Newton's achievements were immense and few possessed the technical knowledge to fathom fully their profundity and magnitude. It has been estimated that around the time the *Principia* (1687) was published, it was read from cover to cover by less than a hundred contemporaries and fully understood by just a fraction of those.[24] Speakers confronted this issue by boiling down Newton's work into more easily digestible versions,[25] while the coffee-houses and academies of Europe provided platforms for the explanation and popularization of these Newtonianism(s). In the eighteenth-century heyday of the popular astronomy lecture, lay audiences were entranced by what the speakers revealed. The exploitation of visual aids – models, orreries, simple demonstrations and illuminated lantern slides – could render such occasions both exciting and sublime, while expositions, dictionaries, handbooks and fictionalized accounts became important channels for the dissemination of key ideas.[26]

Just as Newton's work was reconstituted and sold to the public so, too, were his manuscripts, his scientific instruments and even locks of his hair, fuelling a phenomenon that raised him to the status of secular sainthood. While this thirst for objects – for a piece of Newton – fed burgeoning trends of commemoration and reverence, his image, literal and figurative, was being carved by craftsmen and savants across England and beyond. As sculptures, engravings and death masks were bought and sold, haggled and competed over by collectors and memorabilia hunters on the Continent, authors and lecturers presented Newton as the epitome of the high-principled scientific genius who had provided a fault-less demonstration of how the scientific method of observation and experiment could uncover the truth about the external world.[27] As with Darwinism centuries later, Newtonian astronomy was customized, re-presented and co-opted to support a diverse range of interests. Criticisms were advanced and the dissemination of its ideas was patchy, but the principles and practices it embodied grew to have a massive impact, lasting through the eighteenth century and beyond.

By the 1830s, ideas about the earth and its creation were also shifting rapidly, and the apparent divergence between a literal interpretation of Genesis and scientific findings, especially from the groundbreaking new field of geology, was the subject of extensive discussion and controversy. While medieval theologians had not interpreted Genesis word for word, seventeenth-century churchmen and scholars had begun to promote a literal reading of the Bible. When understood in this way, Genesis seemed to describe the creation as a series of sudden, miraculous, God-inspired events that could have occurred as recently as 4004 BC, according to one well-known estimate. According to the Bible, this episode had supposedly established a fixed and unchanging natural order, a great chain of being from God to man to the lowest life forms. By the nineteenth century, however, this system was being brought into question by the flourishing young science of geology. Within this field, new techniques enabled the surveying and mapping of strata and their fossilized contents in chronological order, an innovation connected with amateur geologist William Smith, the first to work out the correct succession of strata in England and Wales. While establishing boundaries between vari-ous stratigraphical systems (Silurian, Devonian and so on) subse-quently caused much dispute, debate was also simmering about the age of the earth. In his book *Theory of the Earth* (1795), Scottish geologist James Hutton had argued that it was created by a never-ending cycle of slow changes over an extensive period of time. One of many competing theories at the time, the majority favoured an alternative model – catastrophism – which held that the earth and the whole work of creation were produced by violent, sudden change. This theory was challenged in the 1830s, when Hutton's 'steady state' model of the earth's history, known as uniformit-arianism, was fully articulated by barrister-turned-geologist Sir Charles Lyell in his controversial three-volume *Principles of Geology* (1830–33). This book showed how ever-present change operated through agents such as earthquakes and erosion, thus disputing the idea of a relatively recent, sudden creation as described in Genesis. Debates on the issue were heated and ongoing, but the basic tenet of Lyell's theory – that the earth had an extensive history – eventually became the more mainstream view.

Meanwhile, further reassessment of the Genesis story had been prompted by sequential fossil finds that revealed the earth had passed through a number of geological ages, each with its own species of animals and plants. It appeared that creatures such as mammoths and mastodons had once existed, but had become extinct. Such findings, commonly associated with French palaeon-tologist Georges Cuvier, challenged the Biblical image of a fixed and unchanging great chain of being from God to man to the lowest life forms, although Cuvier advanced the idea of immense geological catastrophes followed with new creations by God to explain new species in the fossil record. More radical, however, were the evolutionary ideas of French naturalist, Jean-Baptiste Lamarck. In *Zoological Philosophy* (1809), he advanced a theory that the simplest forms of life had originated in matter as the result of a series of physical and chemical reactions – a type of spontaneous generation. This living matter, forced to exert itself to adapt to its environment, had transmuted (evolved), passing on its acquired characteristics to the next generation, producing higher and more complex organisms until the development of man.

Lamarck, sidelined in his lifetime and after his death, had a limited impact on his peers and the public, and failed to provide an adequate explanation for what made life-forms grow and develop (a mechanism, natural selection, was later discovered by Charles Darwin and Alfred Russel Wallace). However, the idea of trans-mutation central to Lamarck's theory introduced the abominable notion that man, supposedly the highest being created in God's own image, had evolved from the most menial life-forms by a natural, non-divine self-motivating mechanism.[28] Finally, radical challenges to Genesis were joined by cosmology in the shape of the 'nebular hypothesis'. A theory proposed in the eighteenth century by German philosopher Immanuel Kant, and refined by French physicist Pierre-Simon Laplace, the nebular hypothesis taught that the stars and planets originated from immense, diffused revolving clouds of cosmic gas and dust (nebulae), a process that, when taken at its most extreme, dispensed with the need for creation and a creator altogether.[29]

Although some ideas, such as transmutation, were so radical they were deemed taboo, most theologians were able to integrate new geological findings into a liberal reading of the Genesis story. At a time when most geologists were Anglicans, it was common for them to explain the great age of the earth by interpreting the six 'days' of creation as long geological ages (day-age theory) or by incorporating a series of catastrophes and new creations into the earth's history to explain the fossil record, as with Cuvier. While eighteenth-century *philosophes* had promulgated the myth that reason replaced superstition through science, the reality remained one of synthesis rather than substitution. The materialist vision of a mechanistic 'clockwork' universe promoted by Enlightenment thinkers was a concern for the British scientific élite, however, because this was a seemingly random world created by blind natural laws. Such radical attempts to remove God from the equation provoked a revival of 'Natural Theology,' whereby the Anglican clergymen who dominated natural philosophy and monopolized related academic posts sought to collect and present evidence of

God's design in nature. Their vision, rooted in seventeenth-century ideas, emphasized nature as God's creation; thus, the central purpose of their surveys was to provide plentiful proof of the divine architect's wisdom and skill. In keeping with Christianity's pre-dominance in eighteenth-century society, belief and intellectual life, it was held that the existence and nature of God could be demonstrated by the study of his creation; conversely, the world was so perfect in its design that it was evidence of a benevolent higher power at work. In 1802, the 'argument from design' found its most celebrated expression in clergyman William Paley's *Natural Theology; or Evidence of the Existence and Attributes of the Deity, collected from the Appearance of Nature*, which famously used the complexity of the human eye as an example of deliberate, intricate invention by God. Within this system it was assumed that each creature had been specifically designed for its place in nature and society; thus natural theology was used as an ideological prop for the existing social order and the political status quo – fine for the parsons in their cosy country parishes, less satisfactory for child workers in mills.[30] Whatever the case, the argument from design dominated popular scientific culture; it was an immensely power-ful and appealing idea. In the late 1820s, the eccentric Earl of Bridgewater, Francis Henry Egerton, was so inspired by Paley's *Natural Theology* and the desire to atone for an eventful life, that he bequeathed £8000 to the Royal Society to finance a similar work.[31] The subsequent eight-volume survey of the 'divine watch-maker's' handiwork, *The Bridgewater Treatises* (1833–40), was hefty, expensive and, as a consequence, not widely read yet it indicated the scientific status quo in the 1830s, the continuing influence of the argument from design, the cross-fertilization and interconnection between science and Christianity, and the might of creationist (although not literalist) views.[32]

So this survey has come full circle. From the myths of modern writers to those of ancient civilizations, from images of medieval flat-earth belief to nineteenth-century creationist views, the history of ideas about the earth is more than a straightforward narrative about the establishment of a scientific fact. Yet amid a mass of shifting perceptions – of science, religion and history alongside changing views of the physical world – a globular earth has been the educated consensus since at least the fourth century BC.[33] Despite this, the flat-earth idea, last commonly believed by educated people in the far reaches of antiquity, has been transformed into a political tool by modern storywriters hunting for an angle, or by polemicists seeking to argue their case for science against supposed Christian closed-mindedness. It is here that history takes a strange twist indeed, defying expectations about the progress of knowledge from varying points in the human past. For the flat-earth idea has undergone a revival through the nineteenth and twentieth centuries – the work of a movement of true believers determined to prove that conventional scientific knowledge is a delusion, the Bible is literally correct and the earth is flat.

Notwithstanding its status as the world's most infamous alternative idea, flat-earth belief is important in several respects. Because believers claim the Bible teaches that the earth is flat, it provides vital insights into the rise of modern creationism, generally traced to the work of Seventh Day Adventist geologist George McCready Price (1870–1963), and The Genesis Flood, a book published by John C. Whitcomb Junior and Henry M. Morris in 1961.[34] In fact, in adhering to an exceptionally strict literal interpretation of the Bible, flat-earth believers might be classed as extreme creationists, a salient point in the light of contemporary debates about the teaching of evolution in schools and other ongoing Christian fundamentalist campaigns, such as those opposed to same-sex marriage and abortion. While flat-earth belief is widely ridiculed, it remains a central point of departure between a literal interpretation of the Bible and the findings of science; indeed, flat-earth campaigns provide interesting comparisons with the modern creationist movement in a variety of ways. For this reason the flat-earth idea is a pivotal concept: it is one of the most radical and readily refutable Bible-based truth claims about the natural world, while the rotundity of the earth is one of the most fundamental scientific facts, a cornerstone of received knowledge. Despite its apparent absurdity, flat-earth belief therefore occupies a unique place in the relationship between science and Christianity. It raises issues central to science education, the uses and abuses of information, the making of knowledge about the natural world and the psychology of human faith, while the extraordinary history of flat-earth campaigns involves a plethora of conspiracies, counter-cultural critiques and subversive discourses ranging from the moon-landing hoax to the end of the world.

Chapter Two

A PUBLIC SENSATION

TIME WAS, they said the Earth was flat, but now they say it's round;
But strange enough, though true, it is, no PROOF has yet been found.
Astronomers – will tell you, if you ask them, o'er and o'er,
Proofs are by no means wanting, by the dozen or the score.
COPERNICUS has told us this, and NEWTON, and the rest;
And people say 'These are the men who surely, should know best':
HERSCHEL indeed says, in his book, 'We'll take it all for granted;'
COMMON SENSE says, now-a-days, that something else is wanted.

'COMMON SENSE' [William Carpenter],
The Earth Not a Globe (1864)

THE MODERN PUBLIC REVIVAL of the flat-earth idea was the brainchild of a travelling lecturer and quack doctor known by the pseudonym 'Parallax'. Born in Stockport in 1816 and christened Samuel Birley Rowbotham, by the late 1830s he was managing a radical socialist commune, allied to Welsh cotton manufacturer and social reformer Robert Owen (1771–1858), set deep in the Cambridgeshire fens. Here, in a vast, flat landscape criss-crossed by a network of watercourses and dykes, Parallax undertook various experiments to discover the shape of the earth centred on one simple question: what was the shape of the surface of water? Parallax deduced that if the earth was truly a globe, water must have a degree of convexity and this was the point he investigated with a series of experiments on a six-mile stretch of the Old Bedford Canal. During the winter when the canal was frozen, he had apparently lain flat on the ice with a good telescope and spotted skaters at Welney, six miles away, while in the summer he claimed to have seen village folk running in and out of the water, and even those who were swimming. He also made observations on boats sailing along the canal with similar results, or so he said, leading to a conclusion that the canal and the earth were flat.

All that remained was for Parallax to integrate his findings into a coherent theory about the earth and its position in the universe, alongside demolishing the multitude of proofs for rotundity and revolution. Utilizing experimental results, mathematical calculations and various Biblical passages, he proceeded to argue that the earth lay at the centre of the universe, was less than six thousand years old, was created in six twenty-four-hour days and was rapidly approaching destruction by fire. Besides advocating a literalist reading of the Bible, young-earth creationism (as it is now known) and an apocalyptic vision of earth history, Parallax contended that the earth was a flat disc with the North Pole at its centre. The South Pole was naturally non-existent in this scheme, and the circular plane was bordered by an immense barrier of ice. How far the ice wall extended, how it ended, and what existed beyond it were questions that Parallax believed no human being could answer.[1]

He was certain, however, that the disc-earth was stationary with neither axial nor orbital motion, while the sun spiral-circuited overhead once every twenty-four hours at a distance no greater than seven hundred miles. Within this geocentric system, the sun was four hundred miles from London, and the moon and stars were absolutely no further than a thousand miles from earth. When it came to the heavenly bodies, the moon was a self-luminous and semi-transparent body rather than a reflector of the sun's light, as commonly supposed, while the stars were mere 'centres of action' that threw down light and 'chemical products' on the plane earth. This earth, as the subject of God's special creation, was the only material world in existence, for Parallax insisted that the sun, moon and stars were only referred to as 'lights' in the scriptures and that anyone who believed the reverse was 'the victim of an arrogant and false astronomy; of an equally false and presumptuous geology; and a suicidal method of reasoning . . . contrary to nature, to fact, and human experience, and to the direct teaching of God's Word'.[2]

When it came to explaining phenomena generally associated with the earth's rotundity and revolution, Parallax was similarly forceful. The seasons were the result of the sun's 'peculiar concentric path', a circular course that expanded and contracted over the disc on a six-monthly cycle, comparable to the movement of a needle on a record. According to Parallax's reasoning, the sun circled nearest to the central North Pole in British summertime and furthest away from it, towards the impenetrable ice barrier, in the depths of winter. Day and night were likewise the result of the expansion and contraction of the solar path over the circular plane, with the sun acting like a flashlight moving over a table, only able to provide light in places where its rays beamed vertically downwards. As for sunrise and sunset, impossibilities if the sun remained constantly moving above the stationary earth, Parallax dismissed them as results of a special law of perspective and a type of optical illusion. Meanwhile, he explained solar eclipses as the result of the moon passing between the sun and the observer on earth.

Naturally, accounting for lunar eclipses was more problematic because the earth was a stationary plane constantly underneath the sun and the moon, so, as Parallax later put it, 'to speak of its intercepting the light of the sun, and thus casting its own shadow on the moon, is to say that which is physically impossible'.[3] Having considered the matter, he decided that the only logical cause was the movement of a mysterious dark body or 'non-luminous satellite'. As for tides, they were not the result of the moon's gravitational pull, as usually assumed, but the consequence of the rising and falling of the plane earth as it floated on the primordial waters, the 'illimitable fathomless deep'. Everyday proofs of the rotundity of the earth, such as the disappearance of ships over the horizon as they sailed out to sea, were merely the result of the laws of perspective and refraction, while 'circumnavigation' was possible if sailors travelled square to the compass and traced a course round the edge of the disc. In fact, for every phenomenon generally taken as proof of the rotundity and revolution of the earth, Parallax developed a counter-argument backed by Biblical quotations, experimental proof and intricate mathematical calculations. With 'evidence' mustered, he decided to exploit his skills in public oratory to win converts to his Bible-based world-view.

Following a scandal at the commune, where Parallax and others had been unfairly accused of sponsoring 'the traffic in human flesh' and participating in free love, he reinvented himself as an itinerant socialist lecturer and then as 'Dr Birley Ph.D.', practising in Manchester, Sheffield and other northern towns. When it came to medical research, Parallax's major interest was in investigating ways to make mankind immortal, or capable of living for a thousand years or more, and he outlined his ideas on the topic in a number of original tracts, including *Biology: An Inquiry into the Cause of Natural Death, Showing it not to Arise from Old Age, but from a Gradual Process of Consolidation* (1845). Published under the pseudonym Tryon, the book's central argument was that death resulted from a general ossification or choking up of the body by 'earthy matter', principally phosphate and sulphate of lime. The way to live a long life was thus to avoid consuming food and drink containing a high proportion of those substances, and Parallax said he had experimental proof from tests and dissections to back up these novel claims.

In the mid-1840s, having expounded his views on life, death and longevity, Parallax turned his back on alternative medicine and radical politics, and threw himself into public dissemination of the flat-earth idea. The momentous quest was to become his life's work and he employed two specific tactics to render his campaign more effective. The first was to set the idea within an anti-scientific thought system, which he christened 'zetetic astronomy', after the Greek verb *zētein*, meaning to 'seek' or 'enquire'. Although it was original in its application to astronomy, the label 'zetetic' was far from new. It had first been adopted by followers of the Greek sceptic-philosopher Pyrrhon (*c.* 365–275 BC), to denote their belief that all perceptions are of doubtful validity, and because there is no certainty, we can only know things as they appear to us.[4] Highlighting Parallax's radical socialist connections, the label 'zetetic' had also been employed more recently by the London free-thought movement around the time of the Napoleonic Wars. Led by radical atheist publisher Richard Carlile, the zetetic network had campaigned against a spate of prosecutions for the publication of supposedly obscene, blasphemous and seditious material at a time when the government was determined to quash any potential revolutionary threat. These zetetic societies, based mainly in Scotland and the North, were the first free-thought organizations, although radicals had been employing the terms 'free enquirer' and 'truth-seeker' in their journals since the late-eighteenth century.[5]

In all probability, this radical connection provided the inspiration for Parallax's use of the term 'zetetic' although, unlike his historical predecessors, he decided to employ it as the foundation for an attack on orthodox science and the idea that the earth was a globe. Here Parallax was particularly astute: his zetetic system was based on empiricism (observed evidence) and induction (inference of general principles from observed evidence), an approach known as the Baconian method after its originator, the sixteenth-century English philosopher Sir Francis Bacon. The Baconian method had become central to investigations of the natural world during the seventeenth century, but Parallax decided to claim it as his own, alleging that conventional science was beset by theoretical assumptions and the laws of nature could only be revealed by zetetics or 'truth-seekers', determined enquirers who considered all of the available evidence and refused to take accepted theories for granted. In fact, Parallax went further, arguing that orthodox ideas were precisely the ones that should be targeted, questioned, examined and challenged through free and open-minded data-collection by zetetics, who listened meekly to Nature's revelations and did not twist results to fit theories. This was the way of conventional science, he claimed: speculation and imagination had ruled the day, spawning the widespread misconception that the earth was a globe. As a result, he contended, the Victorian public could be compared to 'squirrels in a roundabout', trapped in a whirl of inconsistency and delusion by the mumbling pretensions of arrogant astronomers who bowed to the fashionable assumptions of their age. Parallax continued:

> It is . . . candidly admitted that there is no direct and positive evidence that the earth is round, that it is only 'imagined' or assumed to be so in order to afford an explanation of 'scores of phenomena'. This is precisely the language of Copernicus, of Newton, and of all astronomers who have laboured to prove the rotundity of the earth. It is pitiful in the extreme that after so many ages of almost unopposed indulgence, philosophers instead of beginning to seek, before everything else, the true constitution of the physical world, are still to be seen labouring only to frame hypotheses, and to reconcile phenomena with imaginary and ever-shifting foundations. Their labour is simply to repeat and perpetuate the self-deception of their predecessors.[6]

He believed that Newtonian astronomy was a 'juggle and jumble' of fancies and falsehoods; an elaborate theoretical trick 'enough to make the unprejudiced observer revolt with horror from the terrible conjuration which has been practised upon him'.

In the face of this élitist conspiracy, the only solution, Parallax declared, was to replace conventional science with a true and practical free-thought method. This was promoted as a back-to-basics approach to knowledge, in which experiments were tried and facts collected not to corroborate any existing theory but to start from scratch to uncover the great universal and primary truths. In many ways, Parallax framed his crusade as one of knowledge and power to the people, for at the heart of his do-it-yourself doctrine was the idea that anyone could accumulate facts to reach the truth if they investigated for themselves with an open mind. It was the Victorian ideal of 'self-help' emphasized; the common man could literally help himself to knowledge by sidestepping textbooks and élites and looking only to nature and 'common sense' as a guide.[7] The value of taking instruction directly from nature, experience and direct observation rather than 'book-learning' was a central element in Baconian thought and seventeenth-century natural philosophy, while the Scottish 'common-sense' realism school of the eighteenth century had argued that ordinary common-sense assumptions are true and can form a solid foundation for further philosophical enquiry.[8]

In addition, as the well-travelled, widely read Parallax knew, anti-élitist undercurrents were prevalent in the working-class scientific culture of his own time.[9] In an era when it was common for skilled working men to educate themselves courtesy of Mechanics' Institutes, Owenite Halls of Science, mechanics' magazines and the penny press, Parallax calculated that his democratic zetetic ideology could have a powerful appeal.[10] It was useful, accessible, practical knowledge at a time when the upper reaches of working-class culture were awash with the drive to self-improve. Of course, this approach enabled him to claim an altruistic motivation for his campaign. In Baconian spirit, he declared that his aim was to 'kindle the flame of free inquiry and hold out the beacon of knowledge and truth' by promoting a straightforward system of enquiry where nothing was taken for granted – not even the 'fact' that the earth was a globe. From the beginning, therefore, Parallax's attack was not merely directed at the end product of science – the facts regarding the shape of the world – for he framed his 'knowledge for all' campaign in popularist terms as an assault on the validity and élitism of conventional science and the method it supposedly entailed. Parallax doubtless realized that such a broad-based philosophical approach would have more power than a straightforward claim about the earth's shape, and his primary aim was to establish credibility for himself and his unorthodox campaign.

Armed with his democratic emphasis on Baconian fact collection and analysis for oneself, Parallax felt able to claim the moral high ground. His methodology was an intellectually and morally improving pastime, the zetetics were free men and women, unrestrained by dogma and willing to be guided first and foremost by facts. They were the true objective investigators, the ideal practitioners of science whose work would lead to the salvation and redemption of humankind. For this reason, Parallax declared, 'Bigots may howl; tyrants may frown; hypocrites may sneer; the coward may quail; the indifferent may marvel but the champion of truth, armed in honesty of purpose, pursues his festal path, invulnerable and victorious.' This mythological image of the zetetic 'truth-seeker', the alternative hero of Baconian science, was to become central to Parallax's work. With such laudable sentiments underpinning his philosophy, he decided he needed a new identity to reflect his alternative perspective on knowledge and the shape of the earth. To be truly memorable, the pseudonym should also relate to astronomy, and for this reason Samuel Birley Rowbotham decided that he would present himself to the world as 'Parallax' (meaning the apparent change in the position of an object when viewed from two different points). The public phase of his campaign was about to begin.

Parallax's reasons for seeking to convince the British public that the earth was not a globe were simple. Evidently an ingenious character, who delighted in controversy and dispute, he could not resist the ultimate challenge of toppling orthodox ideas and a fact so established as the earth's rotundity. During his time as a socialist lecturer he had seen the passions that scientific and religious topics could evoke and, moreover, the money that people would pay to listen to a feisty debate on these themes. This was especially so if the issue was eye-catching or outrageous in any way; evening lectures were a major form of entertainment and instruction for the Victorians, and astronomy was one of the most popular subjects on which travelling lecturers could base their talks. Since the eighteenth century, audiences had marvelled at the star charts, planetaria, orreries and lantern slides displayed by those who made their living educating lay people on the lecture circuit, a phenomenon that could not have gone unnoticed by Parallax.[11] Furthermore, astronomy was not the only topic to capture the public's imagination at that time, for alternative sciences, from mesmerism (a form of hypnosis) to phrenology (diagnostic-cum-personality reading from the bumps of the head), were also drawing curious crowds at demonstrations in Mechanics' Institutes and lecture halls nationwide. Through popular scientific culture, Parallax had seen at first hand the venues, audiences and ideologies he would be able to exploit.

Meanwhile, a flat-earth campaign had another more obvious appeal. Ideas about the earth were changing rapidly, and the apparent divergence between a literal interpretation of Genesis and scientific findings, especially from the new field of geology, was contentious. This was especially the case in the wake of Lyell's *Principles of Geology* (1833) and Edinburgh publisher Robert Chambers's widely read *Vestiges of the Natural History of Creation*, which he published anonymously in 1844. Influenced by phrenology, Chambers argued for the progressive 'development' (evolution) of species including humankind, although he was careful to distance himself from the radical French evolutionary thinker Jean-Baptiste Lamarck by emphasizing that this process was divinely guided by laws that God had deliberately built into nature. Nevertheless, the 'development' thesis posed problems for those who interpreted Genesis literally: one churchman went so far as to lambast *Vestiges* as a 'filthy abortion' capable of corrupting the nation's 'glorious maidens and matrons'.[12] Within this context, one suspects that Parallax saw the potential to feed off controversies about science and the Bible and exploit popular scientific culture in a single campaign. By teaching that the earth was flat, and emphasizing the scriptural basis of the idea, he could confirm the Bible as the ultimate authority on the earth and its creation, a critical and controversial point. By claiming to defend the Bible against science, and by setting the two in irresolvable conflict, he was bound to attract an audience willing to pay to hear such ideas. In addition, his emphasis on democratic fact-finding could potentially attract self-educated working men, while experimental proofs would gain further authority and credibility for his cause. So, under a banner of power to the people and to the Bible, Parallax sought to mobilize public support.

Questions remain about whether Parallax truly believed his own theory, for his time as a socialist lecturer and commune member throws serious doubt on this claim.[13] With divergent ideological, social and political agendas, it would be highly unusual for an extreme Biblical literalist to become entangled with radicals and, moreover, to dedicate himself to their secular cause. Socialists tended to subscribe to the Enlightenment ideal that progress would transpire through reason and empirical investigation rather than orthodox religion, and to attack Biblical authority in line with the privileged 'priestcraft' suited their ideological and secular goals.[14] On balance, it seems most probable that Parallax constructed a myth of a lifelong quest to prove the literal truth of the Bible pinned on the most erroneous of facts.[15] However, it appeared that he had found his niche: the use of arch-debating skills developed in the most intimidating circumstances to test the boundaries of science and popular belief. In a colourful existence that defied convention, only two strict rules applied: his real identity was to remain secret, and he would always maintain that he truly believed the earth to be flat, despite later revelations to the contrary.

Scheme devised, Parallax embarked on his first public engagement, a lecture on zetetic astronomy at Trowbridge Mechanics' Institute in early 1849, followed by a taxing appearance in Burnley. His first lecture in the town went off without a hitch, but the *Blackburn Standard* reported that Parallax ran away from a subsequent meeting when he could not explain why the hulls of ships disappeared before their masts when sailing out to sea. According to the paper, the audience had hung around to see if he would return, and when he did not 'they assuaged their disappointment by concluding that the lecturer had slipped off the ice edge of his flat disc, and that he would not be seen again till he peeped up on the opposite side'.[16] Albeit quick on his feet, Parallax clearly had some work to do in strengthening his arguments, and he published two pamphlets on zetetic astronomy later that same year. The first was a sixpenny tract entitled *Zetetic Astronomy: A Description of Several Experiments which Prove that the Surface of the Sea is a Perfect Plane and that the Earth is not a Globe* and this was swiftly followed by *The Inconsistency of Modern Astronomy and its Opposition to the Scriptures!!* With one title focusing on experimental proofs, the other on scriptural evidence, the publications captured Parallax's tactics to a T. From that point forward, his campaign would be characterized by a two-pronged assault on the globe idea; like modern-day 'scientific' creationists, he used Biblical literalism and experimental proof, one of the hallmarks of science, to appeal to the truth.

Both pamphlets were highly unorthodox, but the second was most original: it was Parallax's first presentation of Biblical arguments against the system developed by Pythagoras, Copernicus, Galileo, Kepler, Newton *et al*. He declared that those thinkers displayed a 'spirit of perseverance, industry and ingenuity', but they had laboured only to support a theory and their data was assumed. Furthermore, their vast, romantic system had been foisted on the public – an outrageous state of affairs. He complained that the Bible, alongside the evidence of our senses, supported the idea that the earth was flat and immovable, and this essential truth should not be set aside for a system based solely on human conjecture. And so he continued, offering scriptural proofs and quotations to back up the thesis that astronomers were mistaken: the earth was not a globe, the sun and moon were lights, not solid masses or worlds, the moon was self-illuminating rather than a reflector, and much more besides.[17]

As *The Inconsistency of Modern Astronomy* went on sale to the public, Parallax continued to work the lecture circuits in England and abroad. In May 1851, the same month as an enormous globe was erected in Leicester Square in honour of the Great Exhibition, Parallax travelled to Ireland to disseminate his alternative view. His first stop was Athlone Court House in Westmeath, where he addressed a large and respectable audience who were astonished by his extraordinary lecture and the startling 'facts' it contained. According to the local paper, at the end of the evening 'several gentlemen entered the lists with "Parallax", and a lively and interesting discussion ensued'.[18] On this occasion Parallax stood his ground, and he moved on to speak at the Mechanics' Institute and the Rotunda in Dublin, evoking a similar response. This pattern of events continued throughout the 1850s, with Parallax travelling from town to town on a relentless campaign to spread the zetetic word. In 1854 he lectured in Leicester, and must have seemed convincing to some for the *Advertiser* reported that his statements seemed 'seriously to invalidate some of the most important conclusions of modern astronomy'.[19] Thus encouraged, he decided to write to the Astronomer Royal, George Biddell Airy, asking him if he would be so kind as to review some zetetic literature and offer his opinions. That Airy replied was unlikely, but Parallax persisted with his campaign. In 1856, he provoked a furore in Great Yarmouth, where his striking lectures caused such irritation that several audience members, including an engineer, a tobacconist and a Baptist minister, challenged him to a public experiment at the River Yare. Despite the drama, the *Norfolk Herald* gave Parallax a glowing review:

> ... inasmuch as the system of the lecturer differs in every point of view from our own study of astronomy, and from all previous teachings on the subject, there must be a great error on one side or the other. 'Parallax', as a lecturer, is a sound logician, a clear, lucid reasoner, calm and self-possessed, we have never seen surpassed.[20]

Parallax had honed his debating talents to an impressive degree and was ready to take on any disputant who crossed his path. On repeated occasions he tried to provoke an argument to create a buzz about upcoming lectures, and established astronomers were naturally the best targets for such publicity-seeking attempts. It was an approach that he employed many times throughout his career, and his visit to Aylesbury in November 1857 was but one example. On this occasion, he chose to target Admiral William Henry Smyth, a well-known local astronomer and friend of Airy, the Astronomer Royal, with a letter requesting support for his public appearances. To incite a reaction, he also enclosed a poster advertising his talks on 'Zetetic Astronomy: Earth not a Globe!' at Castle Street Hall, which had been pasted up around the town. The poster informed the public that tickets cost a shilling or sixpence from Mr Gunn the Grocer, a standard price that would probably only be paid by skilled workers and the middle classes. Also in keeping with lecture conventions of the day, Parallax framed his talks in educational terms, as a course with a syllabus and a single overriding theme:

> The lectures will explain a number of experiments which prove that the Earth is a plane and not a globe; that it has no motion, either on axes or in an orbit round the sun; that the sun, moon and stars are lights only and not worlds; and shewing [sic] the true cause of day and night, seasons, tides, eclipses and other phenomena. After each lecture 'Parallax' will hold a public discussion on the subjects advanced to which philosophers and Scientific men are invited.[21]

Looking to attract controversy, the poster also called the attention of people from all religious denominations to Parallax's ideas about the incompatibility of science and the Bible. It even claimed that Sir John Herschel, the eminent astronomer, had declared himself dissatisfied with Newtonian philosophy and the idea that the earth was a globe. On reading the comments, William

Henry Smyth was most amused and forwarded the advertisement to George Biddell Airy with a humorous note attached:

> Here's a pretty go. A man who calls himself 'Parallax' – though I am told Rowbotham would do just as well, solicits my patronage on the enclosed plea. Did you ever?[22]

Unfortunately Airy did. Tongue-in-cheek, he joked back to Smyth that 'Mr. Parallax is a hero, decidedly', and although he had received two or three such flyers, he had not posted them on his wall. But seriously, Airy continued, some good might still come of Parallax's quest. Although zetetics had clearly had the doctrine of the earth's sphericity and rotation driven into them at some point and rebelled, the crusade illustrated the 'urgent need' for the 'rational and argumentative instruction' of the people that he himself proposed.[23]

While the Astronomer Royal ruminated over the popularization of science in his offices at Greenwich Royal Observatory, Parallax continued to tread the boards of halls and institutions in villages and towns across Britain. By the summer of 1858, he had moved north to the Midlands and was taking Northampton by storm with experiments on the laws of perspective at Bridge Street station, and lectures at the recently established Mechanics' Institute in the Market Square. Complete with its own library, museum and reading rooms, it had been set up to disseminate scientific, mechanical and other useful knowledge to skilled manual workers (although such places were also frequented by the middle classes), and usually played host to lectures on subjects such as natural history, geometry and French. Thus it was with some curiosity that the *South Midlands Free Press* reported on Parallax's astronomy course:

> No doubt many of our readers have been mystified and surprised within the last week by the announcement that, in three lectures, at the Northampton Mechanics' Institute, a gentleman, who calls himself 'Parallax', would undertake to

prove the earth not a globe ... We were highly gratified by
the manner in which this important subject was handled
by 'Parallax' – a pseudonym which the lecturer informed his
audience he had adopted in order to avert the effect of an
insinuation that his startling announcement is but the morbid
desire of an individual to be known as the propounder of a
philosophy boldly at variance with that of the great astronomers of the past and present. His subject was handled in a
plain and easy manner, his language and allusions proving him
a man of education and thought, and certainly not a pedant.
The experiments mentioned, divested of technicality in their
recital, and understandable by all, were of such a nature as to
cause a start of surprise at their simplicity and truthfulness ...
It is not for us to pronounce a verdict upon so important an
issue; 'Parallax' may be in error, but as far as his reasonings
from fact and experiment go, there is much to set scientific
men thinking. His arguments consist of facts, and such as are
patent to all degrees of mental capacity ... In the discussions
which followed, 'Parallax' certainly lost no ground. either in
answers to questions or to some broad assertions quoted from
learned authorities.[24]

The lecture tour continued, and eventually Parallax established
himself in London with a new wife. He had been married before,
to a woman called Nancy, but in July 1861, at the age of forty-
five, he settled for a second time with Caroline Elizabeth West,
the sixteen-year-old daughter of his laundress. Their first baby,
Samuel, was born a year later while they were living in Kentish
Town; they went on to have a total of fourteen children, although
only four survived childhood. A man of many interests, Parallax
also branched out into the world of technological innovation around
this time with patents for a variety of original ideas.[25] These
included a new type of caustic soda, a liquid india-rubber, a process
for soap manufacture, a 'fire proof starch' to make material non-
flammable and, most intriguing of all, a 'life preserving cylindrical
railway carriage', described at length in the *Mechanics' Magazine*.

Accidents were a frequent occurrence on the nation's rail network, established two decades before, and Parallax was terrified of train travel; leading to a determination to invent a revolutionary piece of rolling stock to solve the problem for ever. His subsequent design was ground-breaking: allegedly a 'perfect safety' railway train, it consisted of a series of cylindrical carriages that rolled along the tracks, like a succession of miniature big-wheels with the passengers suspended inside.[26]

Despite his concern about train travel, Parallax remained fearless when it came to lecturing on flat-earth theory, and he continued to petition societies and institutions for rooms to speak on what he liked to call 'the first principles of science'. In April 1861 he had some success at Greenwich and managed to get a booking for a run of lectures near to Britain's best-known observatory at the Royal Institution on Royal Hill. As usual he divided his syllabus into three distinct sections, outlined on posters and flyers for prospective audience members. The first lecture would consider how the earth was circumnavigated, why a ship's hull disappeared before its masts when outward bound, why the Pole Star set when travelling southward, why a pendulum vibrated with less velocity at the Equator than at the Pole and other famous 'proofs' of the rotundity and revolution of the earth.[27] At the second meeting, he would explain how the sun, which was now, he argued, 4028 miles from London, orbited the earth. This would be followed by the causes of day, night, winter, summer, sunrise and sunset under these circumstances, concluding with a challenge to mathematicians and an open discussion. In the final lecture, he planned to deal with the cause of tides, the self-illuminating moon, the reason for solar and lunar eclipses, the position of the earth in the universe, and finish with a 'scientific' explanation for its ultimate destruction by fire. In reaction to such unconventional subject matter, a journalist at the *Greenwich Free Press* felt driven to comment:

> To say that these lectures are extraordinary in their character
> is but saying the least that can possibly be said concerning

them. The exceedingly gifted lecturer, who apparently prefers to be known as 'Parallax', demonstrates the Newtonian theory of astronomy to be in opposition to facts; and in so doing demonstrates that the Bible is literally true in its philosophical teachings. From this, the groundwork of his philosophy, spring teachings and doctrines which cause us to hold our breath in the contemplation of them, and compel us, as public journalists, to withhold our opinion on subjects so vast, so important to man, and so utterly at variance with the commonly received notions of the day. Is it for us to say that a greater than a Newton shall not arise? No! we wait the issue. If 'Parallax' be wrong there can be nothing easier than for our taverns of Greenwich to overthrow his doctrines; but if our readers think they would have an easy task so to do, we can only say be present at his concluding lectures, and judge for yourselves ... It is urged that this 'somebody or other' who has the audacity to come right into Greenwich, above all places in the enlightened world, is very strong – strong in his facts, strong in his arguments, and appears after all to get on the right side of his audiences.[28]

Among Parallax's converts at Greenwich was William Carpenter, a journeyman printer, mesmerist and spiritualist. Thirty-one years old, with 'auburn hair, bright, restless eyes and an animated manner', Carpenter had been born and bred in Greenwich, within shouting distance of the Astronomer Royal's residence, Flamsteed House, and by 1861 had established his own printing business in Lewisham. Then an ordinary man about town, or so he thought, he later remembered that when he first heard of Parallax's lectures on zetetic astronomy, he had laughed at the very thought of a lecturer 'going about' spreading such ideas. But that was soon to change. After spending an hour and a half listening to Parallax's talk at Greenwich, Carpenter said he had never doubted that the earth was flat. He recognized that his belief was 'not popular and keeps a man back in the world' but maintained that he 'cared nothing' for that.[29] As a mesmerist and spiritualist, he had already been involved in alternative forms of science and belief, and had an established interest in experimenting with physical and meta-physical reality and challenging orthodox ideas. Although he was not motivated by a particular desire to defend the Bible, the do-it-yourself zetetic philosophy and its explicit challenge to scientific authority held an intoxicating appeal. With its egalitarian emphasis on Baconian fact-finding and practical experimentation, the call to 'find out for yourself', zetetic astronomy embodied one attraction of the alternative sciences alongside an appeal to scriptural truth. Despite its foundation on an erroneous 'fact', Parallax's two-pronged assault on science was beginning to reap rewards, for it appears that Carpenter was attracted by the anti-élitist overtones of his zetetic campaign. He decided that he would do what he could to assist Parallax.

Meanwhile, Parallax repeated his lectures in Greenwich, lead-ing the *Free Press* to report with growing concern:

> It seems impossible for any one to battle with him, so power-ful are the weapons he uses. Mathematicians argue with him at the conclusion of his lectures, but it would seem as though they held their weapons by the blade and fought with the handle, for sure enough they put the handle straight into the lecturer's hand, to their own utter discomfiture and chagrin. It remains yet to be seen whether any of our Royal Astronomers will have courage enough to meet him in dis-cussion, or whether they will quietly allow him to give the death-blow to the Newtonian theory, and make converts of our townspeople to his own Zetetic philosophy. If 'Paral-lax' be wrong, for Heaven's sake let some of our Greenwich stars twinkle at the Hall, and dazzle, confound, or eclipse altogether this wandering one, who is turning men, all over England, out of the Newtonian path. 'Parallax' is making his hearers disgusted with the Newtonian and every other theory, and turning them to a consideration of facts and first prin-ciples, from which they know not how to escape. Again we beg and trust our Royal Observatory gentlemen will try to

save us, and prevent anything like a Zetetic epidemic prevailing amongst us.[30]

Such statements could only act as further inducement, and Parallax continued to wield a battleaxe against conventional astronomy, with lectures in Portsmouth and Gosport. According to one paper, 'hot words' had ensued after the lectures and continued long into the night, but Parallax was unperturbed: having caused a sensation in Gosport, he moved on to lecture in Plymouth in late September 1864.[31] Following one particularly intense debate in the Athenaeum there, one onlooker, a navy staff commander, complained that there were many gentlemen present who had sailed the Southern Seas, yet they had not attempted to dispute Parallax's statement that the circumference of the disc-shaped earth from Cape Horn to the Cape of Good Hope, from the Cape of Good Hope to Port Phillip, and from Port Phillip to Cape Horn was about 30,000 statute miles, an impossibility if the earth was a globe with a diameter of 25,000 miles. Such incidents led the *Western Daily Mercury*, a local paper, to conclude that Parallax 'treats his subject in a very clever and ingenuous manner, and succeeds in drawing many over to agree with him'.[32]

Such coverage was bad news to the better informed, and several surveyors and naval men arrived at the next lecture armed with facts in the hope of winning a final victory. According to the local newspapers, the discussion was the fiercest yet, but despite spending 'three hours on his feet in an atmosphere of 120 degrees', Parallax managed to counter every argument with ingenuity, wit and consummate skill. There was nothing for it, the men decided, but to challenge him to an experiment to decide publicly the truth or falsity of his ideas. Seeing the opportunity for more publicity and a chance to gather experimental proof, Parallax was only too happy to oblige. Co-opting experiments, a fundamental characteristic of science, would help to persuade people that he was telling the truth, and by such means he would make more impact than by speaking in a noisy lecture hall.[33]

Once he had agreed to the test, Plymouth Hoe was named as the best site to make the necessary observations. Meanwhile, the *Mercury* expressed certainty that the experiment would prove to be a 'very interesting' event.

In the interim, the paper's correspondence pages were crammed with letters from irate citizens of Plymouth, many of whom were disgusted by zetetic exploits in their town. Among those most appalled were amateur astronomers and local seamen, who wrote in droves complaining about Parallax's 'foolish assertions' and his attempts to mislead the public about the most fundamental scientific facts. Keen to make amends, they offered a series of proofs for rotundity, from circumnavigation to the curved shadow of the earth during an eclipse of the moon. One sailor, from a naval and nautical school, even felt it necessary to add that during twenty years of voyaging he had never seen the ice barrier that was supposed to surround the disc-shaped earth, and Parallax's claims to have observed boats at great distances on rivers and seas were impossible unless his eye had been elevated far above water level. Amateur astronomer James Willis agreed, declaring that as Parallax had posted himself as a teacher, he should be willing to replicate his experiments openly for all to see. This drew a response from Parallax who declared, on 6 October, that he was ready, willing and able to 'do battle, inch by inch' with his Newtonian opponents 'and upon their own ground'.

His challenge immediately found some support among the *Mercury*'s readers. Several correspondents praised Parallax's attempts to defend the scriptures against the 'vain devices' of science, arguing that human beings should not attempt to penetrate infinite mysteries but should accept the Bible as literal fact. For them, the recent controversy was just the result of 'the rancour which is stirred up in the human mind when its presumptuous inventions and darling dogmas are assailed', while there could be little doubt that navigators warped their calculations to fit their preferred theory of a globular earth. Inflaming the religious significance of the argument, globularists retorted that Parallax was a Jesuit missionary who was trying to persuade people that the earth was flat on behalf of the Catholic Church. The comment was an interesting take on the myth that the Church had oppressed science by insisting that the earth was flat, and it also served to reflect anti-Catholic feeling common in England at that time. One correspondent 'Veri tas' claimed that he had been watching the 'Jesuitical mission-ary' for the last fifteen years and come to the conclusion that his motives were 'simply to excite wonderment at his strange doctrines, blind the ignorant with his false statements, attract an audience and pocket the cash'. As Parallax had no scientific reputation at stake he hid behind a made-up name, and when lecturing in Somerset he had turned out the gas when unable to answer a question and escaped under cover of darkness to try his luck elsewhere. It was a pity, the correspondent continued, that Parallax had not been set upon disproving the multiplication table, rather than the shape and motions of the earth, because everyone would be in a position to make their own judgement, which was patently not so in the present case. It was a sad reflection on the education system, but those who were ignorant of science 'may just as well try to decide a dispute between two Frenchmen' as make sense of the conflicting arguments of Parallax and a Newtonian. Personally, he concluded, he would take a sailor's calculations over Parallax's trickery on any day of the week.[34]

While residents of Plymouth tussled over Biblical interpreta-tion and science education, Parallax was lecturing at the Mechanics' Institute in Devonport, another navy town nearby. Some locals were disgusted that the Institute, established for the dissemination of useful knowledge, should be used for such a purpose, and Parallax's appearances led to the usual confusion of recriminations and abuse. One correspondent, 'Theta', from Devonport's famous dockyard, wrote that Parallax had been careful to use the time allotted to present his 'facts' in a wholesale style, reserving proofs for the short discussion and only if challenged directly. Even then, Theta continued, many found his arguments difficult to disprove without doubt, because it appeared that Parallax had an answer for every refutation. When asked why the pendulum vibrated faster towards the Poles, a well-known proof of the earth's rotation, he said that in the case of the North Pole this was due to the expansion and contraction of the pendulum caused by a difference in temperature; tides, meanwhile, were apparently the result of the disc-shaped earth shifting on the primordial waters rather than the effect of gravity, and when asked to account for the curved shadow of the earth cast during a lunar eclipse, Parallax had retorted simply, 'What proof was there that it was the shadow of the earth at all?' It was a disgrace, Theta declared, that in the nineteenth century a man should have the audacity to stand up before an intelligent audience and contradict the established axioms of nature's laws and call the nation's most renowned men of science impostors. Even worse, he complained, when Parallax had hit a snag in his flimsy defence, rather than admit defeat, which would have been the gentlemanly thing to do, he simply snatched up his hat and stalked out of the venue, leaving his questioners fuming. Yet despite Parallax's dodging and Theta's scathing review, the *Devonport Independent* felt able to conclude that those who tried to debate with Parallax often 'became excited and lost their command', and to an unprejudiced observer it looked as if the zetetic had maintained the upper hand.[35]

Amid all of this, broader questions were being asked in Plymouth's *Western Daily Mercury*: just who was Parallax, and had he created such uproar before? Rumours flew that he might be the oilcloth-seller from South Street who had lectured on flat-earth theory several years previously, while one letter-writer was reminded of an eighteenth-century 'Parallax', who had tried to convince a London crowd that the bronze lion on top of Northumberland House was able to wag its tail. Although a more enlightened number of the assembled multitude had remained unconvinced, it was said that others, unaccustomed to zoological phenomena of any kind, went home with serious doubts about the matter. History seemed to be repeating itself in a vague way, while more revelations surfaced about Parallax's past in a letter from one

George Breeze. An acquaintance of those entangled with Parallax during his visit to Great Yarmouth in 1856, Breeze claimed that he had bailed out of experiments on the convexity of water at the River Yare, and had refused to speak at the local hall unless the audience bought tickets for the event. As a consequence, he concluded that Parallax was an extortionist whose only true belief was in money.

At this the ever-polite Parallax lost his temper and fired off a stinging rebuke. Tired of the 'scurrilous, vulgar and unmanly assertions' that were polluting the *Mercury*'s pages, he again challenged his 'foul-mouthed' detractors to meet him for a public experiment. If they dared refuse, Parallax warned that they would be classed as 'cowards and assassins' by the good people of Plymouth, who had seen for themselves his labours to seek the truth for its own sake.[36] As for George Breeze, Parallax said that he would have no hesitation in resorting to legal action in retaliation for the libellous claims, which were so outrageous that he now felt bound to give a detailed reply for the benefit of faithful followers. His investigations into the shape of the earth had involved nearly twenty years of 'overwhelming labour', he explained, during which time he had borne endless abuse and lost, rather than gained, money. Through it all he had relied on his heartfelt Christian faith, and belief in his mission to serve God, the truth and his fellow men. His reward, he claimed, had been the good wishes of thousands of worthy and intelligent people worldwide. In fact, Parallax continued, he had been invited to the royal courts of Europe and been offered the favour of monarchs, while learned societies had striven to enrol him in the lists of the great and the wise. Nevertheless he had preferred to labour alone, keeping zetetic astronomy free of all worldliness and grandeur.

As for the experiments on the River Yare, Parallax claimed they were a set-up and that he was the victim of a similar conspiracy in the south-west. Somebody had even taken up a knobbed stick under cover of darkness and dealt him 'two successive blows of a very dangerous character' after his lecture at

Devonport Mechanics' Institute, behaviour that could never be condoned.[37] The only solution, Parallax reiterated, was to meet his opponents for the ultimate test, and with that, he waited for them to name their day.

He soon received an answer: 'Theta' announced that a committee from Devonport Mechanics' Institute would meet him to take up the repeated challenge. 'The public ought to be made aware of the extravagancies [of] this strange planist,' Theta continued in the local paper, so that 'they may not be beguiled by his loquacious twaddle and milk-and-water moonshine'. On that note he demanded that Parallax take part in the practical test for all to see at Plymouth Hoe.[38]

Two days later Parallax replied, saying that, despite Theta's 'peevishness of temper', he would happily accept the offer, suggesting 'let us be neither Newtonians nor "Parallaxians" but simply "Zetetics" – seekers for practical truth'. And so it was decided. The scene where Sir Francis Drake had been warned of the Spanish Armada's advance while playing bowls nearly three hundred years previously would be the site of an alternative set of observations to 'decide' whether the earth was a globe.

Weeks passed and rumours abounded that Parallax would slink out of the challenge, but he arrived at the Hoe as arranged on Monday, 24 October, at noon. His arrival was witnessed by a man who was later to become one of the nation's most famous astronomy writers, Richard Anthony Proctor, who happened to live nearby. Born in 1837 to wealthy parents, Proctor had graduated from Cambridge in 1860 with a degree in theology and mathematics, and had studied law thereafter. By the time of Parallax's visit, however, he had turned to the study of astronomy to assuage his grief over the death of his first child and, having published his first article a year before, was now preparing a scholarly book, *Saturn and its System* (1866). Thus it was with some interest that he watched Parallax attempt to prove by observations on Eddystone Lighthouse, first from the Hoe and then from the beach, that the surfaces of water and of the earth were flat. He reported that weather conditions were favourable as the party prepared. A telescope was directed from the cliff towards the lighthouse, which, as both Newtonians and flat-earthers expected, could be seen in full from the lantern down to the treacherous Eddystone Rocks below.

The scene was set for the real test. The party proceeded to the beach and set up the telescope as near as possible to water level. Newtonians anticipated that due to the curvature of the earth (which was overridden by the height of the cliff), only the lantern at the top of the lighthouse would now be visible, while Parallax claimed that the entire structure, fourteen miles out to sea, would still be discernible because such curvature did not exist. With these findings in mind, both parties stepped forward to make their observations. Then came the twist: instead of the lantern being visible in its entirety, as it usually was, only half of it could be seen through the telescope. This was due to a lessening of the air's refractive power on that particular day. The phenomenon of refraction usually reduced the earth's downward curvature by about a sixth, flattening its surface and making distant objects more visible than they would be in an airless environment. However, this time, due to weakened refraction the downward dip of the globe – its curvature away from an observer – was exaggerated even more. The result was an even greater triumph for globular theory than expected. But before the Newtonians could celebrate, Parallax turned to the crowd. Expert in the art of tailoring information to suit his campaign, he claimed the peculiarity was actually an argument in favour of the idea that the earth was flat, and victory belonged to the zetetics. For hadn't the globularists said that the whole lantern should be visible from the beach, when they themselves had just shown that only half of the light could be seen? There must be something wrong with the accepted theory, he crowed.[39] The argument was unbelievable but, according to Richard Proctor, it worked. Placing the blame on inadequate science teaching and an ignorance of the laws of optics, he later regretted that many Plymouth folk left the Hoe that morning agreeing with the *Leicester Advertiser* that 'some of the most important conclusions of modern astronomy had been seriously invalidated'.[40] It seemed that the dynamics of experiment and what constituted proof could be affected as much by spin and rhetoric as by what was actually true or false.[41]

Naturally Parallax's claim to victory only deepened the controversy, and his opponents were outraged by his behaviour during the recent experiment. Over the next few weeks, letters flooded in to the *Western Daily Mercury* denouncing the so-called 'shallow conjuror' and demanding that he stage a rematch to decide the issue again. But Parallax had no intention of accepting such offers, and he threatened to sue the libellous letter-writers if he did not receive a public apology within three days. Yet still the criticisms continued: if Parallax believed his theory, why did he hide his identity? If he really was dedicated to saving the world from the darkness of the Newtonian system, why was he ashamed? The answer must be that he was a liar, who could only take refuge in bluster and prevarication when the whole of the lighthouse could not be seen from the beach. Added to this, critics offered words of advice: 'If Parallax intends to proceed against every individual who believes him to be an extorter of money, he will have to prosecute nearly every inhabitant' of the surrounding towns.[42] In reaction, Parallax retorted that he was the commander of an ever-victorious army, battling to protect Biblical truth, while Newtonianism was responsible for sinking the public into a mire of atheism by directly contradicting the word of God.[43]

As the sensation continued in the south-west, new convert William Carpenter was toiling over his printing press compiling numerous pamphlets and flyers. Eager to express his flat-earth faith, the stream of literature that issued from his newly established zetetic printing office was ample illustration of his dedication to the cause. Carpenter's first foray into the world of zetetic propaganda was *Theoretical Astronomy Examined and Exposed – Proving the Earth not a Globe*, which he published in eight parts from 1864 under the pseudonym 'Common Sense'. The name highlighted the zetetics' appeal to practical first-hand experience and down-to-earth knowledge, a brand of no-nonsense realism that led them to state that the flat-earth idea was 'common sense', whatever the claims of an 'expert' élite. Carpenter's commitment to these popularist principles was further illustrated by his efforts in compiling the work. He later claimed that night after night for three years he 'deprived himself of the enjoyment of the society of his unselfish wife and children' while he slaved over the 128-page collection, composing it in his print shop and committing it directly to type. Finally the pamphlets were finished, and reviewers could not fail to notice the 'striking arguments' they contained. For the price of a penny, readers could discover that the earth was as 'flat as a halfpenny', how astronomy was 'an old wives' fable' based on a web of meaningless equations and random guesses, and why the Copernican solar system was the 'most profound piece of presumptive idealism ever formed by man'. Worse still, Carpenter revealed, the leading lights of modern astronomy seemed incapable of fixing on any figure for planetary distances, velocities and dimensions. On the earth's distance from the sun, for example, Copernicus computed it as 3,391,200 miles, Kepler contradicted him with an estimate of 12,376,800 miles, while Newton had asserted that it did not matter whether it was 28 million or 54 million miles 'for either would do well'.

Such lack of certitude was beyond Carpenter, mere evidence to support his argument that madmen-astronomers were 'seized with a frenzy' of strange theories and estimations that 'steam printing presses could hardly keep pace with'. Yet despite such technological advance, Carpenter argued, the Astronomer Royal with 'all his mechanical appliances and skilful Assistants' could not 'prove the earth to be a globe' or that 'it moves'. With such assertions, his book went on summarily to dismiss all apparent proofs of rotundity and revolution: sailing round the earth, for instance, was in his view the equivalent of walking round London – the latter did not prove that London was a globe, so why should the earth be? In fact, Carpenter claimed, there was plenty of evidence to the contrary, such as the Allegheny mountains of Virginia, USA, which were apparently easily seen from Tenerife. Despite such proofs of the flatness of earth, Carpenter grumbled that the gullible public were still enveloped in a 'fog of mystery'. As victims of received knowledge, he supposed that had they been 'taught that the Earth was a cone or a cylinder' they would probably believe it and teach their children the same. Insisting, however, that 'working men have brains, and though they frequently seem to be out of use, they never get rusty', Carpenter had resolved to re-educate the people, fly the 'flag of reason' and 'sweep the present outrageous system to the wind and the waves'. The time for 'shilly-shallying' was past. With help from 'Common Sense', British artisans, mechanics and labourers would be forced to form their own views about the shape of the earth.[44]

Predictably the British press had mixed opinions about *Theoretical Astronomy Examined and Exposed*, for the unorthodox book seemed to defy definition, lying somewhere between Jules Verne's science-fiction classic *Journey to the Centre of the Earth* and the seventh edition of Sir John Herschel's standard textbook *Outlines of Astronomy*, both published in the same year as the first part of Carpenter's work. The reviews reflected the confusion. Bewildered, the *Spiritual Times* reported that the collection contained 'not a few ideas that cause us to ask ourselves do we stand on our head or our feet?' The *News of the World* was equally unsure:

This is a novel attempt to show that astronomers are all wrong, not only in their calculations of distances, but also about the shape of the earth. The author takes great pains to show he is right; and his readers will admire, at least, his own confidence in the new theory which he promulgates and the vigorous manner of his onslaught upon the astronomers.[45]

Carpenter's prose style certainly went down well at the *Era*, which commended 'Common Sense' for writing 'boldly' with 'sarcastic power not often met with in these calmly polite times'.[46] The reviewer must have overlooked the *Morning Advertiser*'s

assessment of Carpenter's arguments, which, it concluded, clearly
arose from 'a peculiar and exceptional mental bias'. Meanwhile, the
Observer was similarly cutting:

> The brochure attempts to show that the earth is not round,
> even though ships may sail to the west by Cape Horn and
> return by the east round the Cape of Good Hope; because
> no 'round world' is spoken of in the Bible; because the stand-
> ing order of the House of Commons has put a stop to the
> allowance of curvature in railway surveys; and for sundry other
> reasons of an equally weighty description.[47]

The *Army and Navy Gazette* was also unimpressed, pointing
out that 'the author has made a great mistake in the choice of his
nom de plume; he should have styled himself "Uncommon Sense",
for his remarks and opinions are somewhat beyond the scope of
ordinary intellects'.[48] Yet, convinced his version of truth would
reign supreme, Carpenter gloried in 'virulent attacks' from the
'venal journalist'. In response, critics received a stark warning, for
he was 'willing to submit to many sacrifices and many losses in
promoting a cause that deeply concerns the whole civilized world'.
And make a loss he did. According to zetetic legend 'all the
Scientific Societies of London' combined to crush *Theoretical
Astronomy* by threatening to ruin the trade of the publisher, Job
Caudwell, and bullying and bribing booksellers 'daring to sell such
a stinging exposure' of modern astronomy. But still Carpenter
persisted, reissuing the whole set at a price of five shillings in late
1866, thus provoking yet more mixed reviews such as the following
in the *Weekly Times*:

> There must be some reason why none of our astronomers
> have attempted an answer to this sweeping denunciation of
> the accepted astronomical theories. Either it is unanswerable
> or it is thought not worth while to answer it. If the latter,
> we think they are wrong, for anyone reading this volume of
> 'common sense' will find his faith wonderfully shaken unless

he knows considerably more on the subject than the generality of people. 'Common Sense' takes sentences from the works of our most celebrated astronomers and pins them most mercifully to facts. Facts are the things he wants . . . We are almost afraid to give an opinion on the subject, for fear we should ourselves come under his merciless logic. That the book is highly amusing we can vouch, and whoever reads it will have enough food for reflection for a long time to come.[49]

Parallax, meanwhile, was developing publication plans of his own. Believing the time was ripe for a full written exposition of his theory, he published *Zetetic Astronomy: Earth not a Globe!* in 1865. The 221-page book was subtitled *An Experimental Inquiry into the True Figure of the Earth proving it to be an Immovable plane*, highlighting the centrality of mathematical evidence and practical demonstrations to Parallax's unorthodox public campaign. In fact, his extensive compilation of scriptural proofs for the flatness of the earth was reserved for the concluding chapter, where he held forth in warfare mode on the rise of infidelity (or unbelief) in England and his intention to defend the word of God as literally revealed by the Bible.[50] The tactical approach highlights his assumption that science would possess more power to persuade than Biblical literalism; scriptural proofs upfront or alone would not have sold the subject so effectively to readers. Although no information remains about the reception or sales of the first edition of *Zetetic Astronomy*, the eminent London publisher Day & Sons issued a second revised and enlarged edition in 1873. Again including in-depth descriptions of various experiments at different locations, Parallax clearly intended to further his reputation as a trustworthy investigator who was willing for readers to replicate his tests if they wanted to gather direct proof for themselves.[51] Building on these efforts to establish legitimacy, his 1873 edition was an extended collage of quotes and calculations from a vast range of authorities, cut and reconstructed to create a potentially convincing and seemingly well-researched case, further supported by lecture reviews from the local and national press with criticism edited out. As a whole, the *Bookseller* reflected, the subject was 'treated with considerable ability' in the second edition of *Zetetic Astronomy*. Whether readers purchased the book for its humour value, and what they made of it, are unknown; nevertheless, demand was sufficient to generate a further edition in 1881 and, one imagines, a sizeable additional income for Parallax.

While Parallax disseminated his ideas in print, his lectures continued apace – most notably with a visit to Gloucester in October 1865. As usual, he was greeted with anger and disdain by scientifically minded locals, and several citizens wrote panic-stricken letters to the Astronomer Royal reporting their concern about the public understanding of science. Amid warnings from the *Stroud Journal* that no ordinary man was a match for Parallax, several Gloucester men decided to get the full facts direct from George Biddell Airy to use to their advantage in the upcoming debate.[52] They realized that a letter from a distinguished authority would have more impact in a public forum than a textbook quotation, and a few complained of experiencing difficulties in locating or understanding the details needed in scholarly books anyway. Among such correspondents was Westgate Street resident William Pumfrey, who wrote 'in haste' on 14 October, apologizing for troubling the Astronomer Royal 'in the cause of astronomical science':

> Parallax is here attempting to prove that the *earth is a plane* and other such like absurdities. I mentioned the variations of the pendulum in different latitudes as being fatal to his theory. He replied by stating that the variations are entirely caused by temperature and that a compensating pendulum is absolutely uninfluenced by any change of position on the earth's surface. I have no means of contradicting him, as I should like to do on Monday evening next . . . If you can kindly place me in a position to do so, many others in the room besides myself will feel very grateful; otherwise he will certainly have the advan-

tage over us. He also states that on the 23rd January 1862 the Pole Star was seen from a part of the earth 23¹/₂ degrees beyond the equator. If you can contradict this statement I shall feel greatly obliged.⁵³

With some impatience, Airy told his secretary to give Pumfrey a 'simple hint' but warned 'we cannot waste our time thus'. Pumfrey was duly informed that the changes in length of a pendulum's rotation in different latitudes had been most carefully demonstrated, providing ample proof that the earth was a rotating globe. As for Parallax's claims that the Pole Star could be seen at 23.5° south of the equator, that was, quite simply, impossible, due to the earth's curvature.

After Gloucester, more lectures followed in Warrington and Liverpool and, by May 1867, Parallax had move on to speak at Leeds Philosophical and Literary Society, the premier venue for intellectual discussions in the surrounding region. An arch-self-publicist, well versed in the art of promotion and propaganda from his days as a political agitator, Parallax posted advertisements in advance according to one local man, which placed 'the scientific in the Town . . . in a state of great excitement'.⁵⁴ As a consequence, on the evening in question, Parallax was greeted by a 'perfect storm of hooting and hissing' intermingled with cries of 'Down with Parallax' and 'Throw him from the window', but he informed the audience that he had been told of a secret plot to 'put him down' and drive him out of town, so the police had already been sent for. A consummate showman, Parallax worked through his syllabus with the aid of diagrams, most notably of his experiments on the convexity of water on the Old Bedford Canal. He was then interrogated by various audience members, including a local builder. One man asked how, if the earth was a plane, mariners could keep sailing westward yet return to the point where they started. Parallax replied that seamen always sailed square to the compass so they had no trouble sailing round the earth as a circular plane. The debate continued but, according to the *Leeds Times*,

Parallax was 'as slippery as an eel' and 'always discovered the strongest confirmation of his doctrines even in their most complete refutation'.[55] Eventually the chairman, a local solicitor, stepped up and asked Parallax for his name. The crowd jeered and heckled, but Parallax refused to answer: he would give his name, he said, if the audience could prove that the earth was a globe. The unruly proceedings then rose to fever-pitch when another audience member got to his feet and read out a letter from the Astronomer Royal. It concerned Parallax's assertions that George Airy had attended his lecture at Greenwich and announced that 'the Newtonian theory was not sufficient, and John Herschel [believed] we were in a "hodge podge" with regard to these things'.[56] Obviously, Airy denied these claims in no uncertain terms.

Yet Parallax remained utterly unflappable and unfailingly polite, whatever the evidence presented.[57] Despite his determination to flout traditions of honour and authority by concealing his true identity, he had still appeared to be a trustworthy source of authoritative knowledge to the ill-informed. To some his persuasive and ingenious arguments had seemed plausible, especially as they were bolstered by a dapper appearance, gentlemanly persona and formidable debating technique. Throughout his career, Parallax was careful to conform to social norms of respectability, and in an era when education was not yet compulsory, his rhetorical strategy could be perceived as second to none.[58] As one correspondent, 'Fairplay', later informed the *Leeds Times*, this situation was exacerbated by the shortcomings of better-informed audience members:

> Without endorsing 'Parallax's' teachings, it must be said that he advanced them, supported them, and fought for them with a skill and intelligence, tact, and good temper which were not at all equalled by his opponents. One thing he did demonstrate was that scientific dabblers unused to platform advocacy are unable to cope with a man, a charlatan if you will (but clever and thoroughly up in his theory), [who is] thoroughly alive to the weakness of his opponents.[59]

Fairplay continued that although audience members 'expressed themselves noisily and indignantly' against Parallax, his theory remained intact. All in all, the controversy illustrated how scientific knowledge was made and contested by a range of people in a broad social context, and there were no guarantees that the globe idea would win through, despite its verifiability and truth.[60] Although audience members were not passive recipients of knowledge, it seemed that less technically minded onlookers did not necessarily reject his highly unorthodox claims out of hand.[61]

Highlighting the collective nature of the scientific enterprise at this time, some better-educated audience members decided to act.[62] Expressing concern that standard textbooks went over the heads of the masses, clergymen, schoolmasters and navy captains published books condemning Parallax's theory and explaining the true astronomical facts.[63] No such publications were forthcoming as a result of Parallax's lecture in Leeds, but his promise to return to the city for a public debate at the Stock Exchange induced several audience members to emulate Gloucester citizens by writing to the Astronomer Royal. One, a local wine merchant George Scott, confided, 'I do not for one moment believe that he believes his own theory. But unfortunately in Leeds he has got many converts – and amongst people one would least expect.'[64] Retaliation was thus in order, and George Airy was faced with a barrage of questions as a consequence – 'Has a vessel ever circumnavigated the globe on a single latitude south as high as Cape Horn, and if so, what was the distance in nautical miles?', 'Have careful attempts been made with very powerful telescopes to view vessels appearing over the horizon?' and so forth. A dutiful man, Airy replied to the public's letters without fail, but could not resist restating his astonishment that 'such questions should be asked in any civilised country'.[65] In this way, the public understanding of science continued to be a cause for concern for people of many different backgrounds and ranks.[66]

This issue was a matter of especial concern to Airy throughout his forty-six-year career as Astronomer Royal. Victorian Britain saw the rise of earth-flatteners, perpetual motionists and the like, who took advantage of improved educational opportunities, affordable reading material and the new penny post to develop and communicate their unorthodox ideas. Sometimes living proof of the adage 'a little Learning is a dang'rous Thing', the misguided could exploit various channels to disseminate their so-called 'wild' notions – and the development of specialist societies and the role of 'public scientist' offered a more visible yet unwilling audience for their seemingly relentless endeavours. Airy, for one, filed letters from flat-earthers, fortune-tellers and their kind on a shelf marked 'My Asylum for Lunatics'.[67]

By contrast, the professor of mathematics at University College London, Augustus de Morgan, delighted in the blunders of so-called 'unhappy enthusiasts'. A lover of puns and puzzles, he coined the term 'paradoxer' in the 1850s to describe proponents of strange ideas that 'deviated from general opinion, either in subject-matter, method or conclusion'. A vague definition, which if used literally would apply as much to Copernicus as to Parallax, de Morgan's term took hold and he became well known for his wrangles with paradoxers in public and in private correspondence. According to astronomy writer Richard Proctor, however, de Morgan 'bore the shower of abuse' from flat-earthers with 'exceeding patience and good nature', and plundered their publications for amusing articles for the *Athenaeum* and his posthumous bestseller, *A Budget of Paradoxes* (1872). De Morgan's work was well received and magazine editors quickly followed suit with similar stories, keen to win readers with entertaining material in the booming and competitive popular-science market. Following de Morgan's lead, new columns, 'corners' and 'braces' of paradoxes, appeared in publications from cheap mass-circulation papers to the more highbrow journals.

Besides humour value, editors and writers had other motives in covering 'alternative' subjects. Some, such as Richard Proctor, saw confuting paradoxes as a useful exercise for those learning basic astronomy, pointing out that 'nonsense-mongers' could act as 'foolometers' for the better informed. He realized, however, that proponents of strange theories could prove a danger to scientific dabblers and general readers, and placed the blame for their success on the authors of badly written books. All it took, Proctor complained, was a half-understood explanation or a carelessly worded account, and the potential 'paradoxer' was formulating a novel theory on any given subject. Driven by over-confidence, once the paradoxer had devised his theory it seemed to take complete possession of his mind, leading him to use any available fact with the least bearing on the topic to fit his theory.

Unfortunately for astronomers, ideas about the solar system seemed to possess an inherent appeal. One reason for this, Proctor believed, was that paradoxers had an over-developed sense of their own importance, matched with a tendency to think big. Another was the way in which the history of astronomy was presented by the authors of scientific books. Usually, he noted, mankind's knowledge of the solar system was simplified to such a degree that the reader was presented with a neat series of theories, from Ptolemy to Copernicus to Kepler, with one superseding another over time. This created the impression that human knowledge of the solar system was entirely provisional and transitory, and that was quite enough for the paradoxer – why then shouldn't he be the next Newton who would revolutionize the scientific world? It was a phenomenon that rankled with Proctor: he resolved to address the paradox problem head-on.

While men of science puzzled over paradoxes and their origins, Parallax continued his quest. In 1867, he descended on Bradford and Dewsbury before crossing the Pennines to cause a commotion with a series of lectures in the North-west. But wherever he went, the same line of argument followed: science and the scriptures were at war and both could not be right. Such rhetoric was a timely propaganda tactic in the wake of Darwin's *On the Origin of Species*, published in 1859. Although the majority of Victorians were able to incorporate the concept of evolution and its challenge to a literal interpretation of the Bible into their world-view without suffering a dramatic crisis of faith, from 1870 secular writers, such as Draper and White, were beginning to promote the idea of bloody warfare between science and Christianity. Their purpose was largely political: to show how religion had stifled science, and the Church's supposed promotion of the flat-earth idea played an important role in this storyline.[68] Yet as Draper rewrote history to create a powerful conflict myth, Parallax was employing the warfare thesis to serve the opposite goal: to promote the idea that the earth was flat and the Bible was literally true in its statements regarding the natural world. Christianity, he claimed, was being destroyed by science and his campaign was an attempt to right the wrong. Thus, over the last decades of the nineteenth century polemicists for and against Christianity were using the flat-earth idea as an ideological weapon to serve very different tactical ends. As far as Parallax was concerned, the people of England had a choice: to side with those who upheld Biblical truth or to believe the man-made fantasies of science.

Such statements were both controversial and powerful, particularly in the intellectual climate of late-nineteenth-century Britain, when traditional religious beliefs and practices were undergoing noticeable shifts. In a period when traditionalists were gravely concerned that society was becoming more secular, Parallax's attempt to restate a literal interpretation of the Bible was designed to have an appeal. Aware of the power of this approach, he also used Biblical authority to promote his ideas about health and longevity in another book published around the same time. Entitled *Zetetic Philosophy: Patriarchal Longevity, its Reality, Causes, Decline and Possible Re-attainment*, it insisted that because the human body was the perfect work of God, and Old Testament figures such as Adam and Abraham had lived for nearly a thousand years, people were capable of living for an equivalent period if they followed his special diet.[69]

Meanwhile, the ever loyal Carpenter was trying to raise interest in his own publications and had decided to send an extract from *Theoretical Astronomy* to the Astronomer Royal.[70] By this time, he was still in possession of a large stock of unsold copies of his book, although the situation was to alter dramatically when it came to the attention of Christian polemicist John Hampden. Born in 1819, Hampden had been blessed by circumstance: he was the eldest son of the Protestant rector of Hinton Martell in Dorset, and with a substantial private fortune to match his privileged background, his prospects had looked particularly promising.[71] Following family tradition, on St Valentine's Day 1839 he had gone up to Oxford to partake of the standard undergraduate fare – a mix of classics and divinity – at St Mary Hall. The college was the obvious choice: once a medieval hall of learning, his father's cousin, Renn Dickson Hampden, was principal, as well as regius professor of divinity and canon of Christ Church.[72] With such family connections, admission was straightforward, and Hampden entered as a 'gentleman commoner', a privileged undergraduate who paid higher fees in return for a silk gown, immunity from certain lectures and a better seat in the dining hall.

Yet despite his advantages, two years into his studies Hampden had left the university without finishing his degree, thus abandoning an education seen as essential for a gentleman and a prospective clergyman in mid-Victorian England. Parish, pulpit and prayer book rejected, Hampden decided instead to spend his time publishing an assortment of tracts calling for reform of the Church of England on strict Protestant lines. When it came to financing his writing career, he faced no constraints. Various relatives had made substantial bequests to his family and when his father died in 1845 he inherited everything with his brother, also Renn. There was just one hitch. His father's will specified that if Hampden were ever to degrade the family name, by making an unsuitable marriage or by any other act, he would forfeit his inheritance and be forced to survive on a bare subsistence allowance of fifty pounds a year. It was an issue that would later come to the fore.

By the time Hampden read *Theoretical Astronomy Examined and Exposed*, he was living on private means. Crucially, he was so impressed with Carpenter's sweeping denunciation, believing it to be 'one of the most able displays of genius, perseverance, and intelligent acquaintance with the scientific literature of the day that Europe or America could boast of', that he paid him £100 for the copyright and hatched plans to reissue the pamphlet. Through 1869, Hampden determined to investigate the issue further and read everything he could find on zetetic astronomy, including Parallax's *Zetetic Astronomy: Earth not a Globe!*. The book was a revelation, for it seemed consonant with his interpretation of Biblical descriptions of the earth and his concerns about rising 'infidelity' in British society. In reality, by 1870 most Victorians accepted the Book of Genesis allegorically rather than literally, notwithstanding sporadic public debates about evolutionary theories. But Hampden, an arch-conservative, did not see things that way: he was bent on defending Genesis to the hilt, and the flat-earth campaign seemed the perfect way by which to restate the authority of the Bible in the face of the perceived scientific attack.

As Parallax had hoped, zetetic astronomy had attracted support from an extreme Biblical literalist; in fact, his two most prominent converts emphasized the two-pronged nature of the zetetic campaign. Hampden, a Christian polemicist, focused on scriptural proofs of the earth's flatness, while Carpenter, a mesmerist and spiritualist, was keen to challenge élitism in knowledge with the zetetic doctrine of finding out for oneself. Alongside their willingness to believe that the earth was flat, Hampden and Carpenter shared another characteristic across the class divide: they both had a profound contempt for authority and religious and epistemological élites. Parallax suited their purposes well. The flat-earth 'fact' could be used to underpin a broad programme of social reform, flagging the way to the new type of society they sought.

Under these circumstances, it was not long before Parallax received a letter from Hampden, then living in Swindon, saying that he had become an immediate convert and asking for more information. From that time forward, Parallax received letters on an almost daily basis from his newest disciple, proclaiming that he was so completely convinced of the truth of flat-earth theory that he would do anything within his power to prove, beyond dispute, that the earth was not a globe. Hampden certainly kept to his word. With no further delay, he bought the rights to *Zetetic Astronomy* in late 1869 and launched his own campaign.

His first move was to issue a forty-page pamphlet, *The Popularity of Error and the Unpopularity of Truth: Shewing the World to be a Stationary Plane and not a Revolving Globe*, including extracts from Parallax's work, which he followed with a new series of self-penned tracts. More in line with Carpenter's frenzied rhetoric than Parallax's pseudo-academic style, these broadsides proclaimed that science was a 'medley of fancies and falsehoods', based on the hallucinations of a few crazy enthusiasts who had been telling the most astounding lies for five hundred years or more. Above all, Hampden identified Isaac Newton as the chief culprit, alleging he must have been 'in liquor or insane' to have invented 'such preposterous theories as Rotundity and Revolution, Gravitation and Attraction from the fall of an apple'. Victorian men of science fared little better and Hampden denounced them as a 'parcel of brainless jackasses' full of miserable twaddle. Meanwhile, he branded geology, geography, astronomy and natural history 'bungling absurdities', which should 'be rejected by all lovers of truth in the world' because 'demented star-gazers' and 'moonstruck mathematicians' should not be allowed to dictate. It would be just as easy to believe that the earth had been 'made by fairies' as to accept it was a globe, Hampden contended, for he was convinced that the earth was a disc bordered by an immense barrier of ice. The North Pole lay at the centre, hell festered on its underside and the South Pole was nothing but a vicious myth, the invention of 'half-witted, well-paid journalists and schoolmasters'. In fact, Hampden branded all efforts to circumnavigate the earth 'fools' errands', while the idea that 'ships and bedsteads, elephants and bishops' were all speeding round the sun at a thousand miles an hour, like 'squirrels in a cage or felons on a treadmill', was an assertion of the 'utmost idiocy'. As for common proofs of the earth's rotundity and revolution, such as day, night, sunrises, sunsets, the seasons and circumnavigation, Hampden dismissed them all as the result of a special zetetic law of perspective, 'optical delusion', 'eyesight and eyeglass failing' or lies concocted by 'expert' astronomers.

Despite the unorthodoxy of his ideas, Hampden pinned his faith on these 'grand facts'. In numerous tracts, he branded refusal to accept zetetic ideas a disgrace in an age of supposed reason and progress, blaming public ignorance on men of science who 'possess far less sense than the cattle in our fields'. Hampden even claimed that their 'spurious, bastard theories' had placed the reputation of the nation at stake, believing that when it was revealed that the earth was not a globe, England would suffer a humiliation greater than 'surrendering the keys of her capital to a foreign foe'. From his perspective, 'astronomers and opticians' had 'never made a single experiment the truth of which can be incontestably proved'; they supported their 'insane theory' because it made 'thoughtless blockheads stare with amazement!' In the belief that scientific professionals were exploiting the credulity of a populace that dared not question their expert knowledge, he distributed flyers in the streets of London, Chippenham and Swindon, imploring working men not to listen to the professors who had misled them 'for centuries past'.

Appealing to class interests, his leaflets assured the general public that it was as 'easy for you to count the lions at the base of Nelson's column' as it was to 'determine the shape of the earth beneath your feet'. Set on penetrating all ranks of society with his campaign, he also called on the directors of engineering firms, scientific societies, geographical and literary institutions to support his unorthodox cause. Even the long-suffering Astronomer Royal, George Airy, could not escape his ire, and received demands for a government Royal Commission to investigate the globe question with immediate effect. By 1870, Hampden was panicking that Britain was in the grip of a heinous conspiracy involving the press, the pulpits and the platforms of learned societies, all of whom were in league with science. He was particularly scornful of journalists, identifying them as the most enthusiastic scientific accomplices, bribed in secret to disseminate glaring fallacies about the shape of the earth.

Hampden's views were timely for, in the booming and competitive world of Victorian publishing, writers were only too keen to exploit new scientific developments, much to the delight of a growing readership from all classes of society. In the decades following the Great Exhibition of 1851, British technological innovation was guaranteed to capture the public's imagination while such stories were welcomed by journalists looking for material with which to fill the popular magazines that proliferated following the development of steam-powered printing presses earlier in the century. As a result of high-speed mechanized production and the removal of stamp duty on certain types of publication, periodicals had become a vital channel for the dissemination of scientific discoveries and original concepts. In formats from natural history to mechanics' magazines, journalists and popular writers sought to extend the frontiers of knowledge to embrace general readers, artisans and amateurs alike, and science-related stories filled papers from gentlemen's magazines to penny tracts of varying levels of complexity and tone.[73] Matching the onslaught of so-called 'steam intellect', astronomy had long been a popular topic, due in part to its awe-inspiring charm as well as its status as an accessible, do-it-yourself science that anyone could practise at home.[74] As the market blossomed, media stories about the solar system were widespread and had an inherent appeal. Hampden watched as the media reported gossip from Greenwich and ground-breaking hypotheses, explorations and discoveries, in turn fuelling the national craze for amateur star-gazing. He was appalled by such developments: to him the conduct of journalists was 'sheer impudence'. He protested that reporters were nothing but the 'illiterate sycophants' of science, who had 'as much acquaintance with the Chinese alphabets' as they did with 'the merits of the Newtonian system' – and he wrote to tell them so.

Word of Hampden's campaign soon reached Parallax, and he decided to act. Hampden's behaviour was entirely at variance with Parallax's gentlemanly style, and he believed that the other man's actions were liable to do more harm than good. He recognized that civility was central to establishing credibility and he wrote informing Hampden that it was their duty to treat their detractors 'with respect and consideration', for zetetics must 'seek to uneducate or educate afresh [and] not to denounce and abuse'. 'All men wish to be right in their convictions,' Parallax continued, 'and do not wilfully cling to error'; therefore Hampden's 'style of advocacy' was 'persecutive, unjust, and injurious' to the cause it was intended to serve.[75]

But Hampden ignored his advice and refused to reply to his letters. No longer could he stand aside while the minds of schoolchildren and the public at large were corrupted by the 'evasive subtleties' of scientific men and their conspirators. He swore that he would 'rest not day or night' until he witnessed the overthrow of globular theory. 'Go it shall!' he declared. 'All further resistance is useless.' With this proclamation, he resolved to force men of science to substantiate the idea that the earth was a globe with convincing proof or confess its absurdity once and for all.

Chapter Three

THE INFAMOUS
FLAT-EARTH WAGER

Shall we, then, any longer submit to be fooled by an
infidel science, which has for centuries forced us to
acquiesce in the impious hallucinations of a few crazy
enthusiasts, whose proper asylum would have been a
madhouse had not their dupes been as insane as themselves? Let this groundless fraud be at length resisted,
and let our children no longer be taught that we are
spun through the air like cockchafers, at the rate of
thousands of miles an hour.

JOHN HAMPDEN
The Popularity of Error and the Unpopularity of Truth (1869)

THE FRENZIED TONE of Hampden's campaign was undoubtedly
amplified by the fact that by the 1870s science was professionalizing at a quickening pace and, in a number of ways, was far in
advance of where it had been just half a century before. Even the
word 'science' had not come into use until the 1830s, replacing
'natural philosophy', while 'scientist', albeit coined in the same
decade, was not widely employed as an alternative to 'men of
science' until the end of the century. In the first decades of the

nineteenth century, there had been no such thing as professional science, with an identity, education or culture of its own. Research was dominated by gentlemanly amateurs and the Anglican clergy, based at Oxbridge or in rural parishes, and was guided by the spirit of natural philosophy, the study of nature to reveal God's wisdom, benevolence and skill.

By the 1870s, however, the situation was shifting noticeably. Compulsory state education was introduced in England in 1870, and around the same time provincial universities began to take off as a supplement to Oxford, Cambridge and London, developments that were to exert an impact on technical education, job opportunities and the public understanding of science, much to the zetetics' disdain.[1] While science was now being disseminated more broadly through society by a plethora of books, magazines and specialist journals, along with a growing network of clubs and societies, specialization between subdisciplines or 'departments' was becoming more apparent, as was the status of men of science, who were gradually assuming more authoritative public roles. That said, the impact of these developments became truly evident even later in the century: in 1870 job opportunities were still limited to a few universities, and popular science writing was seen as a distinct activity with special skills, not necessarily the province of men of science.[2] Indeed, what constituted a man of science was still vague, since without clear channels of education and training, job opportunities and occupational advancement, professional boundaries and definitions remained fluid and uncertain. This was an obvious cause of concern for men in the field, particularly those from middle- or lower-class backgrounds: it was almost impossible to practise science to a high level without an income or a job.

Nevertheless, by 1870 aspects of the modern scientific profession were slowly and unevenly becoming established, while barriers to progress were under attack from a coterie of ambitious young scientists who were seeking to wrest influence from the established intellectual élite. Determined to release science from its links with aristocratic interests and the Anglican Church, and win social and intellectual authority for themselves and their peers, the group – represented most notably by Darwin's defender, Thomas Henry Huxley, and his co-conspirators in scientific dining coterie, the 'X-Club' – set about promoting a number of distinct professional goals. Chief among these were moves towards forging connections with government, industry and educational institutions in order to gain paid positions, research facilities and cultural standing for men of science at a time when Britain was transforming into a modern industrial nation, the so-called 'workshop of the world'.

This was a golden opportunity for science and its practitioners to stand at the forefront of important innovations on a national scale, while in retrospect it seems that considerations of cultural and occupational prestige were further underpinned by a certain ideological tendency. Now known as 'scientific naturalism', this was the assumption that science was the most legitimate road to absolute truth, founded as it was on rigorous empirical verification, and that only causes directly observable in nature could be invoked to explain how the world worked. Naturalism meant that God and the supernatural were ruled out of the equation in terms of explaining natural processes, for Huxley and his accomplices were determined to eschew natural theology and its clerical supporters, and establish science on an independent footing as a profession and a body of knowledge distinct from the Bible and the Anglican Church. Focused and shrewd, these men were keen to overthrow the old order and challenge the religious, political and social establishment with a power-base of their own. One way to do this was to separate God from nature by highlighting what Huxley called 'New Science' – the accumulation of hard evidence and an emphasis on natural laws.[3] Yet even though Huxley *et al.* sought to sideline 'Parsondom' with an aggressive campaign to win authority in knowledge, they still did not argue that science and religion were at war. Instead they declared that science and religion were two separate spheres, each with its own role to play in different aspects of human life.[4]

The secular tendency 'scientific naturalism' stood in stark contrast to the ideology proposed by the zetetics, who, like modern creationists, were using the authority of orthodox science to promote a broadly religious programme. In the face of such developments, Parallax and Hampden attempted to counter-attack the emerging professional status of scientists, a phenomenon that only accelerated as the century progressed. For, strange as it may seem, Parallax was also attempting to claim cultural authority with his radical theories, co-opting the rhetoric and prestige of science to bolster his campaign to promote flat-earth theory and the larger cause of Biblical truth. Within this scheme, he recognized that a gentlemanly demeanour was a critical aspect of establishing himself as a trustworthy purveyor of knowledge, and this was a tactic that Hampden had lately and very publicly undermined.[5]

In fact, following his recent complaints about the tone of Hampden's campaign, the stream of letters and questions from Swindon had suddenly dried up. Parallax claimed that he was puzzled by this uncharacteristic silence, never guessing that Hampden might have dared to embark on a new strategy to disseminate zetetic astronomy without his approval, support or advice.[6] But that was precisely what Hampden had done. Inspired by the strength of his new-found convictions, on Wednesday, 12 January 1870, he placed a notice in *Scientific Opinion*, a new, wide-ranging weekly journal that plugged the 'cause of science and the interests of scientific men in England'. The advertisement declared:

What is to be said of the pretended philosophy of the 19th century, when not one educated man in ten thousand knows the shape of the earth on which he dwells? Why, it must be a huge sham! The undersigned is willing to deposit from £50 to £500, on reciprocal terms, and defies all the philosophers, divines and scientific professors in the United Kingdom to prove the rotundity and revolution of the world from Scripture, from reason or from fact. He will acknowledge that he has forfeited his deposit, if his opponent can exhibit, to the satisfaction of any intelligent referee, a convex railway, river, canal or lake. JOHN HAMPDEN.[7]

The notice doubtless raised eyebrows among the paper's educated readership, who were more accustomed to debates about whether the earth was an oblate or a prolate spheroid than whether it was round or flat. It stood in particularly stark contrast to the dry news from government departments and international scientific societies that filled the neighbouring columns. The offer was so unorthodox, in fact, that one anonymous correspondent in the best-selling science magazine, the *English Mechanic*, complained that Hampden 'deserved the horse-whip' for his impertinence, while another letter-writer recalled: 'The challenge was looked upon as a skit; no one supposed that there was anyone sufficiently foolish to have meant it, and few true men of science would condescend to literally plunder the poor creature.'[8] However, in an era of growing professionalization, when codes of conduct were still being carved out by men in the field, Darwin's co-discoverer of natural selection, Alfred Russel Wallace, proved willing to take up the bet.

At the time of Hampden's challenge, the forty-seven-year-old naturalist was busy at his London home, near Regent's Park, compiling a new collection of essays, *Contributions to the Theory of Natural Selection*, which bore testimony to Wallace's most famous moment to date. In February 1858, while in a malaria-induced fever on Ternate, an island in the Dutch East Indies, he had independently hit upon the theory of transmutation (evolution) by natural selection, thus offering a long-sought solution to the question of how species changed.[9] Eager for feedback, he had copied out his account and sent it to the older, better-established naturalist Charles Darwin, who, unknown to Wallace, had been developing similar ideas in his private notebooks over a period of twenty years. When Darwin read the letter, he rightly feared that all his originality would be 'smashed' if he passed on Wallace's paper as suggested. Anxious to act honourably, he consulted his trusted confidants, the wealthy geologist and author of *Principles of Geology* Sir Charles Lyell, and the botanist Director of Kew Gardens, Joseph Hooker. Hurriedly, in Wallace's absence, they agreed a gentlemanly compromise. It was decided that the findings should be announced jointly at the prestigious Linnean Society of London, while Darwin set to work at once on his own short monograph, which would describe the theory and his detailed research. Following thirteen months and ten days of 'hard labour', the result, *On the Origin of Species*, was rushed into print in November 1859. From a hut in the tropical jungle, Wallace wrote to his friend Henry Walter Bates, an entomologist with whom he had travelled, recording his reaction to this chain of events:

> I do not know how or to whom to express fully my admiration of Darwin's book. To him it would seem flattery, to others self praise; but I do honestly believe that with however much patience I had worked up and experimented on the subject I could never have *approached* the completeness of his book – its vast accumulation of evidence, its overwhelming argument, and its admirable tone and spirit. I really feel thankful that it has not been left to me to give the theory to the public. Mr. Darwin has created a new science and a new philosophy, and I believe that never has such a complete illustration of a new branch of human knowledge been due to the labours and researches of a single man. Never have such vast masses of widely scattered and hitherto utterly disconnected facts been combined into a system, and brought to bear upon the establishment of such a grand and new and simple philosophy![10]

Having motivated Darwin to publish his findings, Wallace remained content to have made a contribution to knowledge and acceded complete credit to Darwin for the rest of his life. Following *Origin*, a decade passed before Wallace decided to gather together his own ideas on natural selection in a popular collection of essays. With this volume, he hoped to capitalize on the success of *The Malay Archipelago*, his critically acclaimed account of his eight-year collecting expedition in the area, published a year earlier.

Indeed, by 1870, the success of this next book was a necessity for Wallace was labouring under financial constraints, despite his reputation and well-known research. Because science was still being established as a profession, jobs were scarce outside the confines of a few universities; Wallace had never gained a degree and disliked public speaking in any case. A self-taught naturalist, he had left school at fourteen and worked in a succession of jobs from surveying to teaching, until he had decided to pay his own way as a travelling specimen collector in the Amazon and the East Indies. Although he was from a cultured background, his family had fallen on hard times, but his work as a land surveyor and visits to Halls of Science and Mechanics' Institutes in his youth had brought him into contact with a number of radical working-class ideas, from Owenite socialism to phrenology. His background and circumstances were very different from those of gentlemen-naturalists Darwin and Lyell, and were central both to his intellectual interests and his financial state. Unlike many of his colleagues, Wallace did not have a private income to fall back on, while the returns from book royalties and the sale of his collections of tropical birds and butterflies were modest. In December 1868 Wallace had resorted to taking a part-time job as a schools examiner in geography to make ends meet, and by 1870 he held high hopes for a full-time post as founding director of the new Bethnal Green Museum for art and nature in London's East End. Under such circumstances, Hampden's ample stake of £500, an annual salary for a middle-class Victorian, was tempting and, more to the point, unlike many men of science, Wallace was not discouraged by the seemingly unorthodox nature of the scheme.

In fact, since his co-discovery of natural selection in 1858, it had become increasingly apparent that Wallace was an independent thinker who was not afraid to voice views that might diverge from those of the majority of his peers. Over the past five years, this maverick tendency had become particularly apparent in his public defence of spiritualism, the belief that the soul can survive the death of the physical body and that mediums can contact the deceased.[11] The modern revival of this belief began in Hydesville, upstate New York, in 1848, where two young sisters, Kate and Margaretta Fox, had allegedly developed a system of raps as a way of communicating with the spirit realm. The spiritualist movement spread quickly across the United States and to Britain, where a fascination with clairvoyance, table-turning, levitation and the materialization of spirits kept sitters transfixed at séances nation-wide.

During the 1850s, various phenomena had been investigated and publicly debunked by leading men of science yet, as with mesmerism and phrenology, Wallace found spiritualism intriguing and elected to investigate further when he returned from the tropics. He first attended a séance at a friend's home in 1865, followed by private sittings with London society medium Mrs Mary Marshall. However, his spiritualist beliefs were fully confirmed in autumn 1866, when his sister Fanny discovered that her young lodger, Agnes Nicholl, appeared to be in possession of similar mysterious powers. From this point forward, séances became regular events in Wallace's North London parlour. According to reports, flying objects, quaking furniture, spirit entities, even lush displays of fruit, ferns and flowers summoned fresh from the ether were frequent sights. For Wallace, most fantastic of all, the somewhat stout and 'very heavy' Agnes on one occasion levitated 'noiselessly' up to a chandelier right before his eyes. Astounded, he could not wait to reveal these experiences to the public, and quickly published a magazine article (later a pamphlet), 'The Scientific Aspect of the Supernatural', on his investigations into 'forces and influences not yet recognized by science'. Without doubt, Wallace believed he had amassed hard evidence of the existence of a spiritual plane, 'a new branch of Anthropology' no less, and he was determined to interest his scientific colleagues in further in-depth investigation.[12] While some men of science expressed an interest and continued to undertake experiments themselves, many

– including Huxley and Hooker – were set on ruling the paranormal out of bounds. Nevertheless, Wallace had a different definition of the boundaries of science and was determined to retain a place for religion, despite the disapproval of some, but not all, of his peers.[13]

Through the late 1860s, Wallace continued to earn scorn for his interest in what Huxley labelled 'disembodied gossip', along with his involvement in paranormal investigations, such as the London Dialectical Society's full-scale public inquiry into the matter, which began in 1869. But most outrageous of all for some men of science, in April of that same year Wallace had dared to link spiritualism to biology, publishing a claim that a mysterious 'higher intelligence' had influenced the development of humankind.

Wallace believed that natural selection could not account for our intellectual and moral faculties and certain physiological features: a higher power must be at work. While natural selection was far from being accepted wholesale by his peers (even Darwin had advanced other natural causes such as the principle of sexual selection, as a supplement), Wallace's adherence to religious causation was an anathema to those who were using the theory of evolution by natural selection as an ideological weapon against the Church.[14] For his part, Darwin was astonished by the claim, writing that he would have been sure that the offending lines had been added 'by some other hand' if Wallace had not explained his article in advance. Although Darwin's personal religious views were themselves in flux, spiritualism and related ideas had no place in his definition of science; he saw human beings as the product of natural laws and nothing more. From his perspective, Wallace had tarnished their theory with a supernatural explanation: he had 'sought to murder [their] child'. He continued, in a letter to his co-founder, 'I differ grievously from you and am very sorry for it.'[15] By contrast, in an era when what constituted 'science' was still under negotiation, Wallace took a broad-minded approach. Although he was certainly not alone in finding a place for religion in science, he was already bordering on the 'crank' to some people's minds.[16] Hence, by the time he considered the Hampden wager, Wallace was suspected in certain quarters of bringing 'science' (as defined by an influential few at least) into disrepute.

With controversies over science and spiritualism in mind (and like Darwin before him), Wallace sought advice about the flat-earth wager from his sometime mentor, the worldly wise geologist Sir Charles Lyell. Having given the matter some consideration, Lyell decided in favour of the wager on the grounds that providing plain proof of the rotundity of the earth 'may stop these foolish people'.[17] On reflection, Wallace was inclined to agree: it seemed to him that 'a practical demonstration would be more convincing than the ridicule with which such views are usually met', and public education was undoubtedly a worthy justification for his involvement.[18] A ritual refutation of the flat-earth idea would, he hoped, be a case where actions would speak louder than words, for surely if Hampden could have an opportunity to see the proof for himself he would finally be convinced that the earth was a globe.[19] Doubtless the wager also appealed to Wallace's dry sense of humour, and he decided to write to Hampden picking up the flat-earth gauntlet. Three days later, he composed another letter, to his close friend, the professor of zoology at Cambridge University, Alfred Newton, confiding that he had taken on a 'heavy wager' with 'one of those strange phenomena' who do not believe in the rotundity of the earth and 'who is willing to pay to be enlightened'.[20]

While Wallace contemplated the educational benefits of the flat-earth bet, the popular astronomy writer Richard Proctor had selected a different course of action: dishing out free enlightenment through the pages of the best-selling magazine the *English Mechanic*. Having lost his fortune in 1866, when the bank where he was a major shareholder failed, Proctor was now trying to make a living as a full-time popular science writer, a financially precarious but much-needed role at a time when scientific theories were becoming ever more complex. For Proctor, a débâcle with paradoxers in the *Mechanic*'s correspondence columns had only served to emphasize the need for a clear exposition of the facts and, as an exceptionally lucid writer, he was particularly well suited to that role. Interestingly, he thought the problem lay to some extent with astronomy writers rather than the paradoxers who preached peculiar theories, for the former seemed content to provide general summaries of the facts without precise observational and experimental evidence to back up their sweeping statements. As a result of vague accounts, Proctor suspected that ordinary readers had been left bemused about basic facts, and their ignorance was an open invitation to paradoxers, who gloried in taking advantage of this state of affairs.[21] As a remedy, he proposed to turn his attention to a clear and comprehensive treatment of the question of the earth's shape and motion for the benefit of astronomical novices. He realized some of the technical information involved would have to be accepted on trust because he could hardly ask readers of the *English Mechanic* to set sail for the southern hemisphere to take measurements or master higher mathematics to such a degree that they could follow the brilliant and intricate calculations on which human knowledge of the universe was based.

Proctor was only too aware that these were key problems in the popularization of science and, just as mediators were increasingly essential to explain ever more technical theories, readers were going to have to accept him at his word. But still he felt bound to assure the public that astronomers based their work on clear and exact, if complex, calculations and, despite the allegations of paradoxers, were not undertaking a sophisticated scheme to promote a false system of knowledge. Such claims, he insisted, were not only laughable but illogical. What could modern astronomers possibly gain from engaging in such a swindle? It was not as if the long-dead Newton or Copernicus could provide modern astronomers with a boost to their careers, he commented wryly; indeed, established men of science could earn more kudos by overturning accepted ideas.

So, with his plan to enlighten the ill-informed through the pages of popular science magazines, Proctor set to work on a series of articles on 'Our Earth – Its Figure and Motions' for publication in the *English Mechanic*. Meanwhile, the Astronomer Royal, George Airy, was busy answering unorthodox correspondence at Greenwich Royal Observatory. Much of this, to his disdain, concerned John Hampden, who seemed particularly set on promoting his cause since Wallace's acceptance of his *Scientific Opinion* challenge. On 4 February, Airy received the first in a series of letters from Hampden, keen to know whether a meteor was entangled in the earth's atmosphere. Happy, as always, to back his assertions with money, Hampden offered the Astronomer Royal twenty pounds for an answer to his question and for further information about whether 'there is an atmosphere that can be said to belong to the earth at all'. In Hampden's opinion, it was 'quite time such ignorant theories be exploded'. Clearly not satisfied by having tempted one eminent man of science to take up a wager, he issued a second £500 challenge entitled 'A Nut for the British Scientific Association' addressed to the Astronomer Royal.[22]

The letter and attachment immediately found a place on Airy's shelf marked 'My Asylum for Lunatics', and were joined by Hampden's letter to the editor of the *Weston-super-Mare Gazette*, which appeared in print the next day. Under the headline 'The Bible versus Falsehood', Hampden commenced his latest critique with the assertion that 'English science is a disgrace to the age.' With characteristic aplomb, he defied 'the whole of the engineering skill of the United Kingdom' to disprove just one of his statements about the shape of the earth, and declared that he was willing to stake '£500 or all he had in the world' on such an in-depth inquiry. Stating that he had no patience with 'sneaks and cowards', who dared not 'face an honest man in broad daylight', Hampden announced that the onus of proof lay with those who made the 'preposterous' claim that the earth was a globe. If all the leading engineers in England continued to shirk submitting their arguments to a legitimate test, as suggested, Hampden declared that he would expose them as they deserved. He went on to claim that the globe idea was a 'superstitious delusion of the dark ages' (rather than the flat-earth idea, as commonly supposed), and modern science was of little assistance in rescuing Victorian England from its grip. In fact, the nation was a disgrace to the age of reason and progress, in Hampden's view, dedicated to 'disowning the Creator' and worshipping infidel idols such as Newton and Galileo at the scriptures' expense. In a period of religious diversification, from spiritualism to Eastern religions, when atheism was on the increase and declining church attendance was a cause for concern, Hampden was determined to restate the place of Christianity in social, spiritual and political life. The flat-earth idea was the headline of a much broader agenda: to destroy false knowledge and re-establish the 'truth' for so long resisted and denied. He hoped a 'new theology', based on the literal truth of the Bible, would spawn a 'new science', and the 'idolatry of ages' would be swept from the earth. A master of metaphor, Hampden was reversing traditional imagery – conventional science was superstition, his radical Bible-science was the only road to unquestionable truth. Ever the antagonist, he finished his letter with a dark threat: 1870 might be the year when 'mighty changes and unsuspected disasters' befell those who had seen fit to betray the nation about the shape of the world.[23]

Such warnings were easily disregarded by George Airy but, due to his position as Astronomer Royal, the matter could not rest there. He was honour-bound to deal with all enquiries sent to him by the public and this responsibility put him in the firing line for Hampden's next fracas: a fiery exchange with Woolwich Arsenal clerk Charles Edward Kettle. Kettle's disagreement with Hampden had commenced in early February over the meteor question, for Hampden had made the same offer to Kettle as he had to the Astronomer Royal: twenty pounds for proof that the earth had an atmosphere. Kettle had replied immediately, offering assurances that indeed it did and, moreover, that it was round and revolved. Various proofs were offered: the round shadow the earth cast on the moon during an eclipse, the invisibility of the Pole Star over certain lines of latitude on the earth's surface, the changing seasons, the positions of the fixed stars, and several more besides. With some insight, Kettle ascribed Hampden's disbelief in gravity, and the rotundity, revolution and atmosphere of the earth, to the simple fact that he had not personally observed these phenomena, and tied up this argument with the pertinent question: 'You have not seen your own brains – do you believe you have any?' In conclusion, he suggested that Hampden send twenty pounds to the secretary of the Woolwich Ragged Schools and £500 to the Asylum for Idiots at Earlswood, Surrey, on the grounds that the first housed 'poor boys who know more about the earth than you appear to', and the second was 'a fit place for such as accuse Newton, Herschel and Airy of ignorance, falsehood and fraud'.[24]

Infuriated, Hampden fought back – with particular vigour because he had formed the mistaken impression that Kettle was a notable at Greenwich Royal Observatory, rather than a lowly clerk at the Royal Artillery base. The authority and status of astronomers were a particular irritation to Hampden and he declared that the nation ought to be ashamed to have a 'parcel of blockheads' in such positions as Kettle appeared to occupy, for scientific men were 'brainless jackasses', not one of whom knew the true shape of the earth on which he stood. Newtonian theory was 'not worth the pen' it was written with, Kettle was hallucinating, and Hampden would be happy to offer £500 for evidence from the Bible where the word 'round' was applied to the words 'earth' or 'world'. In answer to Kettle's specific proofs, Hampden was similarly incredulous – 'lose sight of the Pole Star – of course! So you do of the sun when it has gone so many thousand miles from where you stand! Wonderful!' As for the seasons, 'they have nothing to do with the motion of the world', and furthermore 'all the stars do move, together with the sun and moon'. Hampden was convinced: the earth had not revolved once from the day it was created. Finally, Kettle could pass on a message to Professor Airy: the Astronomer Royal was a coward if he ignored Hampden's challenge.[25] Days later, Kettle informed Hampden that in his 'hotheadedness and over-eager desire to get a respectable person to take notice of his absurd theories', he had made yet another mistake: rather than being an employee of the Royal Observatory, he was a Woolwich Arsenal clerk. In his reply, the well-bred and well-to-do Hampden leaped on this fact and swept into a patronizing denunciation of the impertinence of a 'humble' secretary seeking to impart knowledge to his social betters, telling Kettle to 'confine yourself to your employment and not talk about [astronomy] as it is evident you are profoundly ignorant'.[26]

Baiting the flat-earth believer was now becoming tiresome, so Kettle responded with deeper criticism concerning Hampden's twopenny pamphlet 'especially for Christians', *Astronomy as Learnt from the Bible: A Familiar Dialogue upon the Earth, Sun, Moon and Stars, Shewing the Situation the Earth Occupies in Creation* (1870). On reading the booklet, Kettle had been struck by the way Hampden seemed to twist the word of God to fit his own theories, a particularly questionable approach for one who advocated a strict literal interpretation of the scriptures. But not only did Hampden's pamphlet appear to offer a contradictory hotch-potch of ideas and assertions ungrounded in Biblical authority: he also appeared to be guilty of the greatest crime of all – blasphemy. Kettle said he had no doubt that the strain involved in concocting such strange and unique theories had caused Hampden's brain to 'collapse', but emphasized that he was keen to escape 'the muddy sewer' of his adversary's imagination and 'rise to the clear atmosphere of real astronomy'. To return to the facts, he posed another brain-teaser: if the earth does not revolve how does one account for the apparent retrograde motion of Mars, the transits of Venus and Mercury, the conjunction of Jupiter and Venus and the precession of the Equinoxes? With this puzzle, and a proverb, 'The way of a fool is right in his own eyes' (Proverbs 12:15), Kettle closed the correspondence.[27] Evidently satisfied with his arguments and sure that the Astronomer Royal would appreciate them, he forwarded copies to the Greenwich Royal Observatory.

In his office, Airy considered the correspondence before him, and was quick to form a judgement. In reply, he informed Kettle it was not prudent to enter into controversies with 'such persons as this Mr. Hampden'. Airy had quite enough to occupy his time without acting as referee in 'paradoxical' disputes, and as papers on such subjects were of 'no use whatever' to him, Kettle was asked not to send any more.[28] Kettle followed instructions, but Hampden continued regardless. Spurred on by his success in enticing Wallace into a public wager, he now sought to tempt the Astronomer Royal with another printed challenge:

> [I] deprecate mere verbal discussion with those who . . . seem wedded to error and falsehood. The absurdities and delusions of *all* our Astronomers and Geographers have been over and over again brought under their special notice. But as long as the Government is paying large salaries for anything and everything the Professors choose to say, there is little chance of forcing them to abandon their medieval science without the pressure of public opinion. Honest and honourable men are starving by hundreds from Government economy . . . and it becomes us more than ever to see that men already abounding in wealth (made at our expense) really teach truth in their several departments of science and philosophy . . . If you then, individually persist in adhering to the monstrous absurdity of the idea of a revolving globe and a convex sea and earth, I am commissioned to offer to make a mutual deposit of any sum not under £50 and not over £200 that you like to name [for the establishment of] the truth of one fact in connection with the theory of the earth's curvature.[29]

It is highly unlikely that Airy replied.

While the Astronomer Royal warned the public against engaging in paper wars with paradoxers, Wallace, a respected man of science, continued with arrangements for his flat-earth wager with Hampden. With this aim in mind, on 8 February 1870 he met Hampden near Waterloo Bridge, at 346 the Strand, to sign a legal agreement setting the terms of their scientific duel. The £500 challenge was to be decided by an experiment to test the convexity of water, with curvature being proved 'to and fro, on any canal, river or lake by actual demonstration and measurement, to the satisfaction of both referees'. Such a test should reveal without doubt whether the earth was flat or round. Wallace maintained that if the earth was a globe 25,000 miles in circumference, every part of its surface must form an arc of a circle. Hence, all land and water must be convex, curving away from an observer placed at any given point. Owing to this curvature, far-off objects would be elevated slightly by the earth's rise before becoming invisible to a spectator, as ships sailing over the horizon routinely proved. Wallace reckoned the rate of the earth's downward curvature from any given point to be eight inches per mile, and eight inches multiplied by the square of the distance in each succeeding mile. Thus, on his calculations, at a one-mile distance, objects smaller than eight inches would be hidden from an observer by the dip of the earth, even if an observer was looking through a telescope. By six miles off, objects of less than twenty-four feet (eight inches multiplied by the square of the distance) would similarly become invisible to an observer.

Back in Swindon, Hampden dismissed this scientific 'bosh and bunkum', insisting that in a flat world it was possible in theory to observe an object straight across the plane from any given point. He sneered at the very idea of a 'hill of water', convinced it was 'an insane delusion' because 'positive planes, horizontal levels, and indisputable flats' are evident on the surface of 'every ocean, lake, canal and river in the world'. Overjoyed, he signed the wager agreement, gloating to an acquaintance that 'the grand match is now sure to come off – I hope and pray the right may win.' However, a location for the test had yet to be decided, although Parallax's disciple knew an ideal spot: the Old Bedford Canal in Norfolk. It was an apt choice, for the canal cut straight through the black fens of East Anglia, an immense low-lying plain known as 'the land of the three-quarter sky'. Once an enormous eerie swamp dotted with isolated island settlements, thousands of acres of this marshland had been reclaimed during the seventeenth century by the Duke of Bedford and his 'gentleman adventurers'. Spurred on by the promise of land won from the water, they drafted in Dutch engineering experts, who struggled against armed opposition from local gangs of riotous wreckers known as the 'Fen Tigers' to divert the winding watercourses to the sea. By 1652, forty thousand acres of fenland had been successfully drained via a sophisticated network of sluices, canals and cuts that divided the landscape into various 'levels'. The canal identified by Hampden as the ideal setting for the experiment gave its name to a vast flat region of 640 square miles extending into Northamptonshire, Cambridgeshire and Bedfordshire. It was appropriately known as 'The Great Bedford Level'.

Hampden's suggestion met with Wallace's approval, and a six-mile stretch of straight, unobstructed canal between the Welney and Old Bedford bridges was selected as a suitable site for the test. The location was already renowned as the scene of the annual Championship of England Skating Matches, a great custom in the Fens, at which every February throughout the nineteenth century hundreds gathered to watch local heroes, such as Wiles of Welney and Porter of Southery, speed along the canal's frozen surface. But Wallace was unaware that this stretch of water was also celebrated for a second reason: as the 'happy hunting ground of the earth-flatteners'. Parallax and his disciples had experimented here since the 1830s, allegedly gathering plentiful proof that 'Old Bedford Water' was a dead-level strait. Indeed, it was claimed that if one went to Welney on a fine day, climbed into the canal and balanced a telescope just above its surface, then one would be afforded a perfect view of barges and bathers at Old Bedford Bridge six miles north. Hampden nursed this secret, believing such previous investigations guaranteed a famous victory over his adversary. Ironically, Wallace shared Hampden's optimism; both parties believed they were setting up a 'crucial experiment', a test by which they would conclusively prove the true shape of the earth.[30]

Assumptions in place, arrangements continued, with both parties depositing their £500 stake at Coutts Bank in the Strand.

However, in mid-February, Hampden made a sudden about-turn, deciding not to attend the trial, after all, but to send a zetetic substitute in his place. Wallace, keen to educate his flat-earth antagonist, was quick to step in with an alternative offer:

I should very much regret if you did not come down to Norfolk to see the experiment I propose to make yourself. I was going to suggest that you should first meet me there, and only in the case that I should not be able to make the thing sufficiently plain to you, should we call in the referees at all. I firmly believe I can make it plain to you and that it would interest you much to see it, and convince you in a way no report of referees could do.[31]

If he had intended to spare his rival public shame, the plan miscarried in spectacular fashion, for Hampden was outraged by the supposed swindle. He barked that to 'trifle with the trial' would be utter 'madness', and informed a confidant: 'I have written a most emphatic No! to the enclosed proposition. What can the man be thinking of, after it has been talked of far and wide, to turn it into a hole and corner affair – how preposterous!'[32]

Wallace's offer rejected, two referees were duly appointed to observe the experiment and finally decide the result of the wager. John Henry Walsh, the sixty-year-old editor of the gentlemen's sporting paper the *Field*, was selected to act for Wallace. By that time 'The Great Walsh', as he was known, had enjoyed a distinguished, if colourful career. A Worcester doctor, eye specialist and fellow of the prestigious Royal College of Surgeons, he had been appointed joint editor of the *British Medical Journal* (then the *Provincial Medical and Surgical Journal*) in 1849. Three years later he was dismissed suddenly following a dispute with the paper's owners, and while his former co-editor languished in Ticehurst asylum, Walsh went on to combine his passions for sport and journalism in a new post as editor of the *Field*. He was a sporting all-rounder himself, a 'good rider to hounds' who owned a stable, trained hawks, kept greyhounds and coached a rowing club, pursuits that led to his role in founding two great London institutions, the All England Lawn Tennis Club and Battersea Dogs Home. Following a shooting accident, when he lost a thumb and forefinger, Walsh had also become interested in the improvement of safety precautions for firearms and ordered a series of intensive trials that led to the development of the modern shotgun. In addition to this, by 1870 he had an enviable literary reputation, having written the definitive encyclopedia *British Rural Sports*, alongside several books on dogs, guns, cookery, croquet, exercise and household management. All in all, then, Walsh seemed an ideal choice for referee. He was of reputable character, a 'perfect stranger' to both parties, an experienced stakeholder for sporting wagers and, moreover, willing to publish the results of the test in the *Field*. (Wallace's additional endorsement of his well-educated referee as 'not scientific', hence 'not prejudiced' in favour of the idea that the earth was a globe, was less convincing.)

The second referee was a different matter entirely. He was appointed by Hampden with Wallace's permission, and the naturalist had insisted that the additional adjudicator should hold 'some public position as editor, author [or] engineer' and must not be 'a personal acquaintance of your own'. Unfortunately, William Carpenter was unfit on both counts. First, he was not a professional but a skilled artisan and, like Wallace, a séance-attending spiritualist in his leisure time. In fact, Wallace's 'ability to turn from the ranks of popular opinion' in this regard had not gone unnoticed by Carpenter, who calculated that on seeing the experimental results the naturalist would make a particularly 'valuable addition to zetetic ranks'.[33] But in terms of his appointment as referee, the printer's lowly social standing was overshadowed by a second more serious fact. Unknown to Wallace and Walsh (who were unaware of Carpenter's publications), he was Hampden's closest confidant and flat-earth associate. With the exception of Parallax himself, a more problematic selection for the post could not have been made.

At five o'clock sharp on Monday, 28 February 1870, an unwitting Wallace met Carpenter at Bishopsgate station in London's East End to travel to the Old Bedford Canal to undertake the test. The naturalist was unaware that in his pocket Carpenter carried a scrawled note from Hampden. Sent in haste from Swindon, it advised Carpenter:

> As I am not disposed to travel so far, my printer, Mr. Alfred Bull, who is <u>thoroughly</u> with us, will attend for me. It will, I think, be satisfactory to you to have someone to consult with and second any suggestions you wish to make. He is <u>an exceedingly shrewd and clever little man</u>, with heart and soul in the subject. You will feel more confident with him to refer to or consult with, otherwise you will be alone. I do not know how many Mr. W. will bring. <u>Do not let them make it a drawn battle, which they may try and do.</u>[34]

Armed with these instructions, Carpenter boarded the train with Wallace. Their destination was Downham Market, a quaint market town thirteen miles south-east of Wisbech and eight miles north of Parallax's old socialist commune site. They assumed that the peaceful location close to the north end of the Old Bedford Canal would provide a convenient base for their preparations, although it did not appear as such when they arrived. Unexpectedly the town was in chaos, for the annual festivities of St Winnold's Fair, one of the largest horse markets in the country, were in full swing, and the narrow cobbled streets were jammed with clamouring dealers and shambling animals, along with merry revellers enjoying the sideshows and the break from agricultural routine. In this bustling atmosphere, lodgings were difficult to locate, so Wallace was fortunate to find sanctuary at the Crown Hotel in the steep central thoroughfare, Bridge Street. A smart seventeenth-century boarding-house run by William Wayman, it doubled as the local Inland Revenue and Excise Office and also had a colourful history. According to local legend King Charles I had found refuge there disguised as a clergyman after fleeing from his English Civil War defeat at Naseby in 1642, and it had provided a hiding-place for local magistrates during the infamous riots that swept across the Fens in May 1816. In fear for their lives, they had been forced to barricade themselves into the pub to escape the 1500 labourers who were rampaging through Downham Market demanding, 'Bread or blood.'

While Wallace settled in at the Crown, Carpenter made do with less historic lodgings in the humble high-street cottage of the local watchmaker's widow, Mrs Howes. Over the next two days, arrangements for the experiment hurried along. Editor and umpire Walsh arrived from London and joined Wallace at the Crown. Then, much to Carpenter's relief, Hampden made a last-minute appearance; he had suddenly changed his mind, claiming 'everybody' in Swindon was insisting he 'must go' to settle the dispute in person.

Party complete, they gathered on Wednesday morning and rode two and a half miles to the Old Bedford Bridge, laden with six signal posts and a hatchet. From here they paced south along the canal's towpath in a strong head wind for six long miles to Welney Bridge, erecting six-foot poles topped by coloured markers at one-mile intervals along the water's edge. Wallace provided an elegant explanation of the 'simple and conclusive test':

> If water is straight and flat, the tops of the poles will of course be straight and flat too. But if the earth and water has a curvature of 4000 miles radius, then the tops of the poles will be equally convex, and they will be seen <u>rising higher and higher to the middle point, and thence sinking lower and lower to the furthest one</u>. With a good telescope curvature will be easily seen if it exists.[35]

At Welney Bridge a large telescope was hauled into a barge, then steadied with sacks and planks so that it stood six feet above the water level in line with the markers on the bank. As the participants wrestled with apparatus in the boat, a score of Welney rustics assembled, their curiosity piqued by the scientific spectacle. The atmosphere grew tense. Telescope in place, Walsh and Carpenter stepped forward to check the view. Confused, they discovered that the six markers along the canal bank were in disarray; they could not even judge which was which, let alone decide whether the line of sight along them was flat or curved. Heated debate ensued, an angry quarrel followed, and the experiment descended into farce while Hampden apparently relaxed, enjoying the scene from a nearby barge. It appeared that Wallace's 'simple and conclusive test' had been a dismal failure.

Despondent, the party returned to Downham Market to redesign the experiment. Days later, Wallace went by horse and coach to the nearby port of King's Lynn in search of a better telescope, accompanied by Hampden and Carpenter. While they visited the town's impressive Athenaeum to see the local Natural History Society's ornithological exhibition, Wallace borrowed a first-rate Troughton's level telescope with a built-in cross-hair (a line marking the true level of any object viewed at a distance) from a local surveyor.

Meanwhile, in that day's issue of the *English Mechanic*, John Dyer's new book, *The Spherical Form of the Earth: A Reply to 'Parallax'*, was causing quite a stir. A fifty-eight-year-old schoolmaster and member of the Royal College of Physicians, Dyer ran a small private boarding-school in New Cross, with his wife, Ann, and four schoolmistress daughters. Already a published author, Dyer's previous work had consisted of lifestyle tips for the general public, and treatises arguing for the 'laws of health' to be systematically taught in colleges and schools. But with his *Reply to 'Parallax'*, Dyer had branched out from the principles of physiology to those of astronomy, a move inspired by Parallax's lecture at Northampton in August 1858. Sitting in the audience at the Mechanics' Institute in the Market Square, Dyer had been struck by Parallax's cool, clever manner, and the effect of his lecture on onlookers 'not in a position to disprove his statements' and who felt quite 'disposed to agree with him'. So concerned was Dyer by this, that he began his own series of lectures on astronomy, teaching exactly the opposite to Parallax: that 'earth might be a globe, the moon non-luminous, the sun a little further from the earth than 4,000 miles, the stars something else than mere lights at a distance of 6,000 miles, and that there might be other worlds than the one we inhabit'.

Then in the late 1860s, Dyer heard of the publication of *Zetetic Astronomy: Earth not a Globe!* and immediately began work on his lengthy reply in the belief that, as an amateur astronomer, it was his public duty to expose Parallax.[36] It was a decision that was applauded by editorial staff at the *English Mechanic*, who were keen to see an end to the public revival of flat-earth views. Indeed, a journalist joked in his review of Dyer's book, 'We can almost fancy something like a feeling of weariness and disgust overshadow[s] the minds of many readers at the mere mention of the name of the specious originator of the "Parallaxian" Philosophy.'[37] Even so, the reviewer continued, it must be remembered that Parallax's arguments had not only been laid before readers of the *Mechanic* but before many others without adequate information or opportunity to refute his erroneous statements. It was common knowledge that, during more than twenty years lecturing in the principal towns of England, Parallax had successfully disturbed the minds of the public. Therefore 'Mr. Dyer's little book' deserved high praise, for it shattered Parallax's bewildering theories to pieces with a clear statement of the facts simply expressed in language that everyone could understand. Moreover, as was fitting for a book aimed at the less well educated, Dyer's *Reply to 'Parallax'* was cheap enough to be afforded by all. But amid the chorus of approval for the amateur astronomer's efforts, a lone voice chipped in with cutting criticism. Unexpectedly, it belonged to none other than astronomy writer and fellow Parallax opponent Richard Proctor, who carped:

> Mr. Dyer must be a very young man not to know that Parallax and co. are merely types of a class which will always exist. So what if Parallax unsettles the minds of many who hear him; serves them right for leaving the subject unstudied. They have

been settled in ignorance and Parallax is unconsciously doing capital work by unsettling them. More power to him![38]

Proctor seemed to have changed his tune, but Dyer was not willing to accept his criticism. In retaliation, he sniped, 'Mr Proctor must be a very young man not to know that there are thousands who have not had the means at their disposal to study astronomy.' From his perspective, it was a duty incumbent on all who had received the opportunity to acquire such knowledge to use it to educate others. In fact, in an era when being an amateur did not necessarily mean less expertise, Dyer was arguably better placed than most to tackle the issue head-on. As someone who did not earn his living from science, he was certainly unfettered by the professional ties that might have discouraged engagement with paradoxers. Under such circumstances, Proctor's seemingly spiteful and superior reaction to his involvement was peculiar to say the least. Why else did he write his 'admirable' articles on the earth's shape and motions, Dyer asked, if not to upset the notions of Parallax and settle the minds of those he had disturbed? Were amateurs and professionals not fighting on the same side with the same weapons on behalf of the same cause: to develop the public understanding of science? Or was Proctor – the popular astronomy writer – more concerned with safeguarding his professional status in a newly establishing field?[39] Besides raising issues about who had the right to disseminate authoritative knowledge to the public, the debate highlighted that even those who opposed Parallax and Hampden could not agree on which tactics to adopt.

While the paradoxer debate continued in Friday's *English Mechanic*, the Saturday deadline for the *Field* loomed. As editor, Walsh was duty-bound to return to the capital, so he decided to leave Downham, pausing only to select a replacement referee for Wallace. Together they plumped for Martin Wales Bedell Coulcher, a forty-four-year-old surgeon and bachelor who lived above his Lynn Road surgery with his elderly mother. Coulcher was a good, solid choice, a seemingly trustworthy professional man and solicitor's son with a medical degree from University College, London; he was also an old friend of Walsh. More promising still, Coulcher was an accomplished amateur astronomer with in-depth local knowledge from a lifetime spent in the Fens.

Reorganization complete, all the participants needed was good weather. When Saturday, 5 March, dawned a crisp spring day, the party gathered without delay on the banks of the Old Bedford Canal. Keen to exploit the opportunity, they divided two by two to set up the redesigned experiment. A large calico sheet was hung from the Old Bedford Bridge, with a thick black band painted across its centre. Then a telescope was placed six miles south on Welney Bridge at the same distance from the water as the black band. Last, between these two points, about three miles from each bridge, a long red pole topped by a marker disc was set up, designed to fall in line with the black band and the telescope. All three points were thirteen feet three inches above the water. The plan was to view the marker disc and the black band in a line through the surveyor's telescope. If the middle marker disc appeared below the line of sight, this would be taken as proof of the flat surface of water and Hampden would receive £500 from Wallace, plus his original stake of £500. If, on the contrary, the middle marker disc appeared above the line of sight, it would be taken as proof of the earth's curvature and Wallace would receive £500 from Hampden, plus the return of his original £500. Wallace calculated that, even allowing for atmospheric refraction, the central marker should appear at least five feet above the line of sight from the telescope at Welney to the black band on the sheet at Old Bedford Bridge.[40]

At 12.40 p.m., Wallace and Carpenter joined Hampden and Coulcher on Welney Bridge. As before, a group of inquisitive locals had gathered for a glimpse of the intriguing proceedings. According to reports, many were perplexed by the unusual event and one onlooker was said to have exclaimed, 'Oi say, Bill, they want to say the water ain't level. Oi know it is, though. Oi've been 'ere these ten years, and Oi know if 't ain't level there's no level anywhere!'[41]

Clearly for the fenmen of Victorian East Anglia, sense perception and first-hand experience gave them primary expertise. With little, if any, formal education, it appears that their ideas about the shape of the earth were liable to be shaped more by their own observations, and cultural factors such as what their neighbours believed. The conjugations and experiments of scientific professionals were of little relevance or importance in their day-to-day lives.[42]

While the crowd debated, Wallace focused on the important task in hand. He set up the surveyor's telescope on the iron bridge so that its horizontal cross-hair was thirteen feet three inches above water level and fixed on the Old Bedford Bridge, six miles north. He checked the telescope and noted that the middle marker was four or five feet higher than the Old Bedford Bridge marker. As he had suspected, the surface of the Old Bedford Canal was curved. The telescope was levelled and Carpenter stepped forward to take the reluctant Hampden's turn. He observed that the centre marker was somewhat below the cross-hair on the telescope, and the far marker on Old Bedford Bridge the same distance again below that. Under usual circumstances, this would be interpreted as evidence of the earth curving gently away from the observer, but Carpenter claimed that the equal intervals between the cross-hair, the centre marker and the Old Bedford Bridge marker proved that all three points were in a straight line. So, he concluded that the water was flat. Wallace reported later that Carpenter 'actually jumped for joy', cheering 'Beautiful! Beautiful!'[43] Meanwhile, Wallace watched, thunderstruck: he had never witnessed such blatant refusal to accept scientific evidence as fact. He leaped in, remonstrating that the cross-hair had absolutely nothing to do with the matter. But Carpenter was adamant. Wallace demanded that the parson of the parish be called to the spot to make an authoritative final judgement. His plea was overruled. Nothing more could be done. A horse and cart were summoned, and Hampden ferried the party six miles back to the Old Bedford Bridge, where the observations 'to and fro' were completed in sullen silence. It was starting to seem as if the flat-earth believers were not 'willing to pay to be enlightened', as Wallace had initially supposed they might be.

Resolution seemed impossible. That night, Hampden accosted Wallace at the Crown Hotel, challenging the man of science to deny that 'he was beaten' and assuring him that it would be 'much more graceful to admit it'. Wallace, weary of the whole affair, did not utter a word. Early on Monday morning, in a desperate final attempt to reach a decision, the two referees met at Coulcher's home near the Bell Inn on the road to King's Lynn. Under the terms of the wager agreement, Wallace needed to prove curvature of the surface of the water to the satisfaction of both referees. With this in mind, Coulcher produced diagrams, declarations, notes and sketches, but Carpenter could not be convinced. Exasperated, Coulcher snatched back the papers, seized a Bible and swore an oath that on Saturday, 5 March 1870, Mr Wallace had proved the convexity of water beyond dispute. But his protests were futile. Coulcher announced that he 'washed his hands of the whole transaction' and ordered Carpenter to leave at once, but he was determined to stand his ground. Coulcher was left with no alternative: set on ejecting Carpenter from his house, he sent his young maid to Church Lane police station to fetch the local constable. Several minutes later Carpenter was shoved over the doorstep by a 'gentleman in blue', with the blunt statement 'Go, or I'll take you!'[44]

That afternoon, Wallace journeyed back to the capital in a separate railway carriage from his flat-earth adversaries. However, the break from debate was short-lived. Once at home, he received an urgent message from Carpenter, suggesting that 'some other gentleman' than Coulcher be sent to his Lewisham base to discuss the wager in a sensible fashion. By this time, it appears, Wallace's patience had all but evaporated. Writing from North London, he snapped in astonishment at Carpenter's 'utter incapacity to perform his duty' and flatly refused to appoint a replacement referee. No choice remained but to select an umpire to decide the wager beyond dispute and, fortunately, a willing judge was again found in the *Field*'s editor, John Henry Walsh. Over the next few days, Walsh reviewed the referees' documents and diagrams before referring the evidence, on Hampden's insistence, to a high-class Piccadilly optician, Mr Solomons. In the meantime, as a precautionary measure, Hampden dispatched Carpenter to the optician's premises in fashionable Albemarle Street to check the investigation on his behalf. What the referee discovered was most unsatisfactory to his mind, for Solomons had entrusted the papers to a young assistant, who subsequently reported to Carpenter that he had skimmed the notes 'for an hour or two' but had not 'sat up all night over them'. The apprentice continued that, 'taking into account the theory of the earth's rotundity', it was his considered opinion that 'if anything had been proved, it was that water was curved'.[45] From Carpenter's perspective, the outlook was not promising.

While Carpenter oversaw proceedings in London, Hampden was busy in Swindon writing letters to his long-standing correspondent Sir Thomas Phillipps. Widely renowned as somewhat eccentric, Phillipps was the best-known bibliomaniac of his era, renowned for his collection of 100,000 books – a hoard so vast he moved out of his house because the collection took up all of the space. A complete obsessive, according to reports, it was Phillipps's personal quest to own a copy of every book ever printed, while it was said that in his younger days he had badgered his friends to find him a rich wife who would be willing to finance his fixation.

Now in old age, besides stockpiling books, he found amusement through corresponding with John Hampden. At first their discussions had focused on religious matters, but of late they had increasingly turned to Hampden's flat-earth theory. Indeed, it seemed that Hampden could write about nothing else. In February, he had even challenged Phillipps to a £200 wager for a single fact proving the 'monstrous absurdity' of a revolving globe and a convex earth and sea. His ultimate object, Hampden had declared, was to force the government to establish a Royal Commission to investigate the flat-earth question, thus relieving the country of the 'disgrace and reproach' that 'persistent and blind adherence to the infidel science of the Middle Ages' had brought upon it. In reply, Phillipps had politely declined the offer, informing Hampden that his arguments were 'by no means well-grounded'. Nevertheless Phillipps posed some questions: was Hampden's earth square-shaped or a cube and, if so, why did it cast a round shadow on the moon; and furthermore, if the sun and moon were round, could Hampden not see how the earth might be the same shape?[46] Unshaken by criticism, Hampden retorted that the earth was neither a square nor a cube: it was flat, circular and bordered by giant icebergs; Hell lay beyond the outer border, and the North Pole was at the centre of the disc. He continued that the sun's course across the plane was horizontal midway between the northern centre and the southern circumference but not more than two thousand miles from the earth's surface – although he was not quite sure about this, stating 'perhaps 200 would be nearer the mark' because 'heat could not travel so far'. As for the earth's circular shadow on the moon, that phenomenon was in full accordance with his theory that the earth was a disc.[47]

Phillipps immediately spotted the opportunity for some fun:

I am glad to find from you that we have a chance of seeing Hell without going into it. If I were a young man I would propose to you to fit out a ship in which we would sail with some of the Philosophers of the day to see it . . . On reaching your boundary we would [climb] onto the icebergs, and crawl on them till we reached the desired view. We would then return to England and represent to Government how all may be made good. If Government would furnish us with a fleet we [could] then empty all the jails in the kingdom & carry [the convicts] to view this frightful abode. You may depend

upon it – all would become good. By the bye a thought occurs
to me – as Hell is such a dreadfully intense hot place, how is
it that all the icebergs are not melted?[48]

He signed off with some more questions: if mankind truly
inhabited a 'pancake alias plane' world, surely it would always be
day and never night? And what lay on the other side of the disc?
For that he would certainly like to see. Hampden replied that he
had staked £500 on the reality of his 'pancake world', and as for
what lay underneath, 'The other side is hell and I pray, dear sir,
that your "wish to see" it may not be realized.' When he received
this letter, Phillipps elected on a brief response, offering commiser-
ations that Hampden had risked such a large sum on his 'flat world
theory' because he would undoubtedly lose his money.[49]

Following his visit to Solomons, Carpenter shared Phillipps's
pessimism, and when he received an invitation to attend the
London offices of the *Field* on Friday, 11 March, to hear the
umpire's final verdict, he steeled himself for the worst. Carpenter
arrived at Walsh's office, next door to the Gaiety Theatre in the
Strand, at one o'clock prompt on the appointed day. To his dismay
he discovered Walsh and Wallace already deep in huddled discussion about events at the Old Bedford Canal. He concluded that
'Mr Walsh' had undoubtedly 'had the screw on from Mr Wallace',
so was not surprised by the unwelcome news that greeted him, for
Walsh at once declared, 'I have had no difficulty whatsoever in
coming to a decision – in favour of Mr Wallace!'[50] The decision
was put in writing on 18 March, Hampden was informed of his
loss, and a week later the final verdict, along with the referees'
reports and diagrams, was published in the *Field*:

> Mr. A. R. Wallace ... has proved to my satisfaction the
> curvature, to and fro, of the Bedford Level Canal between
> Welney Bridge and Welches Dam (six miles) to the extent of
> five feet more or less. I therefore propose to pay Mr. Wallace
> the sum of £1,000 ... unless I have notice to the contrary
> from Mr. Hampden.[51]

Further revelations were to follow. On 26 March, 'with extreme regret', Walsh announced that Hampden and Carpenter had refused to abide by his decision and, even worse, a zetetic pamphlet had fallen into his hands. Carpenter's name was on the title page, and in it the referee described how the same experiment as that tried by Hampden and Wallace had been run before at the same site by the infamous Parallax, with a result in favour of a flat earth. Finally, the truth dawned: Carpenter was prejudiced, rendering the 'original' experiment on the Old Bedford Canal null and void. It appeared that Wallace and his scientific supporters had been duped, and Walsh appealed to his readers:

> The good faith and perfect fairness of Mr. Carpenter were not quite of the nature we believed them to be, and we have no hesitation in affirming that he was a most improper person to be selected to act as referee. The deception was, to say the least of it, 'unscientific'.[52]

The great irony, Walsh continued, was that Hampden, Carpenter and their 'master', 'Parallax', all professed to be truth-loving zetetics, ardent in the cause of 'true' science. That their scheme had backfired could cause no regret to anyone who sincerely valued the truth, Walsh declared, and he reiterated his certainty that the experiment had proved that the earth was a globe and all 'honourable men' could not fail to agree with his decision.

While the participants argued over who had the monopoly on truth, Hampden's fury at the verdict was now virtually assured. In a rage, he demanded an immediate refund of his £500 stake, telling the editor that he would be held legally responsible if he dared to pay out on the wager. Livid about the charges made against him, he railed at Walsh's sheer audacity:

> I would be the first to denounce the conduct of my own brother throughout the length and breadth of England, if he were to depart one jot from the path of strictest honour and rectitude. No man living ever yet dared charge me with acting unfairly in any matter in which I was engaged.[53]

Naturally, Hampden went on, he knew of Parallax's statements – they were so well documented that who 'but an unborn ass' could pretend otherwise? But this in itself was no evidence of fraud. Rather, it was Walsh who had exposed himself to the 'derision of the whole world' through his 'monstrous decision', for with Wallace he was guilty of 'a gross perversion of the facts'. 'At your peril,' Hampden menaced, 'touch my money and I will serve you both with a writ.' Unsurprisingly, his warning was ignored, and on April Fools Day 1870, Walsh paid Wallace the £500 winnings plus his £500 stake, although he took the precaution of requesting an indemnity against any legal expenses that might arise on account of his actions.

In its next issue the *English Mechanic* applauded the outcome, trusting that Hampden's loss might 'render him a little more inclined to give due credit to the disinterested labours of astronomers'. During the same week, the prestigious new science journal *Nature* also reported on the 'very amusing investigation', baffled by how the flat-earth 'Gentlemen' could 'coolly claim the victory' when the diagrams clearly showed the central marker to be more than five feet above the line of the two extremes. Meanwhile, in the *Field*, the usually chatty letters page, 'The Country House', was crammed with correspondence from Newtonians and zetetics debating the recent débâcle. Most readers were amazed by the affair, regarding Hampden's challenge as an intriguing illustration of 'a very curious psychological phenomenon', while one letter-writer mused that the saga revealed as much about lunacy as it did about astronomy, for it seemed to touch 'on the somewhat undefined line between sanity and insanity'. Science enthusiasts puzzled how Hampden was not only 'incapable of following the simplest inductive reasoning' but was also insistent on 'propounding a theory contrary to facts universally received as proved'. The wager was clearly the result of a bizarre 'mental anomaly', comparable to a colour-blind person offering £500 on the assertion that there was no visible difference between red and blue, and allowing a painter and a dyer to act as umpires. In fact, one correspondent grumbled, it would have been almost as reasonable for the 'flatist to attempt to prove his denial of gravitation by walking out of his bedroom window'.[54] Soon afterwards, the *Scientific Opinion* editor who had published Hampden's irrational challenge chimed in to support Wallace's unconventional decision to accept, supposing that either 'a sublime pity for ignorance or a supreme contempt for charlatanerie' must have spurred the man of science to become embroiled in such a pursuit.[55]

In scientific circles, however, Wallace's peers were more critical of his 'injudicious' involvement in a bet to 'decide' the most fundamental and established of scientific facts. Ironically, in the light of Hampden's challenge to scientific authority, it was Wallace's involvement in the wager that was perceived as truly undermining the developing scientific profession and the body of knowledge it sought to represent. The earth is a globe: this is a non-negotiable fact. Wallace was seen to have threatened the very authority of the body of knowledge he sought to defend by making the earth's rotundity seem debatable. This point aside, astronomy writer Richard Proctor (who had chosen to fight the issue by pen rather than telescope) complained that even if a wager had been deemed appropriate, Wallace's involvement was still out of place: 'If Newtonian astronomy really needed defence, Mr. Wallace would not be our selected champion. It would be as fitting to send the Astronomer Royal, Professor Airy, to defend the theory of natural selection.'[56] Hence, in an increasingly specialized and professional world, scientific expertise was being limited by subdiscipline – even when deciding a fact as basic as the shape of the earth.

Nor did Wallace receive any more support from his associates in the life sciences. The Director of Kew Gardens, Joseph Hooker, sternly disapproved of the naturalist's rash conduct, later reminding their mutual friend Charles Darwin that Wallace had 'lost caste terribly' by 'taking up the Lunatic bet about the sphericity of the earth, & pocketing the money'. This, Hooker asserted, was 'not honourable to a scientific man, who was certain of his ground'.[57]

Along with Huxley and others, the high-principled Hooker was seeking to promote a new professional class of scientists, and ideals of gentlemanly restraint, objectivity and dedicated service to the nation were central to his agenda. Above all, he viewed Wallace's conduct as a misuse of expertise; he had culpably exploited the flat-earther's ignorance, tarnished the disinterestedness central to the pursuit of pure knowledge and undermined the all-important authority of scientific men.[58] In a period when social and scientific standing were inexorably intertwined, Darwin was inclined to agree, and over the next few years, due to a number of complex factors, he became increasingly distanced from Wallace.[59]

By contrast, zetetic ranks across the nation rallied to Hampden's defence. One supporter memorably claimed that Hampden had been 'tricked by the telescope', while in the correspondence columns of the *English Mechanic*, a nameless zetetic from Watford voiced their staunch defiance:

> The Bedford canal farce proves nothing one way or the other
> – only Hampden's folly in risking so large a sum with the
> whole scientific world ranged against him, as well as the public
> press and popular prejudice, and the avidity with which his
> opponent grabbed his £500. No wonder he abuses them – it
> is enough to make a saint swear![60]

The affair certainly raised a storm of curses from Carpenter; his frustration was vented in a pamphlet with the somewhat dubious title *Water Not Convex, The Earth Not A Globe! Demonstrated by Alfred Russel Wallace on the 5th March 1870*. On perusal of the thirty-page booklet, a *Sunday Times* reviewer was appalled to find 'a great deal of abuse and ill-nature and very little reasoning or even statement'. Largely in agreement, the provincial papers chorused their disapproval; the *Derby Mercury*'s gruff warning that 'a shilling will be ill-spent on its purchase' captured their general consensus. Yet, as ever, opposition fuelled Carpenter's dedication to his cause, and a storm of books, flyers and flat-earth poetry thundered off his Lewisham printing press in reply to his critics.

Of the last, the most widely circulated was a lengthy composition called 'The Flying Philosophers', featuring the lines:

> There's Bedford water standing now, as flat as a
>   billiard table;
> And fifty Wallace's to prove it 'convex' are not able,
> I say that HAMPDEN <u>gained the day</u>; and, more, that
>   I can prove it,
> And show that earth's a level plane, and all the world
>   can't move it!
> That sun and moon and stars go round the earth, just
>   as it seems,
> And just as Scripture says they do, and not as science
>   dreams![61]

But Carpenter's reaction to the outcome of the wager was gentle beside Hampden's fury at the 'monstrous decision'. Soon afterwards, he condensed his wrath into a strongly worded tract, *Is Water Level or Convex After All? The Bedford Canal Swindle Detected and Exposed.* Overflowing with 'ungentlemanly language', according to a *Scientific Opinion* reviewer, he continued that the pamphlet was 'the most libelous and disgraceful tirade' the journal's staff had 'ever been pained by reading'.[62] Besides the cacophony of abuse, Hampden's publication argued that the water in the Old Bedford Canal was 'as flat as any billiard table in the metropolis' and that only the 'mad drunk' could pretend ignorance of this fact. So confident was Hampden in the truth of this statement that he was willing to make a second experiment with the stake doubled to £1000 on either side.[63] Not that a second set of proofs was required: Hampden went on to contend that the true result of the first experiment – in his favour – was a historic moment giving the 'lie to every Philosopher on the face of the earth' and marking the 'beginning of a terrible and world-wide revolution in human science'.

While Hampden was an arch-purveyor of warfare imagery, in reality he was clearly undermining his own point in this regard. Like Parallax he was competing for cultural authority by drawing on the status of science – his alternative Bible science was the 'true' science, zetetics were the honest seekers of truth. Moreover, as a result of Walsh's deceitful decision, Hampden believed that he had become a zetetic martyr, the 'blameless victim' of roguery and a sophisticated plot to cheat him of his money. But 'little wonder', he exclaimed, for in a ranking of public reputations for shiftiness and guile 'members of our scientific societies are second only to horse dealers and jockeys'. In this way, Hampden sought to co-opt the status of gentlemanly investigator, the bearer of credible, trustworthy knowledge on which the public could rely without doubt. He therefore paraded the wager as proof of the alleged 'rascality of scientific champions' and the 'despicable subterfuges which British science is forced to resort to, in order to defend the Newtonian theory'.

Set on exposing the scientific ruse, Hampden went on to sling a host of slanderous insults at his Old Bedford Canal opponents. Wallace was branded a 'pitiful dastard' and 'a swindler and impostor, a coward and a liar', who 'grabbed like a starving highwayman on his booty'; Coulcher was ridiculed as the local 'sawbones', a 'half-starved apothecary' only too keen to snatch 'Mr. Wallace's ten-pound note' and 'say just what he told him to', while Walsh was roundly castigated as a 'simpleton' with an 'ignorance of the simplest facts of practical engineering', coupled with a 'despicable cowardice and dread of offending the scientific world'. Such language certainly undermined any claim he might make to be a gentleman of science but, nevertheless, Hampden continued his onslaught with the contention that Walsh's verdict had so sickened the common people that the editor was now forced to 'sneak about' London after dark, 'lest he should be kicked by the very scum of Whitechapel'. But this was just the beginning, for Hampden promised that by the time he had done with this set of 'sharpers', knaves and blacklegs, they would be 'hoisted on their own petards' with reputations worse than 'the meanest felons in Newgate'.

Financial loss was now looming large and Hampden decided to address Walsh's refusal to refund his £500 stake when instructed to do so. In *Is Water Level?* he issued an ominous threat: the editor had crossed him at his peril – the proper penalty was 'deep humiliation and disgrace'. Enraged, Hampden snarled that the umpire had:

> placed a rod in the hands of Mr. Hampden, who, perhaps of all men in the world, is most ready to inflict the severest retaliation on those who dare attempt to force their lying frauds upon his acceptance. Deception and falsehood, meanness and cowardice always excite in his mind an intensity of loathing that few are able to realize.[64]

Hampden believed that, as a man of the 'strictest honour and rectitude', it was his moral duty to exact justice and teach the editor a lesson in gentlemanly conduct. Dedicated to his plan, he decided to serve Walsh with a writ for conspiring to obtain the sum of £500 under false and fraudulent pretences. Meanwhile, a public notice denouncing the perpetrators of the 'Bedford Canal Swindle' arrived at the Greenwich Royal Observatory marked for the attention of George Airy. In it Hampden menaced:

> Messrs Wallace and Walsh will, to their dying day, regret and bewail the precipitate and indecent haste with which they grabbed and pocketed my £500. They must by this time hate the very sight of water: but as long as I live, neither they nor their friends shall be allowed to forget the lies, imposition, and fraud, which, for the sake of a paltry gain, they disgraced themselves . . .[65]

Hampden was determined to keep to his word, for the issue was much greater than a straightforward wager, personal pride or codes of gentlemanly conduct. From his perspective, the globe concept epitomized science's most blatant assault on scriptural authority: Newtonian astronomy, a doctrine that went beyond evolution in challenging the Church. While the majority of theologians, men of science and the public at large were reconciling their religious beliefs with various evolutionary ideas Hampden – a warfare polemicist *extraordinaire* – stood at the extreme. Society appeared to be becoming more secular and Newtonian astronomy was to blame. The globe-earth idea blatantly undercut the Bible; it was the first step on the road to irreligion and atheism, which would lead the British public to reject the Bible and even deny the existence of God. For Hampden, the man-made theories of ortho- dox science were 'treacherous quicksands', leading to imminent social, spiritual, political and religious collapse. With such issues at stake, £500 was a small price to pay, particularly as the wager afforded Hampden much-needed publicity for his marginal cause.

Under such circumstances, Wallace's efforts to clarify the public understanding of science were almost certainly bound to fail. His actions illustrated his dedication to the pursuit of 'truth', which he believed could be revealed by intellectual means of scientific inves- tigation and proof, and were another example of his willingness to step across discipline boundaries (in this particular instance from biology to astronomy) in this pursuit. However, the zetetics like- wise perceived themselves as seekers and defenders of 'scientific' truth. Unlike Wallace, their reality was founded in their religious beliefs, so no amount of experimentation or direct observation would have been sufficient to convince them of the rotundity of the earth.[66] For zetetics, who filtered their reality through theory, it was a case in which seeing was not necessarily believing. With a sweeping social programme rooted in an interpretation of the word of God, Hampden was not 'willing to pay to be enlightened', as Wallace had at first supposed. Although the wager might have seemed amusing, the consequences were to cut far deeper than he had initially assumed.

Chapter Four

TRIALS AND TRIBULATIONS

THE TRUTH SHALL KEEP US FREE!

Rise up and join hands with the truthful and bold;
Be earnest; be candid; – be men:
Fiercer work there is yet! Ye, like slaves are still sold;
And must fight for your freedom again.

Be free! Bid the fetters of '*Science*' be riven!
Bid ASTRONOMY seek her defence!
Let her facts and her principles plainly be given;
Her *theories*, – bid them go hence!

Be men, and permit not your thinking to be
In the hands of a privileged few:
Strike out for yourselves; let philosophers see
What the hard-handed workmen can do!

WILLIAM CARPENTER,
Wallace's Wonderful Water (1875)

AS STORIES ABOUT the wager filled the London press, Parallax was at his Kentish Town home devising his personal response to the débâcle. Unsurprisingly, he was outraged by his disciples' disastrous intervention, declaring that there had never been an instance where it could be more justly said 'save me from my friends'. Set on undoing the damage done to his cause, he decided to publish a pamphlet about his original tests on the Old Bedford Canal, *Experimental Proofs that the Surface of Standing Water is not Convex but Horizontal* (1870). Parallax recognized that the wager had cast doubt on his attempts to present himself as a credible investigator, as well as on the trustworthiness of his experimental results, and he concluded his analysis with a lengthy complaint:

> For the long period of thirty-one years I have laboured singlehanded to bring this important subject before the world: both on the [lecture] platform and in local Journals, and travelling from place to place – never resting longer than a few months in one locality, but like ... a scientific or philosophic gypsy breaking up his tent and pitching it 'here, there and everywhere' in order to ... draw [this great question] to the attention of all classes and degrees of intelligence (and as a matter of course I have had to bear every possible form of opposition, the bitterest denunciations – often amounting to threats of violence and personal danger, the foulest misrepresentations, the most reckless calumny, and the wildest and most desperate efforts to stay my career and counteract my teachings).[1]

Yet despite his expertise in the field of flat-earth theory, Parallax grumbled that Hampden and Carpenter had kept him entirely in the dark about the upcoming wager. Not only was this 'a needless insult', he complained, it was also a foolish decision, considering they had known little or nothing of the scientific instruments that the experiment involved. Consequently, they had rendered themselves 'literally the helpless victims of their more philosophical and practical opponents', a mistake that had been bound to cost Hampden his £500. Yet whatever the damage done by the faulty test, Parallax still held out hope for the future of the campaign. Ever tactful, he said that the controversy seemed to have caught the attention of scientific men and professional societies who might investigate further, and for this he offered the participants his thanks.

While Parallax attempted to twist the wager to his advantage, Hampden sent two sympathizers, the Swindon printer and publisher Alfred Bull, and a Brighton associate, Mr Gutteridge, to the Old Bedford Canal to try some experiments on his behalf. The men arrived on 4 May and, seeking to undercut professional authority, decided to rely on 'common sense evidence' and the help of local witnesses to establish persuasive experimental proof that Hampden had been correct.[2] When *Scientific Opinion* reviewed their report on the experiments it restricted its remarks to a simple plea: 'When shall we hear the last of this foolishness?' The unwritten response was that it would be quite some time. Hampden had already drafted a new batch of handbills and flyers, branding Wallace a fraudster, claiming that globe-manufacturers and professors were in 'an agony of confusion and alarm', accusing the Royal Geographical Society of conspiracy, and scoffing that it was little wonder the missing explorer Dr Livingstone refused to return from Africa in the light of such scandalous behaviour.

Over subsequent months the barrage of invective continued unabated. In July, *Nature* warned its readers, 'Mr. John Hampden has again brought his sophisms and misstatements before the public in the form of a periodical called the *Armourer*,' which, the paper quipped, 'had been previously discontinued [to] the regret of hundreds of its readers'.[3] As the first issue of the paper went on sale, debate about Parallax's lectures and experiments spilled into *Nature*'s correspondence columns, provoking its editor to ask, 'Will nothing stop "Parallax's" mouth?' Parallax responded a week later, boasting that the only thing that would 'stop his mouth' was 'a practical experiment, fairly conducted, honestly reported, and logically applied', and with that he challenged 'friends of the Newtonian system' to repeat the test on the convexity of water at the Old Bedford Canal. Fearing further controversy, the editor elected to add a note to the proposal: readers should be wary of accepting the offer for 'Mr. Wallace's treatment at the hands of these gentry shows us what to expect'. 'Let "Parallax" take a good telescope and a return ticket to some seaside place,' the editor continued, 'and watch the ships travelling to and fro over the horizon', after which time he would be welcome to column space to explain his observations.[4]

In the interim, Wallace had been attempting to disregard the zetetics, and was concentrating mainly on his scientific work. In the autumn he attended the annual gathering of the influential learned society, the British Association for the Advancement of Science, presenting a paper on the eccentricity of the earth's orbits and the precession of the equinoxes, their impact on climate and the development of plants and animals over time. During the same period, the flat-earth campaigners were working on a range of new ideas. Carpenter began a series of 'Anti-Newtonian Papers', insisting they were publications that 'no zetetic philosopher should be without', while Hampden fired off letters denigrating Wallace and denouncing the Royal Astronomical Society as a bunch of 'professional liars'.[5]

As part of his anti-professional crusade, Hampden did not overlook Wallace's friends and associates, or the scientific societies to which he belonged, but still Wallace sought to ignore him. He must have hoped that Hampden would tire eventually of his campaign, and through early 1871 he focused on propounding his contentious views on spiritualism, natural selection and the links between them in the belief that such issues were central to science. From Wallace's perspective, the influence of a higher power on man's mind remained a critical question: not only to advancing human knowledge and uncovering truth, the principles by which he was driven, but also to influencing the direction that scientific research would take at a time when the situation was malleable. The parameters of professional enquiry were still in transition and spiritualism was in its heyday, with seemingly inexplicable phenomena from table-turning to transportation being exhibited by mediums and magicians in parlours and theatres nationwide. The possibilities were endless, Wallace believed, and he could not understand why apparent evidence of 'unexplained powers of the human mind, or the action of minds not in a visible body' was being rejected out of hand by many men of science. That they were placing limits on enquiry to further their professional status was truly extraordinary, Wallace asserted, and he was willing to say so in public. For him, the narrow-minded, 'unscientific and unphilosophical' attitude of some of his peers was 'the most striking instance on record of blind prejudice and unreasoning credulity' and a slur on the name of science.[6]

Increasingly radical, inventive and willing to flout convention in pursuit of the truth, in testing the boundaries and definition of science, Wallace had more in common with the zetetics than one would at first suppose. Meanwhile, his wide-ranging approach to 'truth-seeking' was becoming more evident in his scientific work.[7] By March 1871, he was busy compiling a number of book reviews, including his reflections on Darwin's long-awaited *Descent of Man*, in which evolution by natural selection was finally applied to the thorny issue of the origins and development of humankind. The detailed two-volume study deserved to be applauded as 'one of the most remarkable works in the English language', Wallace gushed, although his review was not a blanket appreciation of Darwin's book. His greatest criticism was reserved for its lack of explanation of mankind's vast superiority to his closest species, which Wallace ascribed to the influence of a 'higher power', and he even implied that 'Mr. Darwin' had suggested the same, for his study seemed to give 'hints of unknown causes which may have aided in the work'.[8]

The review was published in May, by which time Wallace had moved with his pregnant wife, Annie, and two children, Bertie, aged three, and Violet, two, to a stately white house in rural Essex, the Dell. Set in four acres on the outskirts of Grays, Wallace had fallen for the sweeping views the site offered and, after months of negotiations, had acquired the lease and designed a home where his young family could relax and grow. In keeping with his dream of a peaceful rural idyll, Wallace set about creating an enclosed and serene little world at the Dell, one of first British buildings to be constructed of concrete. He even planned to build a well and a windmill in its grounds, which were sheltered by woodland and surrounded by thick-set walls. For Wallace, the house was the perfect retreat from the competitive culture of professional science in the capital and he confided to Darwin that he was so busy with 'road-making, well-digging, garden & house planning' that he had effectively given up all other work for a while.[9]

The same could not have been said of Hampden, who remained fixated on his flat-earth campaign, and while Wallace was preoccupied with domesticity at the Dell, he was issuing flyers at a remarkable rate. His most recent production, in late May, claimed that the recent death of leading astronomer John Herschel had been hastened by attacks on conventional astronomy, while another broadside, in June, declared 'British Science Outlawed' on the grounds that professionals had refused to defend the Newtonian theory or refute his accusations against Wallace. 'The meanest felon in a London jail has a pride of character which an English Professor dares not assume,' Hampden railed, for 'these cowardly dastards are too conscious of their guilt to go into any court in England and attempt to defend their principles or their conduct'.[10] Convinced of his truth, Hampden then turned his attention to the Astronomer Royal, George Airy, informing him by post:

> I hope you may not soon have to follow that other poor deluded man Herschel into the Westminster vaults, but if [your] lying impostures last much longer, you may rest assured that the ultimate exposure which inevitably awaits you, will cause you great discomfort. I have posted ... A. R. Wallace, as a thief and a swindler for many months, and not one of his ... friends has dared to come to his rescue. You must be all tarred with the same brush ... [11]

During this period, Wallace was distracted by building troubles at his new home, but the situation altered when the pregnant Annie received her own letter from Hampden:

Madam, if your infernal thief, of a husband is brought home
some day on a hurdle, with every bone in his head smashed to
pulp, you will know the reason. Do ... tell him from me he
is a lying infernal thief, and as sure as his name is Wallace
he never dies in his bed. You must be a miserable wretch
to be obliged to live with a convicted felon. Do not think or
let him think I have done with him.[12]

Just weeks before, Wallace, then president of the Entomological
Society of London, had informed fellow member Robert Mac-
Lachlan that he had instituted legal proceedings against Hampden
in January but 'as the man is half mad I don't want to indict him
criminally and imprison him, and so I suppose he will continue to
write endless torrents of abuse as long as he lives'.[13] But this latest
death threat was too shocking to overlook and Wallace decided
to report it to the police. Hampden was arrested and taken into
custody at West Ham police station on Saturday, 9 July, charged
with threatening to kill Wallace. Hampden complained that the
accusations were ludicrous – he had never once threatened Wallace
'with any violence which would have killed a canary', and would
not even have mentioned the naturalist's name had he not been
swindled out of £500. The only threats that the prosecution could
rake up, Hampden fumed, were that 'if I ever met him I would
spit in his face' or that 'he would never die in his bed', and as for
those,

I do not deny having said 'I would spit in his face' and perhaps
soil his trousers with the tip of my shoe, if I ever met him,
and I warned his wife that if some of my less discreet friends
should do him any injury, I should not bear the blame, as I
was living and spending my whole time in Wiltshire and he
in Essex, so that the chance of my even getting a spit at him
was rather improbable.[14]

Nevertheless, early on Monday morning, magistrates ordered
Hampden to pay £100, find two people to pay good-behaviour

bonds of thirty pounds each, and bound him over to keep the peace for three months or else spend that period in prison. However, Hampden, who declared he would rather rot in the 'blackest dungeon in the kingdom' than behave as dishonourably as Wallace, could not come up with the money. With no further ado, he was sent to prison until he could meet the demands.

A week later the bonds were paid and Hampden was released, only to face the legal proceedings that Wallace had instituted. The papers subsequently reported that Hampden slipped away from the hearing at London's Guildhall, without even instructing a lawyer, while Wallace had arrived with two solicitors and a clutch of witnesses to ensure that justice was done. Summing up, the judge remarked that it was 'a most curious thing that in the nineteenth century any man should be found to wager £500 that the earth was not round', and he should have thought there was a 'craze' upon Hampden, but such considerations were beyond the scope of the case. After a brief absence from court, the jurors returned with a verdict in Wallace's favour, setting damages at £600.[15]

Yet whatever was said about delusions and crazes, Hampden was sufficiently lucid to plead poverty, transfer his assets to his solicitor son-in-law, and declare himself bankrupt so that he did not have to pay a penny of the damages owed. Wallace, satisfied that his name had been cleared, chose not to enforce the court's decision or to become embroiled in a complicated case to overturn the fraudulent bankruptcy claim. He at least received sympathy from Darwin, who wrote that he had been 'grieved' to read about the 'madman' threatening his life, recognizing 'what an odious trouble this must have been . . .'[16] Meanwhile, in a pamphlet, *After all the Commotion, John Hampden Triumphant, Always Has Been and Always Means to Be* (1871), the 'madman' in question declared:

It is now notorious all over the world that I, John Hampden, have, singlehanded and alone, thrown all the ranks [of British scientific societies] into confusion, by having detected and exposed the shallow subterfuges of their boasted professions and the falsehood and delusion of their most cherished and fundamental theories, till they naturally regard me with mingled feelings of terror and bitter animosity.[17]

For Hampden, the legal decision was evidently an incentive rather than a deterrent: indeed, even as the case was progressing through the courts, he was engaged in another acrimonious debate with well-known sea captain and Royal Geographical Society member George Peacock. The controversy commenced when Hampden noticed that Peacock's company trademark was a ship resting on top of a globe and wrote suggesting that he would do just as well to rest his ship on a gooseberry bush. The company had committed two gross blunders, Hampden continued: they had set the ship where no ship had ever been, on the crest of the North Pole, and they had depicted the world as round 'when it is as flat as the table on which I am writing'.[18] Over subsequent months, Peacock proceeded to offer proofs of the rotundity of the earth and 'the paternal advice of an old sailor who has been knocking about on land and sea for the last fifty years of his life', while Hampden challenged him to 'come out like a man, and show me 50 or 20 or 10 miles of a curve', from which time, he claimed, he would 'forever hold [his] peace'. By September 1871 Peacock decided that the debate had been taken far enough and, eager to expose the flat-earth error, he published the correspondence in a sixpenny pamphlet, *Is the World Flat or Round?*, receiving a mixed response from the critics. Although many joined with the *Gloucestershire News* in praising Peacock's 'racy and satirical rejoinders', many also agreed with an *Exeter and Plymouth Gazette* reviewer, who questioned whether 'Captain Peacock's game has been worth the candle'.[19]

A month later Wallace adopted a similar approach to that of Peacock by writing a letter to the *English Mechanic* to revive the dormant debate. It was an unusual response, considering his position and the extent of the controversy, yet Wallace's letter about proofs of the earth's rotundity was deliberately designed to rile leading zetetics, whom he branded 'persons of a peculiar frame of mind'.[20] In particular, he asked why the sun was not visible over the entire planet at once, as it would be if the earth were flat, and asked the editor to open his columns to receive zetetic replies. In the eyes of Wallace's colleagues, courting further debate was certainly inappropriate and only the ardent controversialist and fellow independent spirit, astronomy writer Richard Proctor, stepped in to offer limited public support.[21] Parallax's theory had already accounted for the apparent rising and setting of the sun, Proctor reminded Wallace, and if he sought to branch out from natural history to astronomy it was vital that he understood related topics in their entirety.[22]

While Proctor advised Wallace, Hampden now set his sights on a national bastion of globular theory, the Royal Geographical Society. Evidently he hoped that harassing a prestigious organization, of which Wallace was a member, would further his unorthodox cause, and this mistaken impression formed the basis for another onslaught of outrageous mail.[23] Meanwhile at the Entomological Society of London, Robert MacLachlan had received a quantity of similar letters, and had written asking Wallace how he should respond. Wallace told him that he was presently in consultation with his lawyer about the best steps to be taken, but in the meantime the most 'effectual check on Hampden's ravings' would be to return his correspondence unopened 'with a short note stating that as you are now sufficiently acquainted with the tenor of his letters all future ones will be burnt unread'.[24]

It is doubtful that Wallace's lawyer advised him to continue goading flat-earthers through the pages of the *English Mechanic*, but by late October his efforts in that direction were beginning to bear fruit. On the twenty-seventh, Parallax wrote to the magazine, praising Wallace's 'courage in his attempts to stay the great zetetic wave which is now setting in from every quarter' and challenging him to a public experiment, while other correspondents pleaded with the editor to 'save us from flat-theorists!' and 'sweep the rubbish into the wastebasket'.[25] But the editor had no intention of following their advice: in a competitive market a controversy involving a famous naturalist was undoubtedly a boon to the magazine's sales, and of all the letters it had received on the topic only Hampden's were deemed unsuitable for publication as violations of good taste. Under such circumstances it is surprising that Wallace accepted Parallax's challenge in the next issue, adding a healthy dose of sarcasm to his reply: 'Surely if the earth were flat,' Wallace jeered, 'it should be possible to see the Alps with a telescope from the East Coast of England,' yet 'how strange that of the hundreds of good telescopes upon our east coasts none have ever shown a glimpse in that direction of anything but sea and sky! . . . Let "Parallax" take his telescope to Southend or Margate,' Wallace continued, 'and exhibit this beautiful sight to an admiring crowd. When he can do so we shall, many of us, become his converts.'[26]

In the weeks that followed, Wallace and Parallax wrangled over the design of their experiment and eventually failed to agree, while Proctor, having ruminated long and hard about ways to deal with flat-earth believers, wrote to the *English Mechanic* to protest about the publication of Parallax's letters. 'No astronomer who values his reputation,' he threatened, 'can possibly write . . . for a journal in which the absurdities of "Parallax" are treated . . . as worthy of serious discussion, or classed with the Newtonian system as though the two were "rival theories".'[27] His arch remarks struck a chord with regular readers and one correspondent sarcastically requested column space to discuss the moon being made of green cheese, and whether its so-called ridges and craters were actually the result of the planet's inhabitation by 'mighty green cheese-mites'. This prompted Parallax to turn his attention to Proctor: if Wallace was unwilling to accept the challenge, then maybe Proctor would be willing to try? Suspecting trouble, the editor stepped in to inform Proctor that if he wanted to accept he would have to use some other publication as a medium for the controversy, and as Wallace and Parallax had failed to agree on terms for their experiment, flat-earth theory was ruled out of bounds.[28] Over subsequent months the paper turned its attention to debates over spiritualism and the supposed evidence of 'psychic force'.

While debate simmered, Hampden had written an article for the *Weston-super-Mare Gazette*, giving twelve reasons as to why the earth could not be a globe. Appropriately the piece appeared in print on 10 November 1871, the day that Hampden was due in the dock at Bow Street Magistrates' Court. Wager umpire John Henry Walsh had now grown tired of suffering 'the grossest vituperation' since his fateful decision, and had decided to have Hampden charged with libel. Following discussion with the judge, Hampden pleaded guilty to sending malicious and defamatory postcards and was committed to trial at the Old Bailey on payment of £100 bail. Twelve days later the case came up before the distinguished old judge Robert Malcolm Kerr, or 'Commissioner Kerr'. The hearing commenced with an announcement from Hampden's lawyer that his client had now drafted an apology to Walsh. According to Kerr, that did not detract from the seriousness of the offence: 'Nothing could be more heinous than to take advantage of a public convenience, such as the postcard system,' he declared. However, considering the signed apology, he was prepared to be lenient: Hampden was released on bail and bound over to keep the peace for a year with a caution that if he dared to reoffend the consequences would be severe.[29]

As Walsh had resorted to the courts, Wallace now opted for a different route to safeguard his reputation, publishing *A Reply to Mr. Hampden's Charges* (1871). The eight-page pamphlet, which was distributed to members of the societies to which Wallace belonged, was a full account of the wager and Hampden's subsequent 'violent abuse', dismissing the perpetrator as 'an ignorant but very foul-mouthed libeller'. In conclusion, Wallace asked colleagues to burn Hampden's letters, stating that although he was unwilling to resort to a criminal prosecution, he would not hesitate to do so if his friends thought such action advisable.

Such actions only fuelled zetetic protests, and through early

1872 they stepped up their campaign. Parallax threatened to sue the *English Mechanic* for not publishing his letters, while Hampden and Carpenter issued a flurry of furious flyers castigating the 'delusive frauds' of the Newtonian system, disseminated by 'Bible-hating' professionals in cahoots with the crooked and pernicious London press. One such flyer, 'An Anticipated Increase in the Waste Paper Department', offered an amnesty for scientific books and globes, contending that such materials would soon be rendered irrelevant in consequence of the upcoming 'entire and unexpected revolution in Geographical Science'. Those who wished to do so could beat the rush by immediately exchanging their 'condemned and pernicious rubbish' for cash at twelve shillings per hundredweight of books and a shilling and ninepence for globes.[30] Simultaneously, in another leaflet, 'Facts or Fiction – Science or Truth – Working Men, Which is it to be?', they sought to mobilize working-class support by warning that the newly established education boards were 'mere clap traps' and that children were being misled by geography lessons 'as false and fictitious as any story in the Arabian Nights'. 'Let the working men of England understand how wide the difference is between . . . Science and Truth,' the zetetics proclaimed, for thousands of pounds of public money was being thrown away on employing professors who misled ordinary people about science and Christianity. The élitist conspiracy was continuing while God-fearing men were starving because they dared to teach and preach God's truth.[31]

Poverty was an issue that rankled especially with Hampden, who appeared to have lost his yearly £50 by this time. Despite his hatred of professional science, by May 1872 he had become so desperate that he wrote a begging letter to the Royal Geographical Society:

Having, as you know, been swindled out of every fraction I possessed by that convicted knave and scoundrel, A. R. Wallace F.R.G.S. . . . perhaps you will be good enough to support an appeal for a subscription on my behalf . . . If the thief himself cannot pay what he owes, let his friends do so for him. For I am in the greatest distress & will die before he shall hear the end of it.[32]

But despite Hampden's imaginings, Wallace was labouring under financial difficulties of his own. Struggling to support his family on a low income, he was still waiting to hear news of his application for the directorship of Bethnal Green Museum, an offshoot of the new South Kensington Natural History Museum, aimed at the working classes. The job seemed ideal, considering Wallace's contempt for class privilege and his dedication to popular education, but in July he was informed that, due to shortage of funds, the museum would not be appointing a director after all. Hopes crushed, Wallace confided to Darwin that this had come as 'a considerable disappointment' for he had 'almost calculated on getting something there'.[33]

Also intent on educating the public, albeit in a much different sense, a Parallax disciple, B. Charles Brough, issued the first number of a halfpenny magazine, the *Zetetic: A Monthly Journal of Cosmographical Science*, in July 1872. In the meantime, the editor of the *English Mechanic* had grown weary of Hampden's 'ribald abuse', and had forwarded one of his offensive letters to Wallace. In mid-September, Wallace also received a supportive note from fellow Royal Geographical Society member Alexander George Findlay, urging him to put a stop to the defamation. Further pressure to prosecute followed in a letter from Walsh. He had recently received a postcard from Hampden, vilifying Wallace as a thief, impostor and swindler, signed with the threat 'Have you any idea how long this is to last?' From Wallace's perspective, following almost three years of abuse, Hampden's campaign had persisted long enough. He contacted the police and had Hampden charged with libel at Bow Street Magistrates' Court. Hampden, who had been bound over in November 1871 to keep the peace for a year, pleaded guilty, adding that, as his year was almost up, he had felt he was free to 'go on' at Wallace and Walsh again. In light of his confession, he was committed to trial at the Old Bailey and allowed out at large on bail.[34]

Three weeks later Hampden was summoned to the Old Bailey, and announced that he was willing to retract his libellous remarks. Following discussion between the judge, Commissioner Kerr, and solicitors for both parties, it was decided that he should draft an apology to be published in twelve newspapers, as chosen by Wallace, and the case was then adjourned. The subsequent hearing opened with a revelation: Wallace's solicitor, William Goodman, informed the court that he had received a malicious document in which Hampden said that he regarded his (now published) apology as a formal matter, and that all he pleaded guilty to was 'calling a spade a spade'. Hampden's lawyer immediately leaped to his client's defence, emphasizing that he wished to apologize again 'in the most humble manner' for the new set of 'illegal expressions'. But this held little sway with Commissioner Kerr. Capable and caustic, he had also seen a flyer in which Hampden said the apology was meaningless and that he did not believe a word it said. 'If you are infatuated enough to do so,' Hampden had told Wallace, in this latest letter, 'you may demand and receive an apology every day in the week, knowing that ... they will only add ten-fold to your disgrace.'[35] Summing up, Commissioner Kerr said he was not concerned about the wager: the simple question was whether a law-abiding citizen should be harassed as Wallace had been.[36] He ordered that Hampden be bound over to keep the peace, draft a second apology to appear in various publications and pay £1000 in bail to secure his release until he was called back to court for sentencing.

As Hampden dealt with the continual court action, Parallax was establishing a monthly magazine of his own. In March 1873 he issued the first number of *Earth Life: A Monthly Journal and Record of all such Facts, Principles and Discoveries as Relate to the Improvement and Preservation of Earthly Existence*, a vehicle for his ideas about health, diet and disease. Under the name 'Dr. Birley', Parallax was now a busy medical doctor with a large practice at

Biblical conception of the world: 1. waters above the firmament; 2. storehouses of snows; 3. storehouses for hail; 4. chambers of winds; 5. firmament; 6. sluice; 7. pillars of the sky; 8. pillars of the earth; 9. fountain of the deep; 10. navel of the earth; 11. waters under the earth; 12. rivers of the nether world.

1. The biblical world view.

2. Diagram of Eratosthenes' experiment in the third century BC to calculate the circumference of earth.

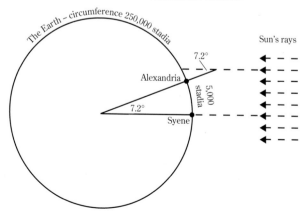

3. Woodcut of the world known to Ptolemy in the second century AD, taken from a 1545 edition of his encyclopaedic *Geography*. The editor, Sebastian Münster, integrated contemporary knowledge into his impressions of Ptolemy's world view.

4. An eleventh-century copy of an eighth-century world map drawn by Spanish priest, Beatus of Liébana, and centred on the Mediterranean Sea. Known as the Turin copy, it portrays the spread of Christianity and features Adam and Eve along with the four sacred rivers of Genesis.

The map provides a contrast to the T and O maps common at the time, for it depicts Europe, Asia and Africa, yet also indicates the existence of a fourth Antipodean continent (to the right).

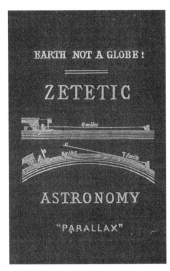

5. Founder of zetetic astronomy, Samuel Birley Rowbotham, aka 'Parallax' (1816–84).

6. Front cover of the second edition of Parallax's classic text presenting arguments for the flatness of the earth.

7. *Above*. George Biddell Airy (1801–92), Astronomer Royal.

8. *Right*. Richard Anthony Proctor (1837–88), leading popular astronomy writer and zetetic critic.

PERFECT SAFETY RAILWAY CARRIAGE.
Fig. 2.

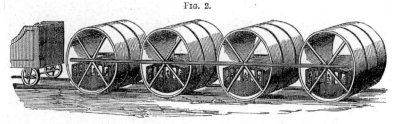

wheel and axle, which require little explanation.
The wheel is made with a cast-iron nave, the inner
surface of which is case-hardened. The front
reservoir or cup for containing the oil has a spherical surface, and will bear any amount of concussion
to which vehicles are liable. The cup is attached
to the nave by means of bolts, with an india-rubber
washer between. This invention possesses many
advantages over the ordinary wheel. The bearing
is shorter, thereby reducing the friction on the
axle, the wheel does not project so far from the
vehicle and has an exceedingly compact and neat
appearance, and does not require to be taken off
until the tire is worn out. These wheels are well
adapted to all sorts of carriages, and have been
extensively used by Messrs. Barclay and Perkins,
Whitbread and Co., Sir Henry Meux and Co.,
Elliot, Watney and Co., Combe, Delafield and Co.,
Goding, Jenkins and Co., Sir Morton Peto, and
many other extensive firms and contractors both
abroad and at home.

Fig. 1.

Correspondence.

PERFECT SAFETY RAILWAY CARRIAGE.
TO THE EDITOR OF THE "MECHANICS' MAGAZINE."

MR. EDITOR,—I beg to forward you a description
of a new principle in the construction of railway
carriages, which will render travelling thereby almost free from danger. The plan here given will not
require the slightest alteration in the present structure of railways; but, as will be seen by the drawings,
the carriages will be of a description totally different
to those now in use.

Thus it will be seen that in every journey by rail
the risk is incurred of serious or perhaps fatal accident from 17 different causes! And the value of
every suggested improvement must be estimated by
the number of causes or chances which its adoption
will take from this fearful list! That which reduces
the chances of accident to the least number will
necessarily be the greatest improvement. It will be
obvious, on carefully reading the description, that the
plan here given will reduce the chances of injury to
the lowest possible number. For example—

9. Master of invention: Parallax's Life Preserving Cylindrical Railway Carriage
as seen in *Mechanics' Magazine*, 29 March 1861.

10. Diagram of the controversial Parallax experiment on Plymouth Hoe
in October 1864.

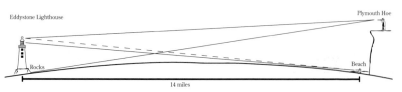

11. Alfred Russel Wallace, with a spirit form materialized by well-known London medium Mrs Guppy, in the 1870s.

12. A diagram of the Wallace–Hampden experiment to test the rotundity of the earth. It was drawn by Lincolnshire surveyor and sculptor, Thomas Wallis, who visited Alfred Russel Wallace at home to discuss the controversy.

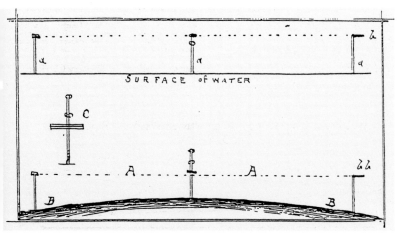

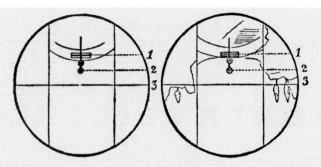

These two views, as seen by means of the *inverting* telescope, are exact representations of the sketches taken by Mr. Hampden's Referee, and attested by Dr. Coulcher as being correct in both cases: first, from Welney Bridge; and secondly, from the Old Bedford Bridge.

13. *Above.* The wager result: referees' sketches of contrasting views through the telescope.

14. *Below left.* Parallax's contentious account of the Bedford Canal experiments, published soon after the Hampden–Wallace wager in 1870.

15. *Below right.* A libellous Hampden broadside lambasting his opponents Wallace, Walsh and Coulcher in the wake of the experiment.

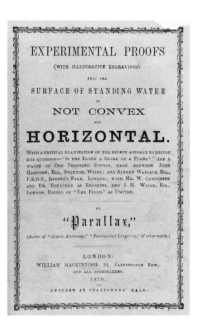

EXPERIMENTAL PROOFS

(WITH ILLUSTRATIVE ENGRAVINGS)

THAT THE

SURFACE OF STANDING WATER

IS

NOT CONVEX

BUT

HORIZONTAL.

WITH A CRITICAL EXAMINATION OF THE RECENT ATTEMPT TO DECIDE THE QUESTION—"IS THE EARTH A GLOBE OR A PLANE?" FOR A WAGER OF ONE THOUSAND POUNDS, MADE BETWEEN JOHN HAMPDEN, ESQ., SWINDON, WILTS; AND ALFRED WALLACE, ESQ., F.R.G.S., REGENT'S PARK, LONDON; WITH MR. W. CARPENTER AND DR. COULCHER AS REFEREES, AND J. H. WALSH, ESQ., LONDON, EDITOR OF "THE FIELD," AS UMPIRE.

BY

"Parallax,"

(Author of "Zetetic Astronomy," "Patriarchal Longevity," & other works.)

LONDON:
WILLIAM MACKINTOSH, 24, PATERNOSTER ROW,
AND ALL BOOKSELLERS.

1870.

ENTERED AT STATIONERS' HALL.

THE HOUSE IN CONVULSIONS
AT THE
LAST NEW PANTOMINE.

THE FIGHT WITH A TARTAR.

FOR THE SPECIAL BENEFIT OF THE ROYAL GEOGRAPHICAL SOCIETY,

Whose funds, it is feared, are considerably reduced in consequence of the refusal of the several members to engage in any more CANAL SPECULATIONS, and to have their names posted all over the Kingdom as Defrauders and Defaulters; also, from the difficulty of inducing fresh members to join the Association, till it has publicly cleared itself from the imputation of harbouring and abetting certain Swindlers whose conduct has caused them a notoriety which few care to possess, and is usually attended with "more plague than profit."

SCENE.—A CANAL SIDE IN NORFOLK.

The Scientific Wager;
The Unscientific Welsher;
The Scotchman who Grabbed what Burnt his Fingers;
The Umpire who did not know Up from Down;
Three Rogues Hoisted on their own Petard;
"Old Scratch" Screaming—"It serves them Right;"
And the Earth proved a PLANE after All!

16. *Above*. A poster advertising a Parallax lecture just one week after the Bedford canal debacle in March 1870.

17. *Below left*. Hampden continued to set public challenges for proof of the rotundity of earth, although the value of associated cash prizes gradually decreased.

18. *Below right*. An early twentieth-century advertisement for Parallax's cure-all Syrup of Free Phosphorus [*sic*], which was manufactured under his alter-ego, Birley, long after his death.

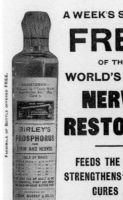

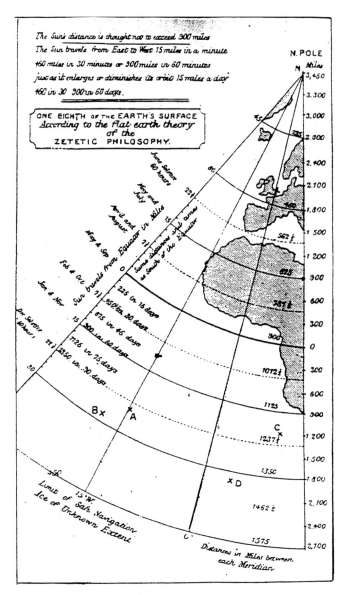

19. Hampden was certain that this diagram, depicting one eighth of the earth's surface according to flat-earth theory, would soon be the 'alone standard of reference when discussing the configuration of the earth-plane on which we live'. Its publication by Richard Proctor in the 'Our Paradox Corner' column of *Knowledge*, on 23 March 1883, triggered heated arguments to the contrary.

his beautiful, twelve-roomed home on Haverstock Hill, North London, and had also embarked on the manufacture of a patent medicine, the 'syrup of free phosphorus', to ensure long life for his patients.[37] The medicinal elixir was the result of his experiments on human brains, and was sold as a cure-all for every ailment from anxiety, hysteria and St Vitus' Dance to impotence, poor eyesight and heart disease. In an era when the marketplace was bursting with patent medicines, elixirs and pills, where pharmacists and quacks were vying with the rapidly emerging medical profession for the authority to cure, the syrup of free phosphorus soon became Parallax's major concern. Now, it appeared, he was set on challenging both the medical and scientific professions, and the knowledge they represented, with radical experimental evidence and a widening range of unconventional ideas.

In the same month that Parallax established *Earth Life*, he wrested the editorship of the *Zetetic* from Charles Brough and renamed the paper the *Zetetic and Anti-Theorist*. To herald the new era, Parallax issued a new challenge to 'scientific men of the whole world' on the paper's front page. In the April issue, he reported that his challenge to an experiment on the convexity of water at the Old Bedford Canal had not yet been accepted but 'we shall continue to refer to it until some worthy champion of the Copernican theory takes up the gauntlet'. Until that time, Parallax declared, all who were interested in the question should make a scientific pilgrimage to the Cambridgeshire fens at least once in their lifetime to undertake the necessary experiments for themselves. In the meantime, he continued his usual round of lectures, including a vociferous debate at Penge town hall with John Dyer, author of *The Spherical Form of the Earth*.

Naturally, the *Zetetic* reported a victory for Parallax in glowing terms, and went on to inform readers that a Zetetic Society had been formed in New York, with Parallax as president, and the United States consul for China, George Henry Colton Salter, the surgeon-general of the Grand Army of the Republic, Hans Powell, and the superintendent of Baltimore public schools, John Nelson

McJilton, among others, on its council. Despite these advances, Parallax's challenge to scientific men had not been taken up and in the July issue of the magazine he decided to switch tactics, addressing it directly to Wallace. Days later, contrary to all expectation, Wallace decided to accept.

The subsequent chain of events was not mentioned in Wallace's otherwise comprehensive two-volume autobiography, *My Life*, in which he referred to his involvement with flat-earthers as 'the most regrettable incident in my life'.[38] Thus one may only surmise his motives for accepting another challenge from Parallax in the wake of the Hampden débâcle, although one clue is provided by his comments in a letter to a Lincolnshire carpenter, Thomas Wilkinson Wallis, written a year after the event:

> Parallax makes the boldest false statements and as the number
> of those who can contradict him from actual experiment is
> small his assertions are believed by thousands. I wonder at
> [his] lectures his gross false statements are not exposed; but I
> suppose there are always a majority present so ignorant as to
> believe that his assertions are as good as anyone's.[39]

Clearly Wallace believed that defeating the founder of zetetic astronomy would have more impact than tackling one of his followers and, with protecting the public understanding of science his ultimate goal, he sought to deal a final death-blow to the flat-earth campaign. In addition, more personal considerations might have been at play. Wallace's growing reputation as a crank, and his marginalization in some circles, while self-induced, perhaps encouraged him to persist in a misplaced defence of scientific authority. Such challenges allowed him to stand up for science, shine a light of truth and conquer false knowledge, a personal bugbear, while he might also have viewed them as an opportunity to differentiate himself from paradoxers in the opinion of his peers. Whatever the case, Wallace's decision to take on Parallax illustrated his crusading spirit and dogged determination to win an argument. Yet in an era when science was establishing itself as the primary source of cultural authority, where accepted codes of professional behaviour were being carved out by men in the field, Wallace's continued engagement with the zetetics could only strengthen conventions of appropriate conduct.

Parallax, meanwhile, was delighted by Wallace's acceptance of the challenge, and his agreement to pay half of the surveyor's fee for the upcoming rematch. Wallace's professional reputation and his connection to Darwin and natural selection provided just what Parallax wanted – a hint of legitimacy for an idea generally deemed beyond the limits of sensible debate – and would also generate more publicity for the cause. Initial discussions focused on arrangements for a test to see whether a boat could remain visible through a telescope as it sailed six miles along the Old Bedford Canal. For various reasons it was decided that the experiments would be undertaken by proxy, with Martin Coulcher, the referee in the Hampden wager, representing Wallace and a professional surveyor on either side. However, it was not long before Wallace and Parallax came to loggerheads. Parallax refused to identify his surveyor or provide references testifying to his technical qualifications and impartiality about the earth's shape, while Wallace insisted that conditions in late August were unfavourable for observations along the canal. Vapours and flickering of the air were common at this time, and neither bridge would be visible along the six-mile stretch of canal, especially when a telescope was positioned near to the water's surface. Wallace suggested a postponement, but Parallax persuaded him otherwise, arguing that he had been able to make clear and distinct observations at the Old Bedford Canal throughout the period from April to October. So long as the test was undertaken soon after sunrise, he insisted that the vapours would be avoided and that there should be no further delay: the test should proceed as planned on Tuesday, 26 August.

Then came the last-minute twist. Two days before the experiment, Parallax wrote to his surveyor, Mr Jones, that 'circumstances of a serious nature' would compel him to remain in London. Jones was therefore instructed to go ahead and make the experiment with Coulcher and Wallace's surveyor, Thomas Burton of King's Lynn, according to the description provided in the *Zetetic*. Arrangements made, the party went to the Old Bedford Canal and located a man with a flat-bottomed boat to row the six miles from Downham Market to Welney. However, the *Zetetic* subsequently reported, it was ten a.m. before the boat moved off towards Welney and the vapours obscured their observations by the time it had travelled three miles. Despite the failed test, Parallax was not willing to relinquish his connection with an eminent man of science: he invited Wallace to undertake a rematch in the spring.[40]

Having gained a reprieve as far as Parallax was concerned, Wallace remained determined to put an end to Hampden's abuse. Despite having been bound over to keep the peace and freed, pending sentencing, on payment of £1000 bail, Hampden had continued his campaign of harassment through 1873, and was brought before a London judge on 17 December to hear his sentence. Wallace's solicitor, William Goodman, did not hold back in his efforts and the court was treated to a full review of the wager and the ensuing legal tangle, with Hampden's subsequent literary endeavours detailed at some length. In response, Hampden got to his feet and pleaded that he was starving while his opponent was living in luxury with his ill-gotten gains, and was attempting to 'inveigle' others into similar experiments. Having heard the evidence, the judge said it was clear that Hampden had gone on attacking Wallace and his family in an 'unworthy and disgraceful manner, improper in every sense of the word' and 'if all persons who had quarrels were to act as he had done the country would not be worth living in for a day'. Yet he also felt that Wallace had been 'most unwise' to take a wager with a man who, on the subject of the earth's shape at least, was 'plainly not in his right mind'. It was high time, the judge concluded, that the defendant was made to feel the full force of the law, and he saw no alternative but to sentence Hampden to two months in London's notorious Newgate Gaol.[41]

Two months passed and Hampden was released, only to return

to his campaign. Infuriated and frustrated by his imprisonment, Hampden continued to fire off letters to and about Wallace, who now had many other difficulties besides. In April his only son, six-year-old Bertie, died from unknown causes, an event that Wallace was to mourn privately for the rest of his life. He continued to work, however, and published a major article 'A Defence of Modern Spiritualism' and a book, *Miracles and Modern Spiritualism*, in the months following his son's death. In addition, he was also engrossed in a ground-breaking study of the geographical distribution of animals and plants, arising from his eight-year research expedition around the Far East. He had been the first to note the discontinuous geographical distribution of flora and fauna between Australia and the Orient, a boundary subsequently known as 'Wallace's line', a landmark theory that would later form the basis for the field of biogeography. But plaudits and praise seemed far away in the winter of 1874 and, in the thick of research notes and drafts, Wallace complained to Darwin that 'my horrid book on Geographical Distribution . . . is almost driving me mad with the amount of drudgery required'.[42]

Hampden, meanwhile, was engaged in writing material of a different kind, leading the *English Mechanic* to report that he sent at least one libellous letter a fortnight to the magazine, and it was 'a wonder . . . that he has not tired of writing when so many of his letters are left unnoticed'.[43] But the comment was premature: Wallace, for one, had not left Hampden's letters unnoticed, and a libellous postcard that he received on 31 January 1875 was the final straw. By this point Hampden had been prosecuted three times to conviction, been found liable for £600 in damages, provided two sets of public apologies and served two months in Newgate Gaol, but was clearly set on fulfilling his promise to pursue Wallace for as long he lived. Finally realizing that full criminal proceedings were in order, Wallace had Hampden arrested and brought up in police custody before magistrates at Orsett Petty Sessions, the nearest court to his home in Grays, where the offending postcard was presented as evidence.[44] An open and shut case, the magistrates committed Hampden for trial at Chelmsford's Shire Hall on a charge of libel, and when he could not pay the bail set, he was escorted to the cells to await a court date.[45]

He did not have long to wait. On Thursday, 4 March, he was brought before the Lord Chief Justice of England, Sir Alexander Cockburn, charged with libel and defamation. The judge, a brilliant man known for his fluent oratory and quick wit, listened with interest while the clerk read out the indictment. It covered six counts relating to various postcards and letters that Hampden had sent since his release from Newgate. Among them was a note to the publishers of the *Encyclopaedia Britannica*, to which Wallace was a contributor, alleging that 'the admission of articles from convicted thieves and swindlers cannot possibly do [you] or [your] work any credit'. Besides these, there were several letters to Wallace defying him to prove that the libels were false:

> I make no secret of having written thousands of letters declaring your conduct to be that of a cheat and a swindler; and I mean to continue to do so as long as your conduct justifies my making such charges. No lawyer or Law Court in England shall prevent my doing so. I know my position too well. You swore you had an answer to my charges. Is this a perjured lie or not? Did you ever know a thief who would hesitate to tell a thousand lies? I only wonder your wife and family are not ashamed to live with you. How much longer is this to last? Do not dream of my getting tired.

Hampden was directed to submit his plea of guilty or not guilty by the clerk of the court, to which he stubbornly declared, 'I am prepared to justify every statement I have made.' With some forbearance, the clerk repeated the question. This time Hampden replied, 'Not guilty', but proceeded to appeal to the Lord Chief Justice that he could not afford a lawyer and would like to apply for a postponement. The judge immediately denied the request on the grounds that the case concerned a 'systematic course of libel', to which Hampden responded that it had also been 'a systematic course of misrepresentation and false conduct on the part of Mr. Wallace' and 'If I am not prepared to prove that[,] I should be put in an asylum.' The Lord Chief Justice quipped that his suggestion was certainly possible, and as the onlookers roared with laughter, Hampden continued to protest that he would not have persisted in libelling a man for five years without substantial proof that his charges were true. But the judge was decided: hardly any behaviour could justify one man libelling another for five continuous years, and leniency would be inappropriate in a case of such persistent abuse. Nevertheless, as Hampden did not have a lawyer, and had failed to submit his plea in writing, the Lord Chief Justice was willing to grant him time to draw up his documents and prepare to represent himself in court.[46]

Hampden was called back to Shire Hall on Saturday as a local sensation, the Purfleet murder trial, was drawing to a close: the Lord Chief Justice passed the death sentence on a soldier for strangling a schoolgirl in a water-closet. The judge removed his black cap and a fresh jury was sworn in to hear the flat-earth case. With no further ado, Hampden was called to the dock, where he pleaded not guilty to the six counts of libel and passed his documents to the clerk to be read out to the court. The clerk was forced to summarize that Hampden had declared his comments, though packed with insults, to be true and, furthermore, because Wallace was championing a fraudulent astronomical system, that they had been published for the public benefit. Under law the plea was deemed an aggravation of the original offence, and thus the onus was on Hampden to prove that his statements were justified. As a next step Hampden called Wallace to the stand for cross-examination, much to the amusement of judge and jury, who anticipated a clumsy attempt at questioning on the part of the flat-earth believer. Hampden did not disappoint, and even Wallace could not help but laugh as his opponent stumbled and blustered through questions about the wager and Walsh's controversial decision.

Clearly enjoying the farcical scene, the Lord Chief Justice

decided after a while that he had better intercede with some
questions to clarify the extraordinary situation. He asked for a copy
of the wager agreement to back up allegations of Wallace's roguery,
but Hampden declared that he did not have a copy in his
possession and that his hands were tied. 'I wish they had been tied
when you wrote those letters,' the Lord Chief Justice quipped, to
which Hampden retorted sulkily that he 'might almost as well sit
down'. But that was not possible and, after several more questions,
Hampden picked a row with the Lord Chief Justice about his
intention to sue Walsh to recover his original £500 stake. In the
face of more sideswipes and sarcasm, Hampden finally lost his
temper with Wallace, raging, 'Why have none of your scientific
friends ventured publicly to justify your conduct or your possession
of the money?' continuing, 'When I saw you had no friends—'
at which point His Lordship interrupted: 'Then you hit him the
hardest, eh?' Realizing that his protests were useless, Hampden
decided to change tactic. He turned to the judge pleading that he
was ashamed of his behaviour, explaining that provocation from
Wallace and extreme poverty had driven him to take up his pen
and dash off the 'reckless assertions'. 'It is not only the act of
taking possession of the money that I complain of,' Hampden
moaned. 'What has offended me much more has been Mr.
Wallace's dogged determination not to write or speak to me.' But
in response the judge merely reasoned that 'Calling a man a
swindler and a thief is not likely to make him consent to become
your correspondent,' at which point Hampden finally relented.
'If it will clear Mr. Wallace's character by sending me to gaol,
by all means send me,' he huffed. 'I don't withdraw my charges.
I only say I am tongue-tied and hand-tied.'

Adopting a serious demeanour, and instructing Hampden to be
silent for once, the Lord Chief Justice took the opportunity to begin
his summing up: Hampden had failed to prove that his libels were
justified, and the case involved systematic persecution by means of
letters and 'spicy' postcards that any Post Office employee could
have read. Under such circumstances, there seemed no alternative

but for the jury to return a verdict of guilty, which they did after a moment's consultation. With that, the Lord Chief Justice turned to Hampden and sentenced him to a year in Chelmsford Gaol, binding him over to keep the peace for a further two years on his release, with payment of good-behaviour bonds amounting to £1400. With no further argument, Hampden was taken down to the cells to begin his prison term.[47]

It was the verdict Wallace had hoped for, considering the lack of alternatives, but legal action had come at a price that he simply could not afford. As a self-employed author without a full-time job, his only option was to contact his publisher, Alexander Macmillan, to request an advance on profits from his forthcoming book, *The Geographical Distribution of Animals*, to 'settle [the] disagreeable ... lawsuit'.[48] The advance was subsequently agreed to and, with legal bills paid and Hampden in prison, the matter seemed to have been resolved. The general consensus was summed up by the *Chelmsford Chronicle*, which reflected that 'No one who read the reports [of the case] can doubt that the prisoner is getting deserts richly merited.' Yet this said nothing about the damage the affair had done to Wallace's reputation, and in certain quarters questions were still being raised, not only about the propriety of taking Hampden's money but also about his decision to take up a wager with a flat-earth believer at all. Such considerations were not limited to gossip in scientific circles; the moral dimension was so striking that even the *Chelmsford Chronicle* felt driven to comment that 'We cannot think that this was a wager that an eminent man of science should have touched.' To be ethically sound, the paper continued, such an arrangement should only be made between parties equally confident in an opinion, but equally ignorant of the result and, moreover, the paper noted, it would have been as easy 'to find a needle in a haystack' as it would to find an intelligent umpire to oversee impartially an experiment to decide the earth's shape. Despite Wallace's professed desire to further the public understanding of science, the entire endeavour had been dishonest and futile, as far as the *Chronicle* could see:

The transaction has cost [Wallace] and his family marvellous
annoyance, it has bereft his opponent of all his means, made
him furious, forfeited his liberty over and over again, threatens
to keep him for years in prison, and may very probably, from
all the indications, deprive him of his reason altogether.[49]

In fact, Hampden had pleaded in court that when many of his
'ill-starred' letters were written he had been so consumed by a
sense of injury, annoyance and hopeless poverty that he was hardly
sane – a claim that, on the available evidence, the journalist said
he could well believe. The *Chronicle* concluded that Wallace, on
seeing the type of man Hampden was, should have at least repaid
the stake immediately, put the wager down to experience and
moved on to other concerns. Such a reaction would have been
preferable to bringing himself, his profession and the fact he
represented into disrepute.[50]

While the *Chelmsford Chronicle* took the moral high ground in
its coverage of the case, the leading science journal *Nature* reflected
simply, 'It is to be hoped that an enforced retirement of a
twelvemonth will result in Mr. Hampden's learning wisdom and
the keeping of peace towards Mr. Wallace and others.'[51] Even with
Hampden in prison, legal action did not end. In January 1876,
the case of Hampden v. Walsh finally reached the High Court
in Westminster following several adjourned hearings. Six years
after the fateful wager, Hampden was finally claiming the return
of his £500 stake, which Walsh had paid over to Wallace against
Hampden's strongly worded demands.

The case was heard in Hampden's absence, as he was still
serving time in Chelmsford, but nevertheless the judge, Mr Justice
Mellor, and a special jury considered the evidence in full. Hampden's lawyer contended that under law either party to a wager, at
any time before the money was paid over and even after the end
of the event, could legally claim to recover his original stake from
the stakeholder. In response, Walsh's counsel denied this point
and claimed that the arrangement was not a wager. The court took

time to consider its verdict and returned with a firm decision. The arrangement was undoubtedly a wager and as such was legally void. As Hampden had asked for his £500 back before it had been paid over to Wallace, he was therefore entitled to recover that sum from Walsh. And so, in another unexpected twist, Hampden was awarded the £500 from Walsh, plus £200 to cover his costs.[52] Moreover, as Wallace had signed an indemnity in April 1870 to cover Walsh's expenses in the event of legal proceedings, and as the final recipient of Hampden's £500 stake, he was honour-bound to pay the £700 on Walsh's behalf. As Wallace later put it, 'Mr. Walsh', an experienced stakeholder for wagers, 'made a great mistake which had serious consequences for me'.[53] Although Hampden still owed Wallace more than £600 from the 1871 judgement when he had declared himself bankrupt, it was decided that only a portion of this amount could be offset against the present liability and costs. With his own legal bills to consider, Wallace's involvement with a flat-earth believer was becoming increasingly expensive.

Just a week later, Wallace received yet more bad news: appeals to the home secretary from Hampden's friends had led to his release from Chelmsford Gaol having served nine months of a one-year sentence.[54] Expecting trouble sooner or later, Wallace carried on as usual, publishing his ground-breaking two-volume study, *The Geographical Distribution of Animals*, and causing a furore at the annual British Association for the Advancement of Science meeting in September, where he allowed another delegate to present a paper on thought-transference. The subject was seen as inappropriate for the nation's most prestigious professional forum by many of his peers, and although the paper was cleared in advance by two committees Wallace had held the deciding vote. In letters to *The Times*, he was accused of 'degrading' the British Association with his 'more than questionable' conduct, allegations he quickly refuted. Unrepentant, he further tarnished his reputation in October when he stood as a defence witness for the allegedly fraudulent American medium Henry Slade. On trial at Bow Street

Magistrates' Court for obtaining money under false pretences, the case against Slade was a *cause célèbre*, critical to the complex relationship between science and spiritualism. The case, instigated by Edwin Ray Lankester, zoology professor at University College, London, and secretly bankrolled in part by Darwin, together with Slade's subsequent conviction (albeit later quashed on a technicality), only served to highlight how spiritualism was gradually being classed as an inappropriate area of investigation for professional men.[55]

Hampden, meanwhile, had fallen on hard times since his release from prison and was living at the Temperance Hotel in St Johns Road, Hoxton, where he continued to work on his unconventional schemes. These included designs for artillery, passenger ships, steamers and gunboats, a type of coach, a plan for British coastal defences and a 'Circular Map of the World'. Having lost his inheritance and nearing bankruptcy, he added that he was looking for a well-connected capitalist, 'who will not shrink from sharing the obloquy incurred by all daring to be in advance of their age,' and who would be willing to buy a quarter-share in his inventions.

While Hampden was considering ways to improve the condition of England, both in terms of her technologies and her ideologies, William Carpenter was busy answering Parallax critic John Dyer with a new pamphlet produced on his Lewisham printing press. By this time he was well established in a radical bookshop in Ladywell Park, specializing in publications on teetotalism, vegetarianism, mesmerism, spiritualism and Pitman's new shorthand system, although his zetetic work continued to take priority.[56] Carpenter's latest sixpenny pamphlet, *The Delusion of the Day or Dyer's Reply to Parallax* (1877), was sarcastically dedicated to Dyer's own kind, the schoolmasters of Great Britain, and was written in the style of a make-believe conversation between two schoolboys, Frank and John, to render it accessible to children. The work highlighted Carpenter's continuing desire to disseminate zetetic astronomy at grass-roots level, but it was to be his last publication for a number of years. In the late 1870s he emigrated to America with his wife Annie and six children, leaving Hampden and Parallax to continue the campaign at home.

Wallace was also in search of pastures new, and in March 1878 he moved to Croydon for the sake of his family's health. While at the new house he informed Darwin that he was hard at work on a book about the geography of Australia, 'for the want of anything better to do', and as the months went by he became increasingly desperate to get a job with a fixed income, however moderate. In July he confided to his publisher, Alexander Macmillan, that he had suffered 'a series of losses and misfortunes' that had rendered it 'very hard for him to get on', and he was 'exceedingly anxious' to obtain any regular literary work, even the deputy editorship of *Nature*, if such a post were available.[57]

By September he had set his heart on becoming the new superintendent of Epping Forest, and was writing letters to every-one he knew with power and influence, including Darwin and Macmillan, to rally support for his application. But despite his efforts, along with testimonials from many of the leading lights of Victorian public life, he was unsuccessful and in 1879 the job went to someone else. In part the victim of a professionalizing world, his outsider status was also self-imposed. Wallace was an indepen-dent spirit, untrammelled by convention and careerist consider-ations, and he had no time for the formalities, bureaucracy and power-play that public life could entail. Modest to the point of self-deprecation, he always portrayed himself as an amateur, which he was in the sense that he could not earn a decent living from his scientific work.[58] A self-taught man, exposed to Owenite socialist teachings as a teenager, his anti-authoritarian leanings ran deep and fed a commitment to a science-for-all ideology. Wallace even actively opposed the government funding of research in the fear that investigation would become the province of a closed intellec-tual élite. Best suited to working alone on subjects in which he believed, there were not many posts that would suit his free-thinking attitude and love of the outdoors. As a result, his financial future was bleak. Increasingly concerned, friends discussed his situation among themselves. Despite Joseph Hooker's criticism of Wallace's perceived transgressions, Darwin was persuaded to intervene with the Liberal prime minister, William Gladstone, to secure Wallace an annual civil-service pension in recognition of his services to science.[59] The first payment came through in January 1881, and Wallace was able to build another new home, Nutwood Cottage in Godalming, where he continued with his broad-based scientific work.

By November 1881, the astronomy writer Richard Proctor had embarked on a new venture of his own, establishing a twopenny illustrated magazine, *Knowledge*, intended to cater for the needs of the general reader. With this publication, he was determined to challenge the increasing dominance of back-slapping specialists and the literary channel for their views, Norman Lockyer's élitist *Nature*.[60] Armed with a disdain for privilege and the 'closed shop' mentality that seemed to accompany the professionalization of science, Proctor emphasized that *Knowledge* would avoid the jargon-laden analyses that blocked the participation of amateurs and ordinary people.[61] In terms of ideology and format, he was harking back to the tradition of the mechanics' magazines earlier in the century by championing the self-improving artisan over the professional expert.[62] Notably he had much in common with Wallace in this regard. Anti-élitist, argumentative, with a deep-seated commitment to popularization, they were both self-employed authors who truckled neither to professional authority nor to the opinions of more cautious peers.[63] Undoubtedly, such traits were crucial to their willingness to engage with paradoxers; it is somewhat ironic that while they were vilified as pillars of the establishment by the zetetics, in the diverse world of Victorian science neither Wallace nor Proctor could be taken as representative of a conventional professional élite.

Despite Proctor's professed reluctance to debate with flat-earth believers, from the first issue of *Knowledge* he set out to stir up Parallax and his followers with articles alluding to paradoxers and flat-earth theory alongside the occasional editorial assault. He even allowed Hampden column space in which to 'calmly express his views', an experiment that seemed to end in dismal failure. Nevertheless, Proctor claimed,

> I am a little pleased with my new invention for silencing paradoxists. Reasoning has been tried in vain ... ridicule is ineffective, and a bad example; denunciation is idle. The plan with paradoxists is to ask them to explain their views and to remove the difficulties, which are, of course, in reality fatal. They either give up in despair, like our enthusiastic earth-flattener, Mr. Hampden, or flounder so absurdly in their efforts to explain their preposterous notions, that even the unlearned (for whom alone, of course, the thing is done) see at once how hopelessly at sea the paradoxists are.[64]

Years of experimentation were paying dividends, as far as Proctor could see, and he remained dedicated to the belief that confuting paradoxes could be beneficial to astronomical novices.[65]

However, his actions had also provoked a dramatic reaction from the newly formed English branch of the Zetetic Society, and in late November 1883 he announced that he had received a challenge to participate in a public debate with Parallax. Predictably he declined the invitation, although he had little doubt that zetetics would use this to their advantage. Nevertheless he was not greatly concerned, although rather than sidestepping discussion from that point forward, which would arguably have been the more logical step, he launched into allegations that Parallax did not believe his own theory, based on evidence presented in *Zetetic Astronomy*. Proctor's proof was straightforward: Parallax had stated that, with his eye close to the water level on the Old Bedford Canal, he had seen through a telescope the whole structure of a bridge over the water, down to its surface, six miles away, yet he had deliberately invented a false law of perspective to explain why these observations could not be replicated in case anyone should try. Furthermore Parallax had claimed that the earth was immovable because in fifty trials when he fired a gun tied vertically, the bullet had twice fallen back into the muzzle, which Proctor asserted was clearly a blatant lie. Building on this point, he observed that throughout his book Parallax had utilized garbled quotations, which seemed to support his flat-earth doctrine, but in every case the reference to the original authority or scientific work stopped just short of where the passage began to bear the other way. 'From this evidence,' Proctor concluded, 'I deduced . . . that as Parallax is manifestly not wanting in intelligence or even in ability, he advocates a theory which he knows to be erroneous.'[66]

Naturally Proctor's comments only served to fuel the controversy, and on 14 December he reported the latest development: the Zetetic Society had invited him to repeat Parallax's experiment of firing a gun vertically to test whether two shots from a total of fifty would fall back into the barrel. Proctor responded to the suggestion with characteristic scorn:

> I utterly decline. If I missed killing myself, I should probably kill someone else: I may very literally say therefore, 'I'll be hanged if I do.' Considering, however, that thousands of experiments on falling bodies show that with the most delicate appliances to avoid disturbances, bodies let fall within places carefully shielded from the external air have not in a single instance fallen either exactly in the calculated place, or exactly below the point of suspension, or any two in precisely the same place, I must be excused for utterly declining to believe that two balls out of fifty *fired* from a gun *tied vertically*, fell back into the muzzle.[67]

His willingness to engage in debate with the zetetics, in print if not in person, led to further controversy in the spring of 1884. H. Ossipoff Wolfson, a Russian immigrant, and ex-secretary of the Zetetic Society, had contacted Proctor requesting column space in *Knowledge* for an exposé of Parallax and his 'humbug catchpenny philosophy'. His object, he said, was to protect others from falling foul of the zetetics, and his first article 'The Flat Earth and her Moulder' in the 4 April issue was packed with revelations designed to prevent anyone ever trusting Parallax again. In the article Wolfson described how Parallax, an 'animated, confident and engaging' character, had persuaded him to endorse his teachings with convincing explanations of how the surface of water and the earth itself were flat. The founder of zetetic astronomy could be somewhat vulgar, Wolfson alleged, but at the time he had put this down to the eccentricity of an old man; it was only later that he had decided Parallax was an 'accomplished quack' and a 'many-headed eagle', who did not believe his own theories and 'dodged about' under several different names. At the time of writing, Wolfson revealed, Parallax was going about as 'Dr. Samuel Birley', professing to possess the secret for prolonging human life and claiming the ability to cure every disease imaginable for a charge of a guinea every two months. Flat-earth theory was merely another means by which Parallax sought to enlarge his ample fortune, over which he kept a most careful watch. 'The only wonder', Wolfson marvelled, was that such behaviour was suffered with impunity, but then again, he supposed, there was always a way for those who wished to escape punishment – 'Let them simply change their names.'[68]

Parallax responded to the allegations in less than a week – or, rather, he instructed his solicitor, Mr Howard Rumney, to respond on his behalf. Legal proceedings for libel were duly instituted against H. Ossipoff Wolfson, while Proctor was warned that if he published any more defamatory remarks he would meet with a similar fate. For his part, Proctor said he was happy to comply, but added mischievously that he had firm proof that Parallax and 'Dr. Birley' were one and the same, and he would be willing to publish any counter-statements that 'Dr. Birley' might wish to make. Meanwhile, Wolfson claimed that Hampden had attempted to come to the rescue by advising him to publish an apology before Parallax visited his solicitor. Despite the subsequent writ, Wolfson remained undeterred, saying he was willing to stand or fall by his statements, which would have to be proved libellous in court.

By early May it had become evident that Parallax was not going to answer the allegations through the medium of *Knowledge*. Determined not to let the matter rest, Proctor suggested that Parallax might like to clear his name by suing him or the magazine's publishers as an alternative. His purpose, he explained, was finally to reveal Parallax for what he was, not because of his promulgation of flat-earth theory but as a result of his work on health and longevity. Proctor believed that misleading the public about medicinal matters was apt to be more damaging than spreading erroneous ideas about the shape of the earth.[69] But his challenge was unsuccessful, and although the debate and the reams of paradoxical correspondence made for entertaining copy, Proctor eventually decided that baiting proponents of alternative ideas was a counter-productive pursuit. He had struggled to balance editorial control with a democratic desire to let everyone have their say, but argumentative readers and paradoxers seemed to be hijacking the journal and the direction of discussions therein. In response, he transformed his 'Republic of Knowledge', closing the correspondence, altering the paper's format and increasing its price. In the final analysis, safeguarding the authoritative tone of *Knowledge* – and the body of facts it represented – remained his first priority.[70] Reflecting on the controversy, Proctor concluded that the scientific profession ultimately only 'suffers by such controversy . . . which equalises in the eyes of outsiders the ignorant and well-informed'.[71]

The key question of whether Parallax believed his own theory was to remain unsolved, for his case against Wolfson never reached court.[72] Six months later, aged sixty-eight, Parallax died of a cerebral haemorrhage at the large North London home he had named Welney House for a favourite location in the Cambridgeshire fens. His activities during his lifetime were sufficient to earn him an obituary in *The Times*, while the *Bookseller* passed comment on the somewhat ironic circumstances leading to his death:

For some years past Dr. Rowbotham could never under any consideration be induced to travel by rail. Patients or friends wishing to see him had to send their carriage for him and in the last few months he visited Brighton for the good of his health, travelling to and fro in a private carriage. Curiously enough the mode of conveyance in which he placed his faith accelerated his death. On an occasion, several months past, he slipped and injured his leg when alighting from a cab, and from that time his health gradually failed.[73]

Parallax's family and followers were understandably distressed. On New Year's Eve 1884 they laid him to rest as Samuel Birley Rowbotham M.D., Ph.D. in Crystal Palace District Cemetery under a gravestone bearing testament to his zetetic teachings.[74]

However, Hampden was not willing to let the memory of the first zetetic philosopher fade with his passing, and in March 1885 he issued the first number of a new twopenny magazine, named after Parallax and intended to perpetuate his principles; in subsequent years it was supplemented by a number of frenzied accounts of the supposed battle between science and religion.[75] According to Hampden, Mosaic cosmology was a fact, modern astronomy was a fable and the differences between them were wide, fundamental and irreconcilable – whatever the majority of the population believed. A leading warfare promoter, he established the Biblical Science Defence Association and Christian Philosophical Institute to promote discussion of the supposed conflict between science and the Bible, evolution and creation, Genesis and geology. The argument continued through a stream of flyers, including 'Britons Never Will be Slaves', which warned Englishmen that their minds were in bondage so long as they believed the debasing and revolting doctrines promoted by 'brainless boobies, infidel upstarts, swaggering freethinkers, knavish professors and the scum of the literary world'. In subsequent publications Hampden offered a hundred guineas to any student who could provide an impartial description of the origin, progress and proofs of the globular theory, and an ample justification for its dissemination in the nation's schools. He even requested that the president of the

Royal Geographical Society offer a prize medal to anyone who could discover the true shape of the world, and launched an assault on arch-professionalizer and then president of the Royal Society Thomas Henry Huxley about his evolutionary ideas.[76]

Wallace remained his prime focus, however, and in keeping with his promise, Hampden distributed a thousand libellous leaflets around Godalming and continued to send poison-pen letters to Wallace's cottage in the town. As usual, the letters were straight to the point: Wallace's cook was informed that her employer was a 'degraded cur and a scoundrel', who would suffer until the last hour of his existence, while his wife, Annie, was told that the campaign to disgrace her 'miserable wretch' of a husband would cease only with Hampden's life. But, contrary to expectation, the impact on his target was diminished, rather than intensified, by years of harassment and abuse. By autumn 1887 Wallace was joking in letters to a friend about one of Hampden's public-speaking engagements, which he saw as an opportunity to 'rile' the flat-earth believer and enjoy 'some excellent fooling', and when Hampden turned up at Nutwood Cottage one day in person, Wallace wished in retrospect that he had invited him in to provide entertainment for his lunch guests.

Unfortunately, he quipped later, such 'happy thoughts' were always apt to 'come too late'. With benefit of hindsight, he reflected that 'fifteen years of continued worry, litigation, and persecution with the final loss of several pounds' had all been brought upon him by his wish 'to get money by any kind of wager'. He recognized that his involvement was due to his 'ignorance of the fact . . . that paradoxers . . . can never be convinced'.[77] His last words on the matter, publicly at least, was that his acceptance of the flat-earth wager had been 'an ethical lapse'.[78] Meanwhile, his sometime accomplice, Proctor, had long moved on to other concerns. Having married an American widow, he had settled in the States, where he continued to promote the public understanding of science through a steady stream of publications and talks. In September 1888 Proctor was nearing the completion of his fifty-seventh book and was returning to England for another lecture tour when he was taken ill with yellow fever and admitted to a New York hospital on a cold, rainy night. Hours later he died there, aged fifty-one.[79]

Eighteen years Proctor's senior, Hampden did not long outlive him, and in late January 1891, newspapers around the capital, from the *Standard* to the *London Star*, reported on his death at seventy-one from bronchitis and heart failure. Among them was the *Illustrated London News*, which informed its readers that 'a curious crank has gone the way of all flesh in the person of Mr. John Hampden', infamous for his unusual beliefs and his habit of spreading a huge map of the world as a circular plane over his dining table in order to instruct his dinner guests.[80] No less droll, the *Daily Graphic* commented that Hampden's interpretation of the prophetic books of the Bible had led him to believe that some startling events would occur between the years 1891 and 1894, and 'very likely he was right in this somewhat vague anticipation'.[81]

Meanwhile, at the end of Hampden's life, a correspondent to the *English Mechanic* reflected on its beginnings, informing fellow readers that Hampden's father, the rector at Hinton Martell in Dorset, 'was by no means a fool' but was 'decidedly mad' on his unorthodox interpretation of the Bible and the prophecies that he believed the Book of Daniel contained. Although this pastime might have had an impact on John Hampden's subsequent ideas, the writer puzzled that the Dorset parish in which he was raised was very hilly and not at all suggestive of a flat earth.[82] To the last, commentators, it appeared, were puzzled by the extremity and anachronism of Hampden's Bible-based views.

Chapter Five

LADY BLOUNT AND THE NEW ZETETICS

The Bible meaning of 'The Earth'
Is sensible and right,
And all who fail to see its worth
Are far from Truth and Right.

To teach a child the sea and earth
Are rushing in the sky,
Distorts his Reason from his birth,
And makes his Bible lie.

'ZETEO' [Lady Elizabeth Anne Mould Blount],
*The Secrets of Nature Exhumed: With Hundreds of Proofs
'the Earth' (and Sea) is Not a Whirling 'Globe'!* (1913)

'In Veritate Victoria'
Victory in Truth

Motto of the Universal Zetetic Society (1893–1906)

ON HEARING NEWS of Hampden's death, readers of the popular science press must have hoped they had heard the last of the zetetics' crusade against conventional science. It was not to be the case, however, for two years after his demise, the flat-earth campaign was revived by Lady Elizabeth Anne Mould Blount on a much grander scale than before. An aristocrat on a Christian mission, Lady Blount had been born Elizabeth Williams, the youngest daughter of James Zacharias Williams, a well-to-do architect and land surveyor who had offices in South London. An earnest Christian, with an interest in religion and science, James Williams had philanthropic leanings and a fondness for lecturing on subjects such as electricity and astronomy. His daughter Elizabeth was to develop similar interests, although she chose to adopt a strict literal interpretation of the Bible in all of its statements about the natural world. Little else is known about her early life, but at the age of twenty-three, she married the forty-one-year-old Sir Walter de Sodington Blount Bt, at St Marylebone Parish Church in London. Her situation shifted radically from this point on, while the same could be said of her new husband's.[1]

The son of Sir Edward Blount, former high sheriff of Worcestershire, Sir Walter's family had a long and illustrious history stretching back to the commander of warships for William the Conqueror, Sir Robert le Blount. Meanwhile, the family title dated from the English Civil War when Walter Blount, a supporter of King Charles I, had his lands confiscated by Cromwell's army and was imprisoned in the Tower of London. His loyalty to the King was rewarded with a knighthood in 1642 and his property was eventually returned. But this was not the last scandal that the Blounts endured: a decadent relative later squandered the family fortune (from iron and coal) on drunken gambling, and other family members had to pay off the debts. Yet all was not lost: in the 1870s, the Blounts still owned six thousand acres and an exquisite Georgian mansion, Mawley Hall, set on hills overlooking the small Shropshire town of Cleobury Mortimer.[2] This property was to become Sir Walter's on his father's death – until he married Elizabeth Williams. On hearing the news his parents had been outraged. The issue was not the difference in social class, or even in age, but religion, for the ceremony had been conducted in an Anglican church. This was considered a great betrayal: the Blount family were staunch Roman Catholics who, having survived the Restoration without changing their religion, were not willing to forgive Sir Walter for his marriage, and cut him off completely. While his mother and various relatives lived at Mawley Hall, he had to content himself with a modest home in the London suburbs with his new wife. There, the couple produced a succession of children: an heir to the title, Walter Aston, and a second son, Edward Robert, along with two daughters, Mary Corisande and Eva Apollonia. One suspects that domesticity and child-rearing were not sufficiently fulfilling for Lady Blount, for in the 1890s her commitment to a strictly literal interpretation of the Bible and social reform led her to resuscitate John Hampden's campaign.

Essentially, Lady Blount's views about the shape of the earth were almost identical to those propounded by Parallax and Hampden, although she placed an even heavier emphasis on the Biblical basis of her ideas. With a Leicester associate, Albert Smith, otherwise known as 'Zetetes', she produced a simple summary of this scriptural world view:

1. Heaven is above the earth (not all round), earth beneath, and water under the earth (Exodus 20:1–4).
2. The sun, moon and stars, placed within the vault of heaven, are powerful lights only, some greater, some lesser, electrical and magnetic, intended for 'signs and for seasons' and to give light to this the only world (Genesis 1:16–18; Psalms 136:7–9; and Revelation 6:13).
3. The earth is represented as being 'outstretched' as a plane, with the 'outstretched' heavens everywhere above it, like a circular tent to dwell in (Isaiah 40:22, Proverbs 8:27, Isaiah 44:24–35, Luke 4:5; and I Corinthians 3:19).
4. The earth is firmly fixed on foundations or pillars, having ends and corners jutting out into the sea (Genesis 1:10; Job 38:4–6; I Samuel 2:8; and Psalms 93:1).
5. The sun, moon and stars move around and above the earth (not more than a few thousand miles off) so that

day and night are 'ruled' by the motions of the heavenly
bodies or lights, and not by the supposed axial motion of
the earth which contradicts the Holy Scriptures as well as
our God-given senses. Heaven is nearer to us than we
have imagined (Joshua 10:12–14; Psalms 19:4–6; Luke
24:51; and Daniel 9:21–3)

6. All that exists was created in six days and not slowly
   evolved, as infidels suppose and recklessly affirm, during
   'millions and millions of years'.[3]

In modern terminology, Lady Blount was a creationist, a
believer that the Bible was the unquestionable authority on the
natural world, and while such world-views could vary widely, she
stood on the radical fringe.[4] At a time when many creationists
accepted 'day-age' reading (interpreting the days of Genesis 1 as
vast geological ages) or 'gap' interpretation (a creation 'in the
beginning' as in Genesis 1:1, followed by a later Edenic creation in
six twenty-four-hour days) of Genesis, she held that the universe
was created in six days no more than six thousand years ago.
Meanwhile, like latterday 'scientific' creationists, she backed her
arguments with an appeal to science and human-sense experience,
but went much further in pinning her teaching on one of the most
radical of Bible-based facts. As a zetetic, her creationist campaign
was founded on the idea that the earth was flat and, in common
with her predecessors, she argued that one could not be a Christian
and believe the earth to be a globe. Blount believed that science
and Christianity were utterly irreconcilable and on this point she
was convinced, informing a journalist from an American newspaper:

If the Bible is the word of God it is absolutely true. We must
accept it as a whole or else accept none of it. We cannot
divorce the religion of the Bible from the science of the Bible,
hence the globists cannot be Christians – nor can Bible
Christians be followers of Newton's philosophy.[5]

Lady Blount was a warfare polemicist who stood at the extreme. Through a period when intellectuals were increasingly attempting to reconcile science and Christianity – in contrast to Victorian Darwinians such as Thomas Huxley – she insisted that the two supposedly separate systems were irrevocably opposed.[6] The Bible spoke of the earth as an immovable outstretched plane, resting on foundations in the waters of the deep, and if she was branded a crank for doubting the claims of conventional science she was willing to accept this outcome in line with her beliefs.

Disgusted by 'illogical Christians' and the 'scientific blasphemy' of globular theory, in 1893 Lady Blount established a new organization, the Universal Zetetic Society (UZS). Above all else, its object was 'the propagation of knowledge relating to Natural Cosmogony [the creation of the world] in confirmation of the Holy Scriptures, based upon practical scientific investigation'. There were notable similarities here with another creationist organization then in its heyday, the Victoria Institute, or Philosophical Society of Great Britain. Established in 1865 by flat-earth believer James Reddie, it was intended to defend 'the great truths revealed in Holy Scripture . . . against the opposition of Science, falsely so called' in the wake of Darwin's *On the Origin of Species* (1859) and the *Essays and Reviews* (1861) of seven liberal churchmen, which had argued against the literal accuracy of the Bible in matters concerning science.[7] However, while the two societies adopted similar general aims, along with the same militaristic motto, 'In Veritate Victoria' or 'Victory in Truth', they did not appear to have any members in common and Reddie's flat-earth beliefs were not shared by other members of the Institute.[8]

While Blount's Universal Zetetic Society was evidently extremely radical, she still attracted members throughout the English-speaking world, who paid a not insubstantial six shillings a year for affiliation. Chief among their number was Lady Blount's vice president, Dr Charles Watkyns de Lacy Evans. Once the best friend of Parallax (he had allegedly shed tears when he died), he had been vice president of the original Zetetic Society and it was a role he was happy to take on again. Similarly keen to join the crusade were other old hands, including Birmingham zetetic James Naylor, author of *Bedford Canal not Convex, The Earth not a Globe* (1873); and with members ranging from teachers to market-gardeners and a time-keeper at an iron foundry, the society seemed to possess a cross-class appeal. With the appearance of wealth and a title on her side, Lady Blount was also able to draw support from the English social élite, and the first list of Universal Zetetic Society members included an archbishop, a major-general, several counts and colonels, the eminent Anglican clergyman and Bible scholar Ethelbert William Bullinger, and the senior moderator in natural science at Trinity College, Dublin, Dr Edward Haughton. Such members endeavoured to disseminate zetetic teaching to the public at large, and lectures with titles such as 'Evolution – What does it Mean?', 'Universal Gravitation: A Pure Assumption' and 'Are we Living on a Whirling, Flying Ball of Land and Water?' started up again. Meanwhile Lady Blount established a journal, the *Earth not a Globe Review* to act as a further mouthpiece for zetetic views. Edited by Zetetes, it was circulated worldwide with the help of staunch supporters in Ireland, Canada, America, South Africa, India, Australia and New Zealand. The only major exception was Russia where, for unknown reasons, it was banned.

Unsurprisingly, the unorthodox nature of the zetetic endeavour rendered the twopenny *Earth not a Globe Review* intriguing reading for flat-earth believers and globularists alike. Naturally the latter came in for a regular lambasting in editorials, articles, cartoons, and even poems with titles such as 'Agnostics', '"Globe" on the Brain', 'In Topsy-Turvy Land', 'The Song of the Evolutionist' and 'That Apple!'. In addition, Lady Blount's followers were keen to repeat John Hampden's success in provoking eminent people to engage in public debate, and in 1897 one UZS member went so far as to challenge the editor of the cheap popular journal *Science Siftings* to prove that the earth was a spinning globe in return for £1000. The journalist replied politely:

We cannot think of accepting your challenge. The 'reward' of £1,000 is doubtless a hoax ... Not a cent could be recovered from anybody, upon the strength of such a 'startling offer' as is published upon the hand-bill. Then apart from this, most of our readers have been educated past flat earth hypotheses. And if we devoted to these such a space as would be needed for the rigid demonstration of the motions and form of the earth, *Science Siftings* would be considered uninteresting, and its demonstrations redundant. Then our circulation would be converted from an increasing to a decreasing one. Probably this last consideration has not presented itself to you; but we cannot lose sight of it.[9]

Interestingly, in a shifting market flat-earth theory was now viewed as a curse rather than a boon to magazine sales and, unlike Proctor, the journalist could not see how the subject could be used in a productive way. Naturally the zetetics disagreed and, believing there was still a market for their ideas, one UZS member suggested forming a 'Parallax Company' to issue a cheap revised edition of *Zetetic Astronomy: Earth not a Globe!*. Eager to attract support, he guaranteed that share prices would be set at an easily affordable level 'so that every plane friend may participate in the reproduction of the grandest and truest scientific literature that was ever placed before the world'.[10]

In the interim, existing members bombarded the *Earth not a Globe Review* with numerous articles and letters. Among the most prolific contributors was William Carpenter, who wrote long letters from Baltimore discussing 'that wager', the 'three pole trick', Alfred Russel Wallace and his so-called 'winning ways'. Evidently emigration to America, with his wife and children, had not dampened his enthusiasm; indeed, it was quite the reverse. After six months on a farm in Dorchester county, Carpenter had established a Pitman shorthand school and printing press in a working-class neighbourhood near Baltimore jail and issued a pamphlet, *One Hundred Proofs that the Earth is not a Globe* (1885). With the byline 'upright, downright, straightforward', it was sarcastically dedicated to his old enemy Richard Proctor, 'The Greatest Astronomer of the Age', and like many of Carpenter's publications, was typeset, printed and bound by the author on the spot without being written down first. Of all the zetetic propagandists, Carpenter had a knack of simplifying the issues at stake for the man or woman in the street and this pamphlet was more accessible and striking than many zetetic works. Among the most eye-catching of the proofs it presented were numbers 33, 62 and 77:

33. If earth were a globe, people – except those on the top – would, certainly, have to be 'fastened' to its surface by some means or other, whether by the 'attraction' of astronomers or by some other undiscovered and undiscoverable process! But, as we know that we simply walk on its surface without any other aid than that which is necessary for locomotion on a plane, it follows that we have, herein, a conclusive proof that Earth is not a globe.

62. It is commonly asserted that the earth must be a globe because people have sailed round it. Now since this implies that we can sail round nothing unless it be a globe, and the fact is well known that we can sail round the Earth as a plane, the assertion is ridiculous, and we have another proof that Earth is not a globe.

77. 'Oh but if the Earth is a plane, we could go to the edge and tumble over!' is a very common assertion. This is a conclusion that is formed too hastily, and the facts overthrow it. The Earth certainly is, what man by his observation finds it to be, and what Mr. Proctor himself says it 'seems' to be – flat – and we cannot cross the icy barrier which surrounds it. This is a complete answer to the objection, and, of course, a proof that Earth is not a globe.[11]

On reviewing the pamphlet, the *Baltimore Daily News* commented drily that 'it can only be described as an extraordinary book', while the *Brooklyn Market Journal* reflected that it was 'forcible and striking in the extreme'.[12] To disseminate the message further, Carpenter printed more than twelve thousand copies and began editing a short-lived journal, *Carpenter's Folly: A Magazine of Facts*. Yet his efforts to gain converts extended far beyond disseminating his ideas in literary form. If anything, he gained most infamy for preaching his ideas to unsuspecting members of the public in locations from New York to Pennsylvania and Maryland, and in sixteen years he claimed to have spoken to 'at least a hundred thousand people'. Such activities were subsequently described by the *Washington Post*, which commented that 'Professor' Carpenter had earned widespread notoriety across the nation, and especially in Baltimore, for his habit of hawking copies of his *One Hundred Proofs* in the streets. When Carpenter was in his hometown, the *Post* continued, he could frequently be found loitering outside Johns Hopkins University or the city college, waiting to impress students with the truth of his theories, a habit that earned the 'celebrated "Prof." Carpenter' infamy as 'the most active of his cult'.[13] Carpenter himself admitted that on one occasion he was ejected from the university building by the janitor, although his proselytizing was not merely focused on undergraduates, as one local historian recalled:

> During [one lecture] season a well-known authority of the day – Garrett P. Serviss – discoursed on popular astronomy. At the end of the lecture as the audience left the building it was met on the sidewalk by an astronomical fanatic, one 'Professor' William Carpenter, who vociferously offered for sale, at a quarter a copy, his pamphlet 'One Hundred Proofs that the Earth is not a Globe'. This book was quite a curiosity.[14]

The *Baltimore Sun* was in full agreement, adding that,

> . . . when people refused to buy a copy of 'One Hundred Proofs', Prof. Carpenter would present them with copies in order that they might not be deprived of the opportunity of

learning that the astronomy now being taught is all *upside down*. He was always thoroughly in earnest, patient and diffuse in his method of presenting his views, and angered only when people refused to give him a hearing.[15]

The newspaper had a particular purpose in covering the story: it had received countless letters about zetetic astronomy since Carpenter moved to town. He was constantly attempting to draw people into debates, the more eminent the better, and when his efforts failed he usually resorted to writing an open letter denouncing them to the public, as the president of Johns Hopkins University, Daniel C. Gilmore, and the astronomer Simon Newcomb discovered for themselves. But eventually the pace of his campaign took its toll, and in autumn 1896 the *Earth* reported 'with sadness' that the 'Professor' had died of a 'stroke of apoplexy' on 1 September at the age of sixty-six.[16] Survived by his wife, three daughters and two sons, he bequeathed a twenty-acre farm in Dorchester county and all of his 'books and records of [his] labours in the cause of God's truth' to his son Lewis, who he hoped would continue the work.[17]

Carpenter need not have worried about the future, however, for many American zetetics were willing to continue the crusade. His brand of the theory had first been exported to America when a Zetetic Society was established in New York in 1883, with the ubiquitous Parallax as president. Its mission had begun with a vengeance: a thousand copies of *Zetetic Astronomy* were shipped over from England, balloon expeditions to observe the flat earth were funded, and challenges were published in the *New York Daily Graphic* offering $10,000 to charity if any American scientist could provide proof that the earth revolved on axes.

As Parallax's theory found a new transatlantic audience, Carpenter had been joined in his efforts by ardent campaigners across the United States. They included the great Virginian slave preacher, the Reverend John Jasper, a firm friend of Carpenter's, who became famous across America for his sermon 'De Sun do

Move and the Earth Am Square'. Jasper was said to have preached to more people than any other Southern clergyman of his generation, and his best-known sermon placed him in great demand. He delivered it more than two hundred and fifty times, always by invitation, in Baltimore and Philadelphia, to the entire Virginia General Assembly and on one occasion to a packed audience including a number of congressmen, in Washington's Lincoln Hall.[18] Alexander Gleason, a Buffalo engineer, was from a different background entirely, but no less dedicated to the cause. The author of a weighty tome, *Is the Bible from Heaven? Is the Earth a Globe? Scientifically and Theologically Demonstrated*, first published in 1890, he ran experiments in true zetetic tradition on the surface of Lake Erie and from the roof of Buffalo City Hall to prove the flatness of the earth.

Experiments of a simpler kind were also tried by an infamous flat-earth exponent from Maine, 'Professor' Joe W. Holden. A sawmill owner from the small town of Otisfield, descended from revolutionary stock, 'Prof.' Holden's gravestone also states that 'the old Astronomer, while a boy at school, discovered that the Earth is flat and stationary and that the sun and moon do move'.[19] Born in 1816, Holden was a former justice of the peace, trial justice, candidate for state senator and census enumerator, leading one newspaper to caution that 'If anybody imagines, because he chooses to take the unpopular side of terrestrial, as well as celestial philosophy, that Uncle Joe Holden of East Otisfield is a fool or a lunatic they may at once disabuse their mind of any such delusion.'[20] Having studied the flat-earth question for more than forty years, Holden had begun giving lectures on the subject at the age of seventy-five, always attending the legislative sessions in Augusta to expound on his ideas. He felt that it was a particularly good ground for disseminating his convictions, and it was said he had never missed a session in more than twenty years. He also lectured on flat-earth theory at the Columbia Exposition in Chicago in 1892, and gave numerous lectures in New England and Maine, where he focused on presenting experimental proof for the flatness of the earth. Unlike Parallax's attempts to persuade his audiences with technical information, Holden focused on solid common-sense evidence to highlight the supposed shortcomings of orthodox ideas. He was convinced that the earth did not revolve because he had set a pail of water on top of a pole in his porch before retiring for the night, and it was still there full of water when he checked the next day. Just to be sure, on several occasions Holden had sat up all night to keep watch on his millpond, which likewise did not move an inch as he supposed it would if the earth truly revolved. Generally, it was reported that his talks were agreeable, gentle affairs, despite the intensity of his beliefs. One of the best-documented was a 'novel and unique lecture' entitled 'The World is not Round but Flat! Does not turn on its Axis and does not Revolve about the Sun', at Central Hall in Bridgton, Maine, in March 1895.

The *Bridgton Scrapbook* subsequently gave a glowing report of the proceedings under the headline, 'A Rich Olio of Fun and Philosophy by Prof. Holden at Central Hall: The Cream of the Town were there, and Laughter and Applause Ruled the Hour':

> His lecture was the local sensation of the week, and served to cause a general rally of the people of this village and vicinity to see and hear him. And Prof. Holden – 'he came, he saw, he conquered!' All walks of life were represented in that composite assemblage, including the profession, the leading business men, and Bridgton's 'Four Hundred', and all were hugely entertained by Mr. Holden's quaint wit and philosophy. Everybody hereabouts has heard of Jos. W. Holden, and few Maine men enjoy a wider newspaper notoriety beyond the land of sunrise.[21]

The 'hero of the evening' had apparently been greeted with a roar of applause as he stepped on to the platform, laden with textbooks, manuscripts, a cloth wall chart and a pointer. Holden went on to speak for an hour and a half, and the paper concluded that 'to say that it was a wide-awake, satisfied, smilingly complacent crowd which poured out of Central Hall is putting it mildly'. Unlike the majority of flat-earth lecturers, Holden had mastered the art of avoiding confrontation with his audiences, and he continued his crusade until his death in 1900 at the age of eighty-six. Many were saddened to hear of his passing, including journalists at the *Semi-Weekly Landmark*, eight hundred miles away in Statesville, North Carolina:

> We hold to the doctrine that the earth is flat ourselves and we regret exceedingly to learn that one of our number is dead, because there are few of us and one can ill be spared. But we are not without hope. One of these days the idea that the earth is round and turns over every 24 hours will be relegated to the roar along with other antiquated notions.[22]

While Joe Holden was seeking to prove that the earth was flat to the population of the United States, in England Lady Blount's *Earth not a Globe Review* continued the challenge to scientific authority. This included an attack on Sir Isaac Newton's 'eclectic suppositions', alongside more letters about Alfred Russel Wallace and the Old Bedford Canal. Meanwhile, Zetetes had passed on the editorship on the grounds of ill-health, and as time went by Lady Blount found space in the columns to hold forth on a variety of interests. A vegetarian and anti-vivisectionist, she despised cruelty in all its forms and had the paper cover these subjects in flowing prose and verse, alongside references to her work as president of the Society for the Protection of the Dark Races. In many ways Lady Blount was an exceptionally forward-thinking individual, and alongside her focus on humanitarian causes, developing her creativity was an ongoing concern. She particularly enjoyed composing poetry and music that reflected her romantic notions and deeply felt beliefs. Her subject matter was broad, with musical compositions ranging from 'Beautiful Flower of Love' and 'Je T'Aime' (dedicated to the Marchioness of Carmarthen) to 'Our Enclosed World: For Pianoforte', and 'The Earth Not a Globe', a piece performed at Crystal Palace by Godfrey's Military Band.

Besides using music to disseminate her message, in 1898 Lady Blount found time to write and publish a fantastical novel, *Adrian Galilio, or a Song Writer's Story* (1898). Suffused with poetry, music and flowery prose, the story offers possibly the best indirect insight into Lady Blount's state of mind and her increasingly remarkable efforts to promote the zetetic cause. Intriguingly, it revolves round the adventures of a certain Lady Alma, unhappily married to a cold-hearted baronet who despises her devotion to her convictions and her steadfast Protestant faith. Unfulfilled and unhappy, she embarks on an affair of the heart with a Roman Catholic priest, who is subsequently shot and then kidnapped. Meanwhile Lady Alma finds solace by reinventing herself as Madame Bianka, a world-renowned flat-earth crusader with connections in high society, who tours Europe giving elaborate lectures on cosmology, the creation, true love and hell. The book was evidently an attempt to popularize zetetic astronomy in a more lightweight literary form, as a *Morning Leader* reviewer noted:

> I have received a copy of the most amazing novel ever written. It is called Adrian Galileo; or a Song Writer's Story ... It is both interesting and original. For originality, it beats every-thing and everybody, from William Shakespeare downwards. It is a novel with a purpose; and that purpose is to prove the earth is flat.[23]

Besides reinventing herself as a novelist, and indulging in semi-autobiographical wish-fulfilment, Lady Blount had other schemes in place. By 1901, she had moved to South Wimbledon with her children, while Sir Walter lived alone in the house next door. Whatever the state of her marriage, she evidently had plenty of time to write, and she exploited her opportunities by establishing and editing a new journal for zetetics, the *Earth: A Monthly Magazine of Sense and Science*, at an annual subscription of 1*s.* 6*d.* In terms of content, it was much the same as her previous paper, with the occasional new controversy cropping up.

Meanwhile, at the Glasgow meeting of the British Association for the Advancement of Science in 1901, H. Yule Oldham informed the geography section that his recent experiments on the Old Bedford Canal had successfully confirmed the globular theory.[24] Details of his paper were not provided in the association's report, but its inclusion on the programme of the nation's most prestigious public scientific forum was interesting, considering that the globular fact was so well established that it was generally seen as beyond requiring further experimental proof. During the same period, in the *Earth*, zetetics were discussing a radical plan for the collections held at the British Museum. There had recently been public debate about lack of storage space at the library, and several correspondents believed the problem could be solved by disposing of some books. The censorship scheme was not new, and one letter-writer described how it had first been advanced in the 1870s:

> The suggestion was that a perfectly unbiased committee should be formed of men of all grades of thought, including the late John Hampden and one other well-known planist, [and the] committee should have full powers to go through the library of the British Museum and get together *all* the modern works which have been written *on*, or *have* the 'Newtonian Theory' of the world being a globe as the base. This selection would necessarily include tons of books, on such absurd *theoretical* subjects as Universal Gravitation, Atomic Origins, Evolution, Geology, Astronomy, Pluralities of Worlds, and other wonderful phantasies too numerous to mention; then, after mature unbiased cogitation, these books should be removed to a warehouse in some deer forest or other depopulated stretch of country, or to avoid any more trouble it would be advisable to ship and throw them overboard in the Mid-Atlantic.[25]

Nor was this the last to be said of Hampden *et al.* in Lady Blount's new journal, for correspondents from all schools of thought frequently harked back to the zetetic heyday with both strongly worded criticism and praise of their ideas. In 1902, 'J.W.' wrote an informative letter on past events, commencing with a polite apology to Lady Blount:

> I cannot assimilate your ideas as I was brought up a mathematician, and went in for trigonometry and astronomy. I find that the question of light alone upsets the whole mass of evidence you have so clearly put forth. Please don't be angry with me.[26]

Like many astronomy enthusiasts, 'J.W.' had attended one of Parallax's lectures to hear his rhetoric at first hand, and had fallen into discussion with him afterwards, with interesting results:

> Parallax told me how he became *hypnotized* with the idea [of a flat earth] when seven years old, and, after one hour's discussion with me, he acknowledged that the laws of light completely upset his theory, and he sold out, to [John] Hampden, a very honest *simple minded* man, all his copyrights and stock of books for £150, five days after and came out with a company advertising 'Dr. Birley's Phosphorous' – and I found after that he had been a doctor at Liverpool, and was struck off the list for some illegal practices, and had passed under *seven different* names – including Rowbotham, Parallax and Birley.[27]

Furthermore, Hampden had allegedly complained that Parallax, after selling the copyright to him, had admitted that the sun's distance must be materially augmented to account for certain phenomena without specifying what they were. Hampden was at a loss, apparently, and privately considered Parallax to be 'a renegade to his teachings and an unprincipled man'.[28] Yet Lady Blount remained unconvinced by these revelations, prompting the correspondent to allege again that after a short discussion with Parallax he had given up on the flat-earth idea and disposed of his literature on the subject. He continued:

> I am sorry that you should be so *hypnotized* with the idea, for there is a factor which astronomers are all aware of, viz:

'refraction' of the air, which so to speak, bends the horizon, or rising sun, upwards before its time, and makes the curvature of the sea to a *certain extent level*, and hence we *see the sun seemingly before her real rising*. There can be no question of the *sun being beneath our feet at midnight*, and the moon's light is a reflection of the sun's indisputably, which it could not be if your teaching of a flat and motionless earth were right . . . Dear Lady Blount, I know you to be a genuine *honest*-minded woman and most desirous of promulgating the truth in Nature, and of the most advanced views possible, and thus I feel it my duty to write you, to keep my *conscience clear*, for I feel for you.[29]

Despite being stigmatized as an object of pity, Lady Blount persisted with the dissemination of her views in a nationwide round of public lectures. In zetetic tradition they were often contentious events, with booing, hissing and stifled laughter from the back of halls, alongside the subsequent heated debates. Set on her evangelical mission, she was not discouraged, and soon developed her own presentation style, integrating prayers and Bible readings as well as the occasional song into her lengthy talks. Such occasions were gold dust for local newspaper journalists, but although they subjected Lady Blount to gentle ridicule, she retained a hardcore of loyal supporters who matched her efforts to popularize the cause. Above all, their preferred tactic was writing irate letters to the local press, which, if they were not rejected by the editor in the first instance, could spark off debates that dragged on for months. In addition to letter-writing, several of her working-class followers sought to convert their peers in chapels, church halls and markets, while others threw their not insubstantial energies into publishing a number of polemical tracts.

The titles of such works were eye-catching, although sales probably never lived up to the authors' hopes. Nevertheless, they continued, driven onwards by the strength of their beliefs; as a defence of the Bible, the flat-earth campaign was perceived as a

God-given mission, so they could never give up. One elderly East London zetetic, David Wardlaw Scott, commenced his book, *Terra Firma: Earth not a Planet, Proved from Scripture, Reason and Fact* (1901), with an account of the personal trials and tribulations that compiling it had involved. Suffering with cataracts in both eyes, so debilitating that he could hardly read or write, he still felt compelled to complete the 287-page volume in the belief that he was performing a service for the good of humankind. More than anything, he was disturbed by the declining importance of Christianity in the nation's cultural and spiritual life and, like all zetetics, his protest was pinned on the most extreme of Bible-science 'facts'. For Scott the flatness of the earth was beyond dispute and he informed his readers that he could see no greater achievement than 'exposing the fallacies of Modern Astronomy . . . so contrary to the Word of God, and so conducive to the promotion of infidelity'.[30]

Similarly dedicated were the South African author Thomas Winship, who published *Zetetic Cosmogony; or Conclusive Evidence that the World is not a Rotating-Revolving-Globe but a Stationary Plane Circle* (1899) under the pseudonym Rectangle, and Ebenezer Breach, or 'Uncle Ebenezer', who took Portsmouth by storm with his public lectures, one of which was described in the *Hampshire Telegraph* in late March 1896.[31] Audience expectations were particularly high on this occasion, according to the paper, because the last such event had ended in a riot 'rather put[ting] a damper upon Mr. Breach'. Nevertheless, he agreed to speak in the Albert Hall at the heart of Portsmouth's entertainment district, and on the evening in question he took to the stage surrounded by portraits of noted astronomical authorities, such as Galileo and Copernicus, and home-made models of the solar system that had been 'saved from the wreck' of the last lecture. Also on display were maps of the world as a circular plane, which, the *Telegraph* observed, looked something like 'bicycle wheels after a collision'. It transpired later that they had been drawn by a shepherd, while the paper noted that Breach's models had a tendency to collapse while he was trying to demonstrate his points. Despite such knocks to his authority, Breach remained determined to use rhetoric and empirical proof to substantiate his case, and had just begun to explain why the earth was not a heavenly body when a 'fearful smell' engulfed the hall. Despite a 'remarkably quick migration' by some audience members, Breach reportedly persisted with an analysis of why the earth did not revolve round the sun, but was interrupted again by someone crying, 'Rats!' and blowing a develene.

However, Breach remained focused, simply calling for a police constable and continuing on to inform his audience that the earth was an oblong 30,000 miles long and 10,000 miles across, while the disc-shaped sun was a mere 5000 miles wide. Then the 'fiendish smell' came on again, according to the paper. However, Breach attempted to finish the lecture, among heckling, cheers and random bursts of applause, but eventually elected to leave the stage after the audience repeatedly broke into choruses of 'Up In A Balloon' and 'For He's A Jolly Good Fellow', pausing only to instruct the constable to keep watch on his models. Although the audience had clearly attended the lecture for its entertainment value alone, the *Earth* later reflected, in all seriousness, that Breach's 'withers were unwrung' by the experience. A prominent campaigner, he went on to challenge the then Astronomer Royal, William Christie, to prove that the earth was a globe and publish tracts with titles such as *Dauntless Astronomy, Fifty Scientific Facts, The Greatest Event of the Age: The Downfall of Modern Astronomy,* and *Twenty Reasons against Newtonianism with Geographical Proofs.*

Yet the most prolific of Blount's disciples was Zetetes, otherwise known as Albert Smith. Even after relinquishing the editorship of the *Earth not a Globe Review*, he had continued to keep busy at his Leicester home, composing tract after tract that argued the case for the flatness of the earth. Not the most approachable reading material, although sometimes unintentionally amusing, the dense publications were packed with Biblical quotations that appeared to support his unconventional views. The best-selling was undoubtedly *Is the Earth a Whirling Globe as Assumed and Taught by Modern Astronomical 'Science'?* first published in 1904. In it Zetetes placed the burden of 'irrefragable proof' squarely on the shoulders of scientists:

> It yet remains for the Copernican school of Astronomy to prove that the earth upon which we walk about so complacently, and that the country which on a fine day looks so calm and peaceful, is flying through space at a total aggregate speed of something like 86,000 miles per hour. Shall we blindly believe a theory which in the nature of things is so impracticable, and a theory which directly contradicts the evidences of our God-given senses? We feel no motion; we see no motion; and we hear no motion; while our senses favour the reasonable and demonstrable fact that the earth is stationary.[32]

He considered it to be a curious fact that in standard astronomical works no attempt was ever made to provide proof of the earth's axial or orbital motion; strangely, he commented, the matter seemed to be 'deftly taken for granted!' Interestingly, and unusually for a zetetic, Zetetes then addressed the question of why the shape of the earth should matter so much and whether there were more productive ways that he could spend his time. It appears that this point had been raised with him personally on repeated occasions, for it was one to which he gave particularly short shrift. 'To those who say "what does it matter?",' he retorted, 'we might as well ask, "Does it matter whether we receive the evolutionary theories of Darwin . . . and other infidel philosophers, or the simple but grand teachings of the prophets of Israel, and the Apostles of our Lord, respecting God and His great Creation?' In the warfare style now typical of radical polemicists on religion and science, he continued that the globular theory was the mainstay of the modern absence of religious faith and that one could never reconcile astronomical and evolutionary theories with the Bible, the word of God.[33] After sixty-seven pages of discussion, he reached the predictable conclusion that science was 'one of the most subtle plots of the Devil', while 'THE BIBLE IS ABSOLUTELY TRUE, and MODERN ASTRONOMY IS ABSOLUTELY FALSE'.[34]

The tone of such comments pre-empted that subsequently adopted by creationists and, with publications like these, the loyalty of many UZS members was never in question for Lady Blount. The key issue remained converting the rest of the population to the truth of her ideas. As a woman who relished a good argument or a debate, she was undaunted by the momentous challenge, and remained keen to appear at any venue where she could gain permission to speak. Throughout 1903 she travelled the country, giving talks such as 'Logic Cosmically Considered' and 'Moonshine and Motion' to a wide range of audiences, including a crowd at Chelmsford's Shire Hall, scene of the infamous Wallace–Hampden trial in 1875. According to the *Essex Weekly News*, this meeting was packed to overflowing, with people crammed on to the landing to listen as Lady Blount lectured for two hours in all. It would have been much longer, the paper continued, had it not been time for her train to depart, although the audience seemed to grasp the basic points of her message. In the aftermath, a reviewer commented under the headline 'An Impudent Hussy':

> Lady Blount is a nice little lady with whitening hair and a delightful lisp. And as she describes in graphic fashion the errors and blunders of the scientists, she has a habit of smiling in the most attractive manner imaginable. Nobody could help liking her. But how terribly severe she was on the poor misguided men of science! They have been under the tutelage of the Evil One. Some of Lady Blount's reasons for believing the earth is a plane and not a globe were very amusing. It was said that vessels had sailed round the globe east to west, but nobody had gone round north or south; and how, therefore, could the rotundity of the earth be 'proved'?[35]

With some delight, the journalist concluded the piece by noting the evident embarrassment of the chairman, Major Rasch MP, and his reluctance to second a vote of thanks to the speaker at the end of the evening. As for the speaker, she was outraged by the review, protesting that it would have been better entitled 'A Counterfeit of Truth'.

Other engagements followed the Chelmsford lecture – at Peckham, Wood Green, London's Exeter Hall and more, but by spring 1904 Lady Blount had decided that she needed to bolster her rhetorical assault with firm experimental proof, particularly as previous zetetic tests were somewhat out of date. In the final analysis, she believed that visual proof was needed, and what could be better than photographic evidence showing the flatness of the earth? The natural location for such a picture was the Old Bedford Canal, the 'happy hunting ground of the earth flatteners', and thus inspired, Lady Blount set to work on her plan. She hired a photographer, Mr Clifton, who had access to the latest innovation in long-distance pictures: a Dallmeyer camera with a telescopic lens. With the arrangements to her satisfaction, she made her way to the canal on Tuesday, 10 May. Here she met Clifton, with a group of supportive zetetics, and together they discussed the best tactics to adopt. The first step, they decided, should be to repeat a version of the classic test: six poles topped by flags placed a mile apart along the water from Old Bedford Bridge to Welney. A good telescope was then moved to the lowest point nearest the water's edge at Old Bedford Bridge, and levelled to the altitude of the flags, five feet above the surface of the water. On looking through the telescope, Lady Blount alleged that the flags could be plainly seen, with each intervening flag at the same altitude over the full six miles; an utter impossibility if the earth was a globe.

Success seemed assured, at least to her, but she remained dissatisfied. Determined to gather as much evidence as possible while she had the opportunity, she asked a loyal zetetic, Mr Shackleton, to go to Welney Bridge and try another test. When darkness fell, he climbed into a flat-bottomed boat, lit an acetylene gas lamp showing a naked light, and proceeded to hold it as close to the water as he could. Meanwhile, six miles north at the Old Bedford Bridge, Lady Blount and her party made some more observations. Lying on a platform with telescopes less than eighteen inches from the water, they attempted to spot the light. Not only was it distinctly seen, according to Lady Blount, but the reflection in the water below was also evident, observations believed to be impossible if the earth was round. The results were more impressive than Lady Blount could have hoped: 'The light flashed out straight to the eye of the observer,' she later wrote, 'with no hill of water intervening for the whole of the distance.'[36]

The next day, Lady Blount returned to the canal for a third and final experiment, this time exploiting Mr Clifton's talents. At the Old Bedford Bridge, a fifteen-feet-long white sheet with a black border was draped vertically from the bridge down to the water's surface, while a wooden platform was fixed at Welney Bridge, likewise at water level. Armed with a 'specially prepared' Dallmeyer camera, complete with telescopic lens, Clifton lay flat on the platform and focused on the sheet six miles north. Lady Blount hoped that he would be able to photograph the entire object, which globularists believed would be hidden by the earth's supposed curvature of twenty-four feet over six miles. However, due to 'indistinct atmospheric conditions', the test had to be abandoned. Lady Blount also had to pack up and leave: she had to give a lecture on the Isle of Wight on Thursday evening and never missed an opportunity for debate. Prior to her departure, however, she instructed Clifton to repeat the experiment as soon as possible.

The next day he achieved some interesting results. As Lady Blount stepped on to the platform to address the audience at the Institute Hall in Shanklin, she was handed a telegram informing her of his findings. Apparently the screen had been completely visible at a six-mile distance, and not only this: its shadow had also been seen reflected in the water below. What was more, to his confusion, Clifton had obtained photographic evidence of these 'facts'. He subsequently reported to Lady Blount:

> Referring to the experiments at Downham, Norfolk . . . I have
> much pleasure in testifying to the fairness of the conditions
> under which they were conducted. I arrived on the spot with

the distinct idea that nothing could be seen of the sheet at a distance of six miles, but on arrival at Welney I was surprised to find that with a telescope, placed two feet above the level of the water, I could watch the fixing of the lower edge of the sheet, and afterwards to focus it upon the ground glass of the camera placed in the same position. The atmospheric conditions were very unfavourable, a day of sunshine having succeeded several wet days and thereby caused an aqueous shimmering vapour to float unevenly on the surface of the canal and adjoining fields. This prevented the image from being as sharply defined as it would be under better conditions; but the sheet is very plainly visible nevertheless. This trouble is well known to all who have practised telephotography. With regard to the lens used, I may say that this had an equivalent focal length of between 16 and 17 feet, which ensured an image of appreciable size being obtained at such a distance. I should not like to abandon the globular theory off-hand, but, as far as this particular test is concerned, I am prepared to maintain that (unless rays of light will travel in a curved path) these six miles of water present a level surface.[37]

Lady Blount was delighted; according to her (inaccurate) calculations, if the earth was round, the screen would have been some twenty feet below the line of vision.[38] She concluded in a pamphlet, *Flat or Spherical?*,

As the whole of the screen, and its reflection in the water beneath were observed and photographed, no curvature can possibly exist; the theoretical scientists are wrong and beaten and Parallax, John Hampden, William Carpenter, and the army of Zetetics were, and are, right in their contention that the world is not a globe.[39]

The photograph had exceptional propaganda value and, after years of supporting her ideas with Biblical quotations, Lady Blount was keen to exploit her new-found visual proof. With this aim paramount, the picture was paraded in successive issues of the *Earth*, while she also reprinted a petition from Norfolk locals testifying to their belief that the canal was flat.[40] Meanwhile, in the popular-science journal the *English Mechanic*, a number of correspondents were unimpressed. Among numerous objections, several explanations were advanced to account for the supposed evidence. Much to Lady Blount's disdain, one correspondent claimed that the picture was a mirage, 'an aerial flickering image', while another disputed the size of the sheet she claimed to have used in the test. Lady Blount was affronted: she had personally supervised the preparation of the sheet, by the Misses Watts of Downham Market, and it was fifteen feet square, no more no less, with a black border a foot from the edge.

But her protestations were not sufficient to convince the sceptics, and another correspondent commented that the 'flat earth lunatics' were imagining things again.[41] The challenge to Lady Blount's trustworthiness hit home, and bitter debate ensued, with letters from globularists and zetetics again becoming a regular feature in the pages of the *English Mechanic*. The journal allowed readers to judge for themselves by reproducing the now infamous picture in its 28 October issue. It was accompanied by a letter from Lady Blount protesting that 'The experiments were done openly, and no attempt was made in any way to sophisticate phenomena so as to fit into any theory whatever.' The calico sheet was an ordinary one, of fifteen feet in length, the photographer was a globularist, and he used a powerful camera lens to ensure an image of reasonable size. She also appended two letters from the photographer, Mr Clifton, to add some 'globist' clout to these claims.[42]

Despite advice to the contrary, a few professional scientists could not resist becoming embroiled in the debate. In January 1905 the Irish astronomer, John Ellard Gore, wrote a letter outlining various experiments where the fifteen-foot dip of the horizon over twenty and a half miles had been photographed at sea. His conclusion, 'The earth is therefore spherical, as all experiments prove it to be', was clearly intended to provide the last word on the matter.[43] But attempting to have the last word in a discussion with zetetics was a near-impossible task. If words were futile, maybe direct proof would be indisputable – or so amateur astronomer Clement Stretton believed – so early in 1905 he decided to run a version of Lady Blount's experiment on Leicestershire's Ashby Canal. On 20 January, he reported back to the *English Mechanic* that the trial had been a success. He had instructed a 'perfectly independent' surveyor (although one suspects he was a globist) to undertake a version of Lady Blount's experiment with a theodolite and a boat. The surveyor placed the theodolite four feet nine and a half inches above water level, and watched the boat as it sailed away. At one hundred yards, the vessel fell under the true line of sight, and after a mile, it was eight inches below this level, proving that the earth was a globe.[44]

Naturally Lady Blount was somewhat displeased. She retaliated in the next issue of the *Mechanic*: 'Mr. Stretton's experiment [was] defective in what it was supposed to prove' – namely, the globular theory. According to her, the position and handling of the theodolite had been incorrect, and if the instrument had been properly employed the result would have been much different. In the face of adversity, Lady Blount remained stubborn: she had openly conducted reliable experiments before globists and planists, and they had provided a 'flat contradiction to Mr. Stretton's conclusions'.[45] The tests had been undertaken at considerable expense to herself, not for personal gain but simply to illustrate the truth. It was a pity, she continued, that Newtonians did not display the same reverence for facts as truth-seeking zetetics. Nevertheless, she emphasized, '*We* stand by our experiments of *facts*.'[46] Her patience was wearing thin: she believed she had illustrated beyond doubt that all water was level and her photographs, witnesses that could never be blackmailed or biased, provided incontrovertible proof that the earth was flat.

Never one to rest on her laurels, Lady Blount embarked on a new project while the controversy rumbled on. One of her primary concerns had always been health, both spiritual and physical, and the newspapers reported that even in her mid-fifties she could still cycle fifty miles a day.[47] In common with other leading zetetics, she was extremely inventive and attracted to alternative medicine alongside alternative science. In the early years of the twentieth century she formulated her own version of Dr Birley's syrup of free phosphorus, Lady Blount's Muric Acid, which she sold as a cure for rheumatism and gout. On a more personal level, she had taken to wearing magnetic corsets, vests and gloves for the good of her health, and having experienced beneficial results, she decided to publish something as a mark of her gratitude. The subsequent pamphlet, *Magnetism as a Curative Agency* (1905), was a *tour de force* of advertising propaganda for the company Appareil Magnétique, wherein Lady Blount waxed lyrical about the positive influence of magnetic currents on her stomach ulcers and circulation.

Crammed with illustrations of ungainly garments, from compound vests to special womb shields, the pamphlet made weighty claims for their positive effects. Most notably, Lady Blount argued, when 'Hercules' compound magnets were worn close to the body, the blood was charged with magnetism, thus 'correcting its morbid condition and polarizing the system'. This action, she alleged, could cure everything from nerviness and constipation to 'brain fag' and bladder disease, while wearers need never again be concerned about 'loss of nervo-vital force'.[48] The pamphlet was also notable for being one of the few publications in which she made no mention of flat-earth theory, although this did not indicate a decline in her devotion to the cause. She continued editing the *Earth* and lecturing to the public with such success that in April 1906 the *Portsmouth Daily Herald* warned that 'the zetetics are now numerous enough to be called a cult'.[49] Spurred on by her seeming success, that year Lady Blount published a summary of her ideas in a seventy-one-page booklet, her own version of Parallax's classic, *Zetetic Astronomy: Or the Sun's Motions North and South; With the Moon's Motions; Fancied and Real; Showing the Uselessness of the Gravitational Theory etc* (1906).

In 1907, having given the fullest expression in print to her flat-earth ideas, Lady Blount decided that the time had arrived to make contact with the *bête noire* of the zetetic movement, Alfred Russel Wallace. By now a grand old man of seventy-three, Wallace was still thriving, living and writing at his rambling house in the small town of Broadstone in Dorset. He had not forgotten his unfortunate involvement with Hampden and Parallax, however, and had even kept some of the papers connected with the case. Judging by his annotations, the letters and pamphlets had some humour value, and this was probably his reaction to Lady Blount's note. In typically animated style she asked if Wallace would be 'so very kind' as to read an article she had enclosed, mysteriously adding, 'I somehow felt led on to write it but I feel that you wd not mind reading it first.' In addition, she enclosed a copy of her magazine, the *Earth: A Magazine of Sense and Science*, along with further comments about her work. 'I am the leader of an array of Zetetics or Truth Seekers,' she began, and 'Learned and Scientific Men and women are in our ranks.' Although she admitted that some of these people did not dare confess publicly their disbelief in the 'whirling sea and earth globe', she assured Wallace that 'many accept my teaching'.

Chief among her theories was the idea that the daily revolution of the sun was caused by 'the Ether', which circulated rapidly, like great whirlpools, above and around the North and South Poles. The sun, meanwhile, was 'a comparatively small electric body', whose rays were carried round the earth by two sets of magnetic currents, moving in opposite directions but in perfect harmony. In conclusion, she informed Wallace that she had turned her attention to writing on the motions of light. Bearing in mind his reputation as a free-thinker, with diverse work on evolution, spiritualism and, latterly, radical politics, she assumed that he would be sufficiently open-minded to provide her with support and free editorial advice. In short, Wallace was seen as something of a crank in certain circles and, despite their differences, Lady Blount seemed to perceive him as a kindred spirit, who might be of some use to her in her work. While bumping along on a train, she added a note to her letter, in looping scrawl, saying she was planning to send on more of her publications.[50] Whether Wallace replied to her first letter is unknown, but he printed 'Lady Blount! The Flat-earth Woman' on the envelope and filed it for posterity.

Alongside flat-earth theory, zetetics had other innovations they wished to publicize. Lady Blount had invented a flying machine, the 'perfect aeroplane', so she said, and three months later the *Washington Post* reported on this dubious claim. Designed to solve all of the problems of aviation, the plane was apparently studded with valves that opened when rising and closed during landing, like the feathers of a bird in flight. With evident scepticism, the *Post* reported, 'Lady Blount says anyone strong enough to ride a bicycle will be strong enough to cycle with her new flying machine.'[51]

Filled with energy and enthusiasm for such schemes, Lady Blount paid no heed to ridicule from her numerous detractors. Non-believers could never dampen her indomitable spirit, and she remained unstoppable in her search for the next original idea. An admirable trait, if somewhat misguided, more often than not her drive to realize her often unconventional notions had damaging consequences for her good name. Undoubtedly her social status could open avenues hitherto blocked to zetetics – she was, after all, famous for having a letter printed in *The Times* stating that the earth was flat – but such success could be double-edged. The papers were also eager to report on her failures, albeit more gently than they did with less refined zetetics, and there were other more serious consequences. Most notably, in 1909 her position in society brought her to the attention of an unscrupulous conman and, as a result, the police. The trouble began when Lady Blount met a businessman, Francis Sinclair Kennedy, who proposed a seemingly lucrative idea. In Edwardian England visiting the dentist was costly, and sets of false teeth were often beyond the budget of the less well-to-do. When he saw that there was money to be made by virtue of necessity, Kennedy proposed the formation of a company with Lady Blount at the helm to provide panache and financial backing.

The business, Lady Blount's Medical Aid Society, would undertake extractions and provide false teeth for which clients could pay week by week. The idea appealed to Lady Blount's philanthropic nature in an era when such schemes were not uncommon and she eventually agreed to the plan. A charter describing the aims of the society was subsequently drafted; it would provide medical aid for people of small means, supply dental services and false teeth and, more ambitiously, support medical research and establish a network of local dispensaries and a central institution in London to which members could subscribe. In March 1909, Lady Blount's Medical Aid Society was registered as a company by Universal Zetetic Society member Austin Fryers, with a nominal capital of £500 divided into £1 shares. Naturally Lady Blount was the major shareholder, closely followed by Kennedy and Fryers, and several clerks, authors and gentlemen were persuaded to buy into the company. In a matter of weeks, the society began trading from premises just off Oxford Street, with furniture and dental machinery purchased on credit; and Lady Blount's younger son, Edward, also became involved.[52]

Seemingly, Lady Blount remained unaware that the business was a sham. She was the latest in a line of titled personages, Lord Haldon among them, who had lent their names to crooked companies such as the Artificial Teeth Society and the National Dental Aid Society, run by Kennedy. His *modus operandi* was as follows: advertisements for the business were placed in the London papers stating that false teeth could be obtained at hospital prices on an easy-payment system from a shilling a week, with the additional benefit of free extraction. As a result of the so-called 'easy-payment' scheme, clients bound by contract paid vastly over the market price for sets of teeth they needed to replace those the society had removed. It was surgical extortion and, more painful still, it transpired that the 'dentists' at the practice were entirely unskilled, and injuries, such as a broken jaw, were common. The scandal snowballed until it was brought to the attention of the cheap working-class paper *John Bull*, which published an in-depth exposé of Kennedy as a warning to unsuspecting readers. According to the report, Kennedy was a 'blackguard', who had served a five-year prison sentence for conspiracy, and had been arrested for various misdemeanours including threatening his wife when drunk.[53]

In the interim, the Metropolitan Police had also become interested in Kennedy's business dealings. They launched their investigations in the spring of 1910, in collaboration with the Companies Department of the Board of Trade, and two issues sprang immediately to the fore. Of these, the most crucial was the weakness of the witnesses involved in the case against Kennedy and his scam Medical Aid Society. The most detailed statement in their possession was from a roguish character, Charles Edgar, who had previous convictions for larceny, theft and fraud, and was currently facing other charges. Further to this, the Board of Trade and the Director of Public Prosecutions were concerned about the lack of evidence in general, and whether such a complex and unusual case would stand up in court. Although it was believed that Kennedy's company had been a fraud from its inception, it would be difficult to prove this to a judge and jury.[54] After some consideration, it was decided that it would be best to prosecute Lady Blount as director, which involved establishing that she knowingly and willingly authorized or permitted the crime. Amid all of this, Kennedy was striving to shift the blame by insisting that although he still appeared on file as a director of Lady Blount's Medical Aid Society, he had resigned his position and sold his shares in October 1909.[55]

However, rather than ceasing business following Kennedy's departure, Lady Blount had decided not only to continue but, ill-advisedly, to expand. At Vine Street police station, Sergeant John Prothero was ordered to gather evidence to assess her precise role. He discovered that in November 1909 Lady Blount had secured a lease in her own name for six rooms on the first floor of Imperial Mansions in Charing Cross Road, at an annual cost of £280. For two months the upmarket premises served as both the dental

headquarters of the Medical Aid Society and the editorial offices
of the Universal Zetetic Society. This unusual arrangement had
been brought to an abrupt end in late January 1910, when the
landlord, music-hall impresario Walter de Frece, evicted Lady
Blount after neighbours complained about strange goings-on in
the building. During the same period, the police had also filed
numerous reports from people who had paid and had their teeth
extracted, yet had received no false teeth in return. There were
more than fifty victims and their evidence was damning. According
to their statements, a mysterious dentist, Dr Parker, had taken out
their teeth, in most cases recommending total extraction even
when only a few removals were needed. 'This advice,' Sergeant
Prothero explained, 'was evidently given to bring the patient more
into the power of the society.' Having no teeth would ensure
prompt payment of weekly instalments up to the two guineas
needed for the false set.

Operations had been further complicated when Dr Parker was
sued by the British Dental Association for practising as a dentist
without qualifications or a licence. As an interim measure, Lady
Blount had apparently enlisted the services of de Lacy Evans,
'personal secretary' and Universal Zetetic Society vice president, to
undertake the work in Parker's place. Even though Evans had been
head physician at St Thomas's Hospital and a private doctor to the
nobility, or so he alleged, the arrangement was far from ideal. He
was not a qualified dentist either and, furthermore, as Sergeant
Prothero put it, 'This gentleman has lost his prestige through
intemperate habits and all the patients were nervous as to the
treatment they were to receive at his hands.' Moved by the plight
of the patients, many of whom now had no teeth, Sergeant
Prothero added a personal note to his official report: 'Their
complaints are heartrending in character and several have suffered
loss apart from the money paid over to the company. In one
instance the complainant has lost her situation through the imped-
iment of her speech, brought about through the loss of her teeth.'[56]

By early 1910 the Medical Aid Society was spiralling rapidly into debt. All of its hired help, equipment and goods had been obtained on credit and, despite the guaranteed customer payments, it seemed that the society was unable to pay its bills. Following eviction from its premises in Charing Cross Road, it had moved to the less salubrious Clerkenwell Road, while a printer and book-dealer sued Lady Blount for her failure to pay £350 for books purchased by her societies. Soon afterwards, a former employee resorted to the law courts to obtain £100 owed to him, and in the meantime, the Medical Aid Society was plagued with letters from the Companies Registration Office seeking further informa-tion for its files.[57] By mid-1910, the activities of Lady Blount's Medical Aid Society had been suspended, and in February 1911, at an extraordinary general meeting of the members in Blooms-bury, it was decided that the company should apply for voluntary liquidation.

True to form, Lady Blount soon moved on to other projects. Although the Universal Zetetic Society appeared to have become inactive, in 1913 she published a book on her alternative astronom-ical theories, *The Secrets of Nature Exhumed*, and relocated to Worthing with her now elderly husband. There she set up business premises in the high street, though Sir Walter was becoming increasingly frail. He developed cancer of the bladder at the age of eighty-two, and died in October 1915 after a heart-attack. He left everything he owned to his wife. The difficulty was that he owned very little: by aristocratic standards Sir Walter was impover-ished and after death duties his estate was valued at less than a thousand pounds.[58]

Nevertheless, Lady Blount was solvent enough to move to Brighton, where she lived in the 'Enclosed World Observatory' in Sussex Square, writing musical compositions and pamphlets on an array of serious topics long after most of her original zetetic colleagues had died. In 1921 she published *The Dreamer*, a study of the creation and the origins of mankind and the world, written in rhyme and question–answer form, and this was followed a year later by a book on the origin of the human soul. By this time she was living at The Observatory in New City, a residential complex near Bognor Regis, where she turned her attention to the third publication in her 'origins' trilogy, *The Origins and Nature of Sex* (1923). The pamphlet was so controversial that its publisher felt obliged to include an explanatory note. Essentially Lady Blount had strayed from the subject – or, at least, from contemporary perceptions of it. While the pamphlet covered her ideas on the origins and nature of sex, as interpreted from the Bible, the publisher noted, 'The doubts as to the shape of the earth, the doctrine of the bi-sexual condition of Adam and other points will doubtless not appeal to each reader.' The unwary had been warned: Lady Blount's attempts to weld her versions of science and the scriptures together into a 'universal theology' took her into realms not often explored.

Throughout her study of sex, she digressed on to the evils of science and the fantasies of the 'astronomical novelists', in long sections espousing her flat-earth views. She also had harsh words for the doctrine of evolution: 'Darwin was clever I do not deny, but in this teaching called Darwinism there are some erroneous beliefs, probably formulated before his mentality had sufficiently developed.'[59] And so the pamphlet continued, hurtling through eunuchs, masturbation, hermaphrodites and more; its theme also had a certain resonance in her own life. Never one to follow convention, be it social or scientific, later that year Lady Blount married a forty-year-old builder, evangelist and ex-Royal Navy man from Portsmouth, Stephen Morgan. She was seventy-three.[60] Whether due to this change in her life or simply her advancing age, Lady Blount never produced another book. But as a woman with the courage of her convictions, who could never resist a challenge or a debate, she continued to proclaim the truth of her Bible-based ideas in private until her death in 1935.[61]

Chapter Six

FLAT-EARTH UTOPIA

In the Christian Catholic Apostolic Church in Zion we accept the Bible as the Inspired Word of God. We believe that the Bible, in the original languages it was written, was verbally inspired – that is, that not only the thoughts were inspired but also the words. We accept all that the Bible has to say regarding the earth, the sun, the moon, and the stars, and we do not propose to surrender the Word of God to infidel astronomers.

WILBUR GLENN VOLIVA,
'Which Will You Accept? The Bible, the Inspired Word
of God or the Infidel Theories of Modern Astronomy',
Leaves of Healing, 10 May 1930

BY THE 1930s another flat-earth advocate was attracting widespread publicity across America as a consequence of his unorthodox views. A formidable character, as striking in his personality as he was in his cosmology, Wilbur Glenn Voliva had been born in a simple farmhouse in rural Indiana in March 1870, the son of a Methodist lawyer, James H. Voliva, and his wife Rebecca, a strict Presbyterian of the old school.[1] Raised in such a devout household, Voliva was inspired by his religious convictions from earliest childhood: Methodist Sunday school and teenage preaching led to ordination at the age of nineteen, followed by the study of theology, Latin and Greek at a succession of Bible colleges. In the spring of 1897, he received his Bachelor of Arts degree along with confirmation of a divinity degree from Hiram College near Cleveland, Ohio, and subsequently became pastor of the Christian Disciple Church at Washington Court House in the same state. Yet after years of dedicated study the position proved a disappointment to him; the teachings of the Church seemed to lack divine authority, and his congregation seemed too self-obsessed to attempt to meet his ideal of true Christian standards. Discouraged with his mission, Voliva wanted to resign, and by 1899 he was on the brink of forsaking the Church to retrain as a lawyer. It was at this point that he was handed a copy of *Leaves of Healing*, a magazine packed with extra-ordinary testimonies of the work of the itinerant Scottish faith-healer John Alexander Dowie. According to reports, the blind had received sight, diseases had been cured and whole families saved. All in all, it seemed like a miracle – visible evidence of God's triumph over Satan.

By the time Voliva became aware of Dowie's work, the late-nineteenth-century evangelical revival was in full swing, character-ized by dedication to personal salvation, Holy Ghost baptism, divine healing and the belief in Christ's Second Coming. Numer-ous sects, summer camps, magazines and Bible institutes sprang up across America; meanwhile, Dowie criss-crossed the country like a multitude of other itinerant preachers, proselytizing and minister-ing with his independent mission.[2] In 1893 he arrived in Chicago and set up a 'Little Wooden Hut', as he later called it, outside the World's Columbian Exposition in Jackson Park, just across the street from Buffalo Bill's Wild West Show. Stories of his healings, first of Buffalo Bill's niece, Sadie Cody, then of Abraham Lincoln's cousin, Amanda M. Hicks, soon spread through the city, and within three years Dowie's travelling healing sideshow had grown into the Christian Catholic Apostolic Church with its own imposing tabernacle at 16th and Michigan Streets.[3] Here on stage, Dowie expounded his message of 'salvation, healing, and holy living' surrounded by crutches, plaster casts, liquor flasks and other paraphernalia given up by their owners after he had healed them. Bystanders remembered that his services were electrifying, if foul-mouthed and controversial, but the tone did not detract from their overall effect.

In early 1899, Voliva felt the same inspiration when he read *Leaves of Healing*. He recalled later how he was gripped by Dowie's dynamic interpretation of the scriptures; to him, this was how Christianity was meant to be.[4] He was especially attracted to the simplicity of Dowie's message and its solid base in scripture, so in February 1899, keen to become part of the mission, he travelled to Chicago, attended meetings in the Central Zion Tabernacle and was ordained as an elder in the Christian Catholic Apostolic Church. A devout and capable disciple, he sped up through the church ranks. Over the next two years, he directed the mission in Chicago and Cincinnati, before emigrating to Australia in October 1901 to act as overseer-in-charge.[5]

Throughout this period, Dowie was furthering plans for the mission in his adopted hometown. By the late-nineteenth century, the streets of Chicago were marked by poverty and distress – slums and tenements, hunger and strikes. The city was the melting-pot of the Midwest, where rapid urbanization and industrialization had brewed social dislocation and labour unrest as well as prosperity and economic advance. Chicago was a modernizing urban centre and the signs of change, for better and worse, were everywhere. Amid all of this, in September 1899, Dowie declared a three-month 'Holy War' against the 'hosts of hell' in the city and, determined to fight fire with fire, organized vocal protests wherever sin was found.[6] It was a wide-ranging crusade. Guided by a strict interpretation of Old Testament teachings, Dowie had a broad definition of evil, and condemned the press, the medical profession, political interests, social organizations and other religious denominations. He believed that an individual's loyalty should lie first and foremost with God and the Bible rather than with secular concerns, and he particularly criticized the rapidly emerging medical profession for raking a profit from flouting God's authority to heal.

Headed by the slogan 'Doctors, Drugs and Devils', Dowie's crusade even endangered his life:[7] brawls, riots and repeated arrests for practising medicine without a licence resulted from his 'Holy War' in Chicago, and it was later reported that thugs in Hammond, Indiana, attempted to murder him with spikes when he lectured there.[8] Faced with such hostility from the state, the professions and the public in general, Dowie concluded that holy living was not possible in modern mainstream society and the time had come to create a Christian Utopia for himself and his followers, a religious refuge free from pollution, persecution and sin. In 1899, with the money that had flooded into the church coffers from its faithful followers, Dowie purchased 6,500 acres of Illinois farmland on which to build his dream for $1,250,000. A year later, on the plains and the railroad line between Chicago and Milwaukee, a few miles south of the Wisconsin border, Zion City was born. Over the next three decades, it would become infamous as the modern centre of flat-earth belief.

At the time of its establishment, the Zion experiment was not as unusual as it might seem. People had long sought to bring into reality their dreams of heaven on earth with the establishment of social, religious and political Edens on American soil. From the Civil War period particularly, idealists of all descriptions – anarchists, socialists, spiritualists, co-operators, vegetarians, free-lovers, Mormons and Shakers – had built sanctuaries from the troubles of the world, and bastions of hope for a brighter future in the land of new horizons and opportunity. While the specific strategies of such mystical dreamers, inspired prophets and radical visionaries varied widely, they were united in their desire for social progress and a better way of life, together with a belief that perfection of humankind was possible, given the right circumstances of co-operation and shared interests.[9]

Set on the western shore of Lake Michigan, Zion was not the first such Utopian community; nor would it be the last. Dowie saw the city as the ultimate gesture of his belief, more meaningful than any sermon, for it was established as a true theocracy, a city ruled by God. He saw to it that Zion was wholly owned and controlled by the Christian Catholic Apostolic Church, a place where the divine would influence every aspect of family, commercial, educational, ecclesiastical and political life.[10] All was co-ordinated through Dowie, who adopted the motto 'Where God rules, man prospers' to illustrate his faith that the city would solve the social, moral and economic troubles of its era. From its inception, Zion was designed to be the model Christian colony, a city in which God's people could escape the poverty, disease and sinful ways of the outside world. Importantly, this sense of togetherness was underpinned by a persecution narrative – later issues of the church journal, *Leaves of Healing*, referred to the city as a 'refuge for the oppressed of God's people'.[11] More positively, Zion was advertised as a perfect panacea for worldly ills, a city in which everyone could live a clean life, receive full reward for their labour, invest their earnings and build a happy home.[12]

As it was in ethos, so it was in design. A vast wooden temple, with seating for more than seven thousand, the Shiloh Tabernacle was built on a two-hundred-acre plot at the centre of the city, literally and metaphorically at its heart. Church teachings were further disseminated by Zion's special parochial school system, running from a kindergarten to a four-year college. Education here was founded on the Bible as the revealed word of God although, unlike the zetetics, Dowie did not interpret the Bible to argue that the earth was flat. Meanwhile, the city's religious ideals were translated into its economic life. Competition was banished in favour of co-operation and equal pay, while all industrial and commercial establishments were owned outright by the Church. In the early years, such businesses included a lace factory, a bank, a department store, a candy factory, a printing and publishing house, and the baking industry that produced the Zion Fig Bar, a best-selling snack from the 1920s to the 1950s. At one point the city produced a million a day, and it was fitting that even this should have its roots in the city's Christian ethos: church elders searching for a distinctive product had turned to the Bible for inspiration and therein found the fig.

From 1901, with work on the new city under way, settlers journeyed by wagon, railroad and boat from as far as South Africa, Australia and Switzerland (where the Church had branches), eager to begin a new life in Dowie's Promised Land. As overseer, Dowie was determined to fulfil expectations and introduced a strict legislative code, based on his interpretation of Old Testament teachings. By law 'all diabolical evils of the world' were banned, and this included alcohol, pigs, tobacco, oysters, lobsters, playing cards, medicines, vaccination, drugstores, hospitals, doctors, theatres, sorcerers, dance halls, opera houses, circuses, houses of ill repute, labour unions and masonic lodges. The keeping or selling for human food of anything forbidden in Deuteronomy 14: 7–19 was outlawed, along with 'Any of the Curses or Abominations which Defile the Spirits, Souls and Bodies of Men'. Disobedience invited immediate expulsion from the Church or the city.

Yet for many pioneers Zion in its early years was a beautiful place to be precisely because of this strict code of conduct: citizens at last felt free to live clean lives according to their beliefs. Testimonials from settlers illustrate that many felt a sense of relief to be able to practise their faith in a wholesome community of honesty, peace and understanding, shielded from the pollution, competitiveness, sin and selfishness of urban life. The Utopian dream of contented co-operative life with like-minded people was a perennial human vision, while the simple appeal of Dowie's mission is demonstrated by the fact that between 1897 and 1906 in excess of twenty-three thousand people were baptized into it.[13] Like other Utopian settlements, Zion was rooted in the desire for a perfect world and bolstered by the magnetic appeal of a charismatic leader, although it appears that, as time passed, Dowie became intoxicated with his success. In 1901, with visions of the city becoming the headquarters of a world religion, he announced to the world that he was 'Elijah the Restorer', the third incarnation of the prophet after John the Baptist. He was immediately challenged and denounced by religious leaders worldwide, but nevertheless went on to consecrate himself the 'First Apostle of the Christian Catholic and Apostolic Church in Zion', or the reincarnation of the apostle Paul.[14]

Meanwhile the city, a triumph for progressive town planning, was experiencing boomtown growth; in just five years the site had been transformed from uncultivated wilderness to a modern metropolis worth seven million dollars. But by 1905, the strain of such intense short-term growth was showing: shortages of raw materials, factory shutdowns and unemployment presented warning signs for uneasy citizens and church elders. Often engaged in back-breaking labour from dawn until dusk, citizens had given a tenth of their income and more to the Church, as well as investing in it their hopes and dreams. Under such circumstances, it is unsurprising that disillusionment and confusion swept through Zion in response to mounting problems: as businesses were left without working capital, life savings drained away and poverty increased, so cynicism and re-evaluation took hold. Settlers, once united in their belief, began to drift back to their original hometowns or to rival churches, while close-knit families quarrelled and divided over religion and economics. The situation was aggravated by the fact that Dowie was living in luxurious circumstances before their very eyes. His three-storey, twenty-five-room mansion, Shiloh House, featured the latest European fashions and every modern commodity, and had been built at a cost to Zion residents of $90,000.

The undercurrent of foreboding and instability increased in September 1905 when Dowie suffered a stroke while preaching in the Tabernacle and travelled to Jamaica to recuperate, claiming $2000 a month from his followers' investments to cover his expenses while he was away. Fearing death and realizing that a trustworthy representative was required to oversee the city in his absence, he asked Voliva to return from Australia and granted him power of attorney to enable him to fulfil his role.[15] It was a fateful manoeuvre. Voliva arrived in Zion in February 1906 and, with the support of church elders, promptly deposed him. The charges against Dowie were as follows: extravagance, hypocrisy, misrepresentations, exaggerations, misuse of investments, tyranny and injustice.[16] According to church elders, the once humble Dowie had not been too sick to 'lay his palsied hand' on the people's money, and his taste for 'tomfoolery' had resulted in disaster for the civic economy.[17] He had not only spent huge sums of the community's money on his mansion, but also on a summer residence in White Lake, Michigan; he had made plans for a palatial home for the entertainment of aristocrats and royalty, and taken expensive world tours staying in the best hotels; he had bought $50,000 worth of furniture, and spent $40,000 on books – according to the *Leaves of Healing*, the list went on to the tune of $2,529,765.71.[18]

It appeared that, following his European tour, Dowie had developed a taste for the high life. In addition there had been his 1903 scheme to convert New York City. The Christian Catholic Apostolic Church extravaganza at Madison Square Gardens had involved three thousand Zionites and a thousand bodyguards, cost $300,000 and was a notable failure – yet Dowie had immediately launched another ambitious 'Round the World' salvation campaign.[19] He had covered the soaring cost of these schemes by asking Zion residents to deposit money in the city bank, which, as with every other institution in Zion, he controlled. There was also evidence of corporate irresponsibility, for church officers were aware that citizens' investments in the city's industries had been siphoned off to fund the expensive refurbishment of Dowie's Chicago base. If Voliva was to be believed, the people were poor and Dowie was to blame. Voliva preached that riches and exaltation were not meant for this world but for the life to come, and no man, however visionary, was greater than Zion City – established to serve as God's kingdom on earth.

If the allegations against Dowie had been limited to the misappropriation of funds and poor judgement, he might have been forgiven by his followers. After all, Zion was a theocracy, and Dowie was a divinely appointed overseer and prophet who had made his message manifest with numerous dramatic public healings. Indeed, some believed that Dowie, through his visible ministry of prayer and laying on of hands, had reclaimed their souls from sin and debauchery or literally saved their lives. There were even suggestions that declining health had impaired his judgement and rendered him unwilling to listen to his church elders' advice. But such theories held little sway with Voliva: devoted to Zion's founding ideals, he was determined to oust Dowie and re-establish the precepts on which the city had been established. Supported by leading church elders, he played his winning card: the accusation of polygamy. Over subsequent weeks, the *Leaves of Healing* revealed to Zionites that Dowie had fallen victim to a diabolical trick and succumbed to the 'black lusts of the flesh'. It was even alleged that a harem for concubines had been discovered in Shiloh House. Further details were provided in *Leaves of Healing*, where elders provided evidence of Dowie having glorified the benefits of 'a plurality of wives'.

Whatever the truth of the allegations, the national newspapers went wild. Sex scandal from the moral high ground of Zion City was too much for them to resist, and the columns of sensationalist rags and respectable publications alike were filled with spicy stories. On 6 April 1906, the *Washington Post* reported that Dowie had attempted to divorce his wife of thirty years to marry a Swiss heiress whom he had converted to his 'peculiar creed'. It added that Jane Dowie had not appreciated his alleged change of heart and had passed on love letters and information about her husband's assignations to Voliva and the press.[20] Meanwhile, on the same day, Ohio's *Elyria Chronicle* paraded the headline 'Dowie Planned Harem of Seven New Wives', accompanied by a report claiming that his planned 'Mexican Paradise Project' was a sophisticated cover-up for a polygamous colony. According to Jane Dowie, her husband planned to take seven wives, had already proposed to five and had threatened violence when she dared to protest.[21] The stream of similarly lurid stories continued and the self-styled prophet's fate seemed sealed. In response, Dowie disowned his wife and son and threatened to have Voliva arrested on charges of criminal conspiracy. But his attempts to avenge their alleged treachery through the courts came to nothing, and in March 1907 he died in his bedroom at Shiloh House. Meanwhile, Voliva seized absolute control, backed by members of the Christian Catholic Apostolic Church, and the Zion City coup was complete.

Although he was now general overseer of a whole city, albeit one in the hands of the receiver, Voliva's personal possessions were valued at just eighty-three cents, according to a financial schedule submitted to the county court. Yet he was a shrewd businessman, with exceptional managerial abilities and frugal tastes, and he dedicated himself to repaying the city's five thousand creditors and releasing Zion from bankruptcy. In a series of sermons, during which he publicly forfeited his watch and cufflinks, citizens were encouraged to donate their worldly goods – their jewellery, watches, fur coats, musical instruments, even their houses and life savings – to salvage their city. The majority of church members, looking for strong leadership after past excesses and keen to abide by Dowie's original vision, supported the new overseer. As a result, with donations, cost-cutting measures and sermons on the virtues of sacrifice and 'The Simple Life', by 1910 Voliva was able to buy back Zion's estate on behalf of the Christian Catholic Apostolic Church.[22]

But while Voliva was keen to hail himself the saviour of God's city, outsiders begged to differ. Frequently depicted by journalists as a dour dictator, Voliva lacked the enthusiasm, sense of humour and air of benevolent paternalism that had softened the charismatic Dowie. In contrast to Dowie's emphasis on pacifism, universal redemption and divine healing, journalists noted that Voliva focused on harsh reforms, ruthless discipline and hellfire rhetoric. It was no coincidence, critics archly remarked, that one of his earliest newspapers was entitled the *Battle Axe*.

However, despite Voliva's sometimes questionable tactics, leading to accusations of fanaticism and, later, financial fraud, a streamlined Zion emerged from the chaos surrounding Dowie's fall from grace. Unfortunately, from Voliva's perspective, the disillusionment, loss and re-evaluation that accompanied it led to fragmentation and growing calls for Zion to become an open city with religious freedom, a democratic political system and a loosening of its now infamous lifestyle laws. Faced with escalating dissent, which increased as time passed, Voliva went on the offensive, resolving to reinstate the clean-living ideals on which the city had been founded. In his eyes, the battle was nothing less than one of good versus evil; the sins of modern society would be banished by force, if need be, and his enemies had better beware. This warning applied particularly to the Independents, non-members of the Christian Catholic Apostolic Church, a powerful faction that promoted a more liberal approach to religion, politics and morals. Vote-rigging and church-burning, riots and fights: independent Zionites were allegedly subjected to ongoing persecution from Voliva and members of his Theocratic Party over the years that followed. Denounced as dirty dogs and worse, their shopping street was nicknamed 'Rat Row', their children were forced to attend different schools from those that were church-affiliated and were even banned from playing in Zion's parks, which Voliva personally owned.

As a result of such developments, in the years until 1935, settlers who had envisaged a cosy Christian community arrived to find a city troubled by bitter political and religious divisions. Disturbed by the near-constant animosity, some pioneers left Zion, while others opted for theological alternatives, joining Independent churches or the itinerant Pentecostal preachers who held tent meetings in the town. A principal problem was the loss of divine authority. Dowie, for all his alleged lapses, was widely viewed as God's chosen leader bestowed with visible healing gifts, and many found it difficult to place a similar level of faith in his bombastic successor. Undoubtedly Voliva's gifts were of a more worldly order. A dictatorial preacher, who appeared to revel in conflict and dispute, when he warned his congregation 'the war is on red hot' the threat was truly meant. Sunday sermons became fighting talk: rather than giving demonstrations of divine healing, Voliva declared he would 'skin' Independent political candidates and 'hang their hides out to dry', while rival churches fared little better. Voliva frequently branded the local Grace Missionary Church a 'goat house', the Pentecostal Church a 'monkey house', and when the father of Pentecostalism, Charles Fox Parham, held meetings in Zion seeking converts from 'Dowie-ism', Voliva castigated him as a 'religious vagabond' intent on poaching his flock. Indeed, the Pentecostal movement was burgeoning, with 50,000–100,000 converts across Canada and the United States by 1910, while Zion was heading into decline, and Voliva was determined to crush the competition, despite their similarities.[23]

Such authority contests were common in radical evangelical circles and experiments in communal life, and in this respect Zion was little different. Through the years that followed, invective and directives characterized public life in the city: in the midst of challenges to his authority Voliva focused on keeping control. It was rumoured that spying on neighbours was encouraged, and that any citizen of Zion who forsook the Shiloh Tabernacle for an alternative risked being labelled a traitor and excommunicated by the Church. Social exclusion, if not exile, was likely, and since Voliva owned or controlled many businesses, errant worshippers also risked their livelihoods and homes.

In an era when religious controversies were governed by different conventions of etiquette, participants frequently eschewed courtesy and tolerance in favour of virulent public abuse. This was especially the case in radical evangelical circles, for advocates were dealing in non-negotiable commodities, promoting intransigent positions based on competing interpretations of the will of God.[24] With no holds barred and no room for manoeuvre, head-to-heads were often unbelievably acrimonious, with libel and bitterness taken to an extreme. In such a context, Voliva soon established a reputation as an arch-disputant, known for placing billboards at strategic points around Zion, 'the only place where it is easy to do right and difficult to do wrong', featuring strongly worded pronouncements. Some were lists of outlawed practices, while others

were one-liners aimed at competing churches – 'You're running a
monkey house – get out of here and establish a zoo of your own.'
There were also longer warnings:

> No gentleman, not to mention a Christian would break into a
> church settlement and attempt to hold meetings, or to estab-
> lish a counter-organization. Those who do are nothing more
> than religious bums, with less honour than a gang of highway
> robbers and thugs. Get out of this community if you have a
> drop of honest blood and go establish a settlement of your
> own. If they want to know who is boss around here, let them
> start something else.[25]

The media reported that the Independents grew so tired of
being bullied that they torched and defaced the billboards whenever the opportunity arose. At one point, it was even said that
Voliva considered constructing the signs from asbestos, so frequently were they burned down, while there were rumours that
steel signs charged with electricity sufficient to electrocute tamper-
ers were being prepared.[26] Unsurprisingly the notices became a
source of ongoing controversy in the area, and in nearby Wisconsin
local papers mocked when Voliva was forced to employ 'giant
negro Zionists' armed with Winchesters to guard the placards
round the clock.[27] From Voliva's perspective, however, it was
paramount that the religious convictions of the people who
founded the city be respected, and while Zionites became increasingly marginalized, he was set on safeguarding his spiritual, political and geographical territory, notwithstanding the cost. With
strong distinctions between inside and outside, socially, philosophically and legally defined, the city was established as a safe space, a
place for the divine and his chosen people; any incursion was
viewed as religious persecution and a direct assault on God.[28]

While opposition increased, the edicts of the general overseer
or 'G.O.', as Voliva was known, became more severe. Throughout
the 1920s Zion was renowned as the strictest town in America,
with lifestyle laws so harsh that outsiders sometimes laughed at the

thought of entering its limits. Such derision only hardened Voliva's determination to safeguard the city, and he became known for marking its boundaries with stern warning signs: 'The use of intoxicating liquor, tobacco, profanity and vulgarity are forbidden in Zion. Only the clean may enter.' Dancing, cinemas, theatres and the congregation of more than two people in public were also prohibited, as were chewing gum and driving in excess of five miles per hour. All of these activities could result in immediate arrest by the Zion Guard, Voliva's special police force, who carried Bibles rather than guns in their holsters. At one time the guard numbered eight hundred men, employed to control law-breaking and moral lapses. Visitors frequently spoke of an eerie feeling that they were being followed as they drove slowly through the town, particularly on Sundays, when pleasure pursuits of any kind – including whistling and reading newspapers – were outlawed. Headlines in the Wisconsin papers, such as 'Wake is Picnic Compared with Sunday in Zion City', warned the unwary what to expect, but many outsiders already believed that Zion was not the place to visit on any day of the week if one was looking for a wild time.[29]

Indeed, what Dowie had done with intense passion Voliva took further, and he launched numerous campaigns, introducing the rest of America to Zion's lifestyle rules. Smokers were a particular bugbear, and over the years Voliva launched a number of nationwide crusades against 'stinkpots', complete with missionaries and a specially created anti-tobacco girls' brigade to push the point home. They would pray for the addicted, Voliva told the press in 1922, but 'we will have our fists doubled when we do it'. It was fine to 'chase the devil with prayers,' he continued, 'but a meat axe comes in handy once in a while'.[30] Voliva's off-the-cuff remarks matched his tactics perfectly.[31] In the late 1920s, he published a *Handbook and Guide to Hell* that described in graphic detail how each and every sinner would suffer everlasting punishment in an overdose of his own vice: smokers would spend eternity locked in a vat of tobacco juice, with other punishment for the 'filthy' sinners' who disobeyed Zionite food taboos. In response, mocking journalists made much of the fact that Voliva lived on a strict diet of Brazil nuts and buttermilk, although he did not decree that Zion citizens should do the same.

City ordinances illustrate that he created a host of rules pertaining to women's clothing, however, and among the items he banned were tan shoes, low necklines, short sleeves, dresses without collars, skirts more than three inches above the ankle, open-work stockings, X-ray sleeves and transparent blouses. Cosmetics, bobbed hair and high heels might also result in arrest, and the wearing of an ordinary bathing suit was a crime. In August 1922, Voliva passed a law stating that men and women, boys and girls – even married couples – must bathe fifty feet apart at all times, while females over twelve had to wear stockings and bathing skirts and keep their shoulders, upper arms and collarbone fully covered: even in the sweltering heat of high summer anything less would be deemed 'immodest, indecent and suggestive of low and vicious morals'.[32] In this context, if any church member happened to catch sunstroke or fall ill with any other malady, treatment would be restricted to prayer, in accordance with church doctrine of divine healing. Like Dowie, Voliva viewed medicine as an abomination and ill-health was considered to be God's will – a consequence of sin to be dealt with by prayer and the laying on of hands.

Voliva was dedicated to interpreting the scriptures literally as the inspired word of God and this commitment extended as much to his view of the planet as it did to his view of a proper Christian life. Although Dowie had accepted a globular world, Voliva concluded from scriptural study that the earth must be flat and from 1914 he became increasingly focused on destroying what he called the unholy 'trinity of evils': evolution, modern astronomy and higher criticism (the historical and textual analysis of the Bible).[33] At a time when it was common for fundamentalists to apply day-age (interpreting the days of creation in Genesis 1 as vast geological ages) or gap (a creation 'in the beginning' as in Genesis 1:1, followed by a later Edenic creation in six twenty-four-hour days)

theories to their interpretation of Genesis, Voliva adopted a strict
literal reading, declaring that if one believed the Bible, one had
to accept the relatively recent creation of the earth in six twenty-
four-hour days.[34] Although many Christians probably believed the
same, they did not publish their views; meanwhile most scientists
and churchmen accepted the great antiquity of the earth and the
general idea of evolution in some shape or form.[35] Voliva's creationist views were, therefore, extreme, and he also went beyond
contemporaries in challenging orthodox cosmology, stating that he
took his astronomy from the Bible and conventional astronomy
was an offence to the word of God. In a sermon on 'Modern
Astronomy' on 26 December 1915, Voliva informed his congregation:

> I believe this earth is a stationary plane; that it rests upon
> water; and that there is no such thing as the earth moving, no
> such thing as the earth's axis or the earth's orbit. It is a lot of
> silly rot, born in the egotistical brain of infidels . . . Neither
> do I believe there is any such thing as the law of gravitation.
> I believe that is a lot of rot, too. There is no such thing! I get
> my astronomy from the Bible.[36]

Determined to found his statements firmly on Biblical authority, Voliva instructed his assistant, Anton Darms, to search the
scriptures for evidence of the earth's shape, and although he failed
to find explicit mention of flatness, he discovered many passages
referring to corners or ends.[37] Consequently, words such as 'ball'
were edited from Zion hymn books, while Voliva later published
Darm's findings, William Carpenter's *One Hundred Proofs that the
Earth is not a Globe*, and special issues of *Leaves of Healing* to
promulgate his unconventional world-view.[38] The highlight of the
latter was a photograph of a twelve-mile stretch of the shoreline
of Lake Winnebago, Wisconsin, taken with an eight-by-ten-inch
Eastman view camera with its lens exactly three feet above the
water. 'With a good pair of binoculars,' the caption read, 'one can
see small objects on the opposite shore; anyone can go to Oshkosh

and see this sight for themselves any clear day. The scientific value of this picture is enormous,' the caption continued, for it was incontrovertible proof of the flatness of the lake and the earth itself.[39]

Through the First World War and beyond, Voliva focused on converting his congregation to flat-earth belief but, driven by a desire to convert the world, in the early 1920s he started to promote his ideas more widely through the American media. Influenced by the Bible and, later, zetetic literature, he argued that the earth was as flat as 'a saucer, a pancake or a stove lid', with the North Pole at its centre and the South Pole distributed around its circumference as an immense barrier of ice. This barrier prevented ships from sailing over the edge of the disc, Voliva asserted, and he was so convinced of its existence that at one time he planned to embark on a voyage around the sixty-fifth meridian, just above the Antarctic Circle and near the supposed rim, to prove his theory incontestably. Because he believed he would be travelling around the entire circumference of the flat-disc world, rather than the South Pole as generally known, he informed the papers that his around-the-world voyage would take at least six months to complete.[40]

Voliva found he could milk much publicity from such unorthodox views, especially in the age of pioneering polar exploration, and journalists delighted in his periodic panics that explorers would sail, fly or walk off the edge of the world. According to reports, he was triumphant when the Italian explorer Umberto Nobile and his crew in the airship *Italia* disappeared on an expedition to the North Pole in 1928. Voliva was sure that the airship had sailed over the edge of the world, and announced to the world's press that the international search effort was futile. It later transpired that the airship had crashed near Norway's Svalbard archipelago and Nobile was subsequently rescued from an ice floe in the Arctic Ocean. Yet Voliva remained dedicated to his beliefs, and in 1929, he sent numerous telegrams to the world's press warning that the explorer and aviator Richard E. Byrd should abort his planned expedition to the South Pole because he would undoubtedly fly over the ice barrier at the edge of the earth.[41] Later Byrd became the first man to fly over the South Pole, although for Voliva this achievement was pure fantasy: he countered again that the South Pole was non-existent and Byrd had merely flown round the rim of the disc-shaped world.

Voliva was convinced that the earth was motionless and the concept of a moving atmosphere attached to a rotating globe was as lost on him as it had been on earlier flat-earth believers. He had similar difficulties with the law of gravitation – 'a lot of bunk', with no basis in fact. He announced to a journalist in 1927:

> The law of gravitation is a myth. Pythagoras, Galileo and Copernicus dreamed (probably because of dyspepsia) the theory of a sphere revolving and rotating in space, and Sir Isaac Newton invented something to hold the earth up. Imagine a man going out of his way to call God a liar![42]

Voliva was so convinced there was no such thing as gravity that on 28 September 1921 he gave a special address on the matter in the Shiloh Tabernacle, complete with a simple experiment designed to convince the congregation with direct visual proof.[43] The *Wichita Daily Times* covered the event under the title 'Voliva Seeks to Disprove Law of Gravitation', accompanied by a description of how he had tried to debunk Newton's theory that objects fall because they are pulled by gravity to the centre of the earth. According to the account, Voliva appeared on stage with a book, a balloon, a feather and a brick. He began by asking the congregation,

> How is it that a law of gravitation can pull up a toy balloon and cannot put up a brick? I throw up this book. Why doesn't it go on up? That book went up as far as the force behind it forced it and it fell because it was heavier than the air and that is the only reason. I cut the string of a toy balloon. It rises, gets to a certain height and then it begins to settle.

I take this brick and a feather. I blow the feather. Yonder it
goes. Finally, it begins to settle and comes down. The brick
goes up as far as the force forces it and then it comes down
because it is heavier than the air. That is all.[44]

The demonstration illustrated that Voliva was not anti-science
as such but, rather, opposed to the findings of science: like other
flat-earth believers, he was more than willing to employ the
rhetoric and practices of science in the promotion of his world-
view. Meanwhile, having allegedly disproved gravity, Voliva
believed he had further proof for a flat earth – for surely, he asked,
if the world were spherical, Australians would need 'hooks on their
feet' to prevent them flying off into space.

From a conventional perspective, Voliva's view of the solar
system was as unusual as his ideas about the earth. Based on his
interpretation of traditional Hebrew cosmology, as described in the
Old Testament, he believed that the sky lined a solid dome, the
firmament or vault of heaven, from which the sun, moon and stars
hung like 'a chandelier from a ceiling'. The heavenly bodies moved
round the stationary earth propelled by two great ethereal currents,
one running north, the other south, and the stars were tiny orbs
that rotated the earth at a distance no further than they seemed.
'Astronomers,' Voliva informed a Dakota newspaper in May 1928,
'have removed the stars to immeasurable distances simply for the
credit and convenience of their own theories.'[45] They were guilty
of similar deceit when it came to the sun, he argued; influenced by
zetetic literature, he claimed it was thirty-two miles in diameter,
2700 miles from earth, and spiralled across the stationary plane
once a day at the same height. When it came to 'that business of
the sun rising and setting', Voliva told the *Iowa Recorder* in June
1928, it was 'only an optical illusion that doesn't prove the earth is
round', while the sun's supposed distance from earth was absurd,
as he explained to the *Kingsport Times* on 16 September 1921:

They tell you that the sun is 92 million miles away. I laugh at
that, not only as a mathematician but as a student of God

Almighty's Word. Did God Almighty create the earth and
then create a light to light it up and put it 92 million miles
distant and make it a million times larger than the earth?
What kind of a fool would build a house up in Kenosha and
erect a light a hundred miles from it to light up the parlor?[46]

Like many other advocates of a flat-earth perspective, Voliva
employed a common-sense argument: a globular, revolving earth
was illogical, and if orthodox science was true it would not conflict
with the word of God:

> . . . when it comes to the word of God, [scientists] have not a
> leg to stand on. I will take the word of God and dispose of
> any astronomer on the face of the earth in less than thirty
> minutes.[47]

For Voliva, as for latter-day 'scientific' creationists, the Bible
was the inerrant authority on the natural world, and as science
contradicted a literal reading of the scriptures, it must be a satanic
device. Yet by staking the word of God on so highly falsifiable a
truth claim as a flat earth, and in forcing a choice between two
extremes, Voliva undermined his own cause and attracted intense
ridicule as a result of his radical creationist belief. But he was
unconcerned by the difficulties inherent in seeking to convince the
world that the earth was flat. He claimed that it was impossible to
be a Christian and believe the earth to be round: scientific knowledge and Christianity were irreconcilable and there was no middle
ground. He believed that Genesis should be interpreted literally
on all matters, and declared his disgust at 'so-called fundamentalists' who 'strain out the gnat of evolution yet swallow the camel
of modern astronomy'.[48] Increasingly isolated in the 1920s, he
announced to reporters that he was 'the only man in the United
States who believe[d] the Bible'. In fact, he added, he was the
'only true fundamentalist' in the whole world.[49] As such, he saw
that he had a duty to uphold his flat-earth beliefs and salvage
misguided hearts and minds. For Voliva, evolution, higher criticism

and modern astronomy were 'the devil's triplets', part of the same vicious plot designed by Satan to seduce the unwary and destroy their faith in the Bible as the word of God.

A tireless campaigner, Voliva was determined to combat the 'devil's triplets'. In July 1925, his stand against evolution took him to Dayton, Tennessee, to appear as a prosecution witness in the trial of John T. Scopes, a young athletics coach, and physics, chemistry and mathematics teacher, who was being prosecuted under the state's new Butler Law for teaching human evolution while substituting for the usual biology teacher at the high school where he was employed. While religious responses to Darwinism varied considerably, the teaching of human evolution in public schools had become a prime target for fundamentalists (a diverse cross-denominational coalition of militant conservative Protestants), spurred on by a sense of crisis in the wake of the First World War. The movement took its name from 'The Fundamentals', a series of booklets published between 1910 and 1915 to reinvigorate Christianity and restate the fundamentals of Biblical orthodoxy – an inerrant Bible, personal conversion, the virgin birth and full divinity of Jesus, the literal truth of miracles and the Resurrection, and the literal truth of the Second Coming. While the term 'fundamentalist' was not coined until 1920, the movement developed a firm identity from this point, along with its focus on removing human evolution from school curricula. By the mid-1920s laws banning the teaching of the doctrine in public schools had been passed in several states, including Tennessee.[50]

In his memoirs, Scopes claimed that he was not sure he had taught human evolution during the lessons in question, but nevertheless, following discussion with Dayton school officials opposed to the new law, he accepted the invitation of the American Civil Liberties Union to stand trial in a test case with a view to having the legislation revoked. A national sensation, dubbed at the time 'the trial of the century' and the subject of legend and over-simplification ever since, the so-called 'Monkey Trial' at Rhea County courthouse was a media circus, widely touted as pitting science against religion, fundamentalist against agnostic, conservative against liberal, and more.[51] The prosecution was led by William Jennings Bryan, the long-time Democratic Party leader, three-times presidential candidate and ex-Secretary of State under Woodrow Wilson. A charismatic popularist reformer and outstanding orator, Bryan, like many fundamentalists, was predominantly concerned about the materialist implications of evolutionary theory, which he saw as a menace to Christian moral values and social stability. More specifically, he viewed human evolution, 'a doctrine as deadly as leprosy', and associated concepts of 'survival of the fittest' and 'might versus right' as the primary cause of social ills such as militarism, imperialism and competitive capitalism. Consequently he focused his not insubstantial energy on the public defence of fundamentalist Christian faith and creationist teaching in schools.[52]

Although Bryan was undoubtedly committed to his cause – he was a Presbyterian layman who propounded day-age theory during the trial – the papers joked that even a prosecutor of his calibre could not meet the exacting standards of Tennessee mountain folk. According to journalists set on ridiculing the creationist camp, Bryan, for all his hatred of evolution, remained a heretic in locals' eyes for believing in a spherical earth, and the people wanted 'good Christian books' in schools, which taught that the earth was flat. Voliva was in full agreement with their view, for in his eyes Bryan could not be considered a true fundamentalist if he believed the earth to be a globe. He informed a local Tennessee paper, the *Chattanooga News*, that 'Even William Jennings Bryan is inconsistent. He doesn't go far enough in his fundamentalist belief. He is not as consistent as we members of the Christian Catholic Church are. If Bryan repudiates modern theories of biology, he ought to repudiate modern theories of geology and astronomy as we do.'[53]

Thus the arrival of Voliva, branded 'the most conservative of anti-scientific conservatives' by reporters, was awaited with anticipation. On 11 July, with a twist of sarcasm, the *Haywood Review* suggested that while Voliva was 'fighting the wicked evolutionists'

in Tennessee, he might also like to persuade the state's legislators
to 'outlaw the school geographies'. But despite his enthusiasm for
the anti-evolution cause, Voliva did not appear as a prosecution
witness, as predicted by the papers. Rumours circulated that he
had more ambitious plans, for he reportedly proposed to Bryan
that they run for the presidency of the United States on a joint
platform to eliminate the twin heresies of evolution and a spherical
earth.[54]

The scheme came to nothing (Bryan died a few days after the
trial ended), and while Scopes was initially found guilty and fined
a nominal amount, the conviction was later quashed on technical
grounds.

As the fundamentalist campaign continued elsewhere, Voliva
turned his attention to educating the youth of Zion to accept his
Bible-based world-view. In common with world-makers of varying
types, Voliva saw education as the means by which he could shape
the next generation and transform his vision into reality in the
realm under his control. Since the days of Dowie, the city had had
a system of parochial schools, and by the 1920s eleven hundred
pupils were enrolled in the grade and high schools. While the
public schools adhered to Illinois state law regarding the curriculum and the teaching of science, the parochial schools were
controlled by Voliva's Theocratic Party, which instituted the Bible
as the principal textbook and outlawed the teaching of organic
evolution in keeping with Voliva's stated view: 'Teach a child he is
kin to the ameba [sic] and he will take on the characteristics of the
ameba. Anyway, such teaching is a flat contradiction to the plain
word of God. The Bible doesn't even mention ameba.'[55]

A globular earth was deemed taboo on similar grounds so,
from the early 1920s, flat-earth theory, alongside creationism,
became a central feature of a revised parochial school curriculum,
supported by special textbooks and maps. Interestingly, rather than
opposing Voliva's innovations, the headmistress Mary Thompson
was proud to report to journalists that the children under her
guidance far preferred their new flat world to the 'old-fashioned

round one in which Columbus believed'. 'The students in Zion schools,' Miss Thompson continued, 'grasp the theory of the flat earth readily because their minds are not so full of globular earth teaching such as older folks have had drilled into them.' From her perspective, the children were proving a pleasure to teach, for they accepted the disc-world idea without question. It appeared that children accepted the concept as a rational and logical explanation, and from the entire student body, Miss Thompson claimed that only one pupil had raised any objection to the unorthodox world-view.[56]

The headmistress's positive progress report was echoed by Eva Baker, a geography teacher willing to inform her classes that the sun was a tiny disc a few thousand miles distant, rather than an immense sphere ninety-three million miles from the earth. 'If the sun was so large,' she informed reporters, surely it would 'light up all the world, instead of confining its hottest rays to a 3000-mile belt between the two tropics?' Believing this point to be unanswerable, she went on to describe how she used maps to demonstrate flat-earth theory in class. It seemed that Voliva had provided parochial schoolteachers with two depictions of the world deemed suitable for use in lessons: a four-colour map of the earth as a circular plane created thirty years before by Alexander Gleason, the Buffalo civil engineer and author of *Is the Bible from Heaven? Is the Earth a Globe?* (1890), and a standard map of the world set on Christopher's projection. The map, which was used by scientists and navigators to make time and longitude calculations, differed from the usual Mercator projection, familiar in other schools, for it showed the earth as it would look to an observer directly above the North Pole with the continents projected on to a circular plane. The North Pole lay at the centre of the disc, and the Antarctic regions were indicated by a white ring around the outer circumference – the ice barrier surrounding the edge of the world, according to Voliva.[57]

Alongside revolutionizing the school curriculum, in May 1923 Voliva began disseminating his message through another innovation, the 50,000-watt radio station WCBD. Established in a wooden building next to the Shiloh Tabernacle, it cost $156,000 to set up and was one of the five most powerful in the nation. Courtesy of its strong signal, radio enthusiasts as far distant as Australia and New Zealand were able to receive uninterrupted religious programming live from Zion, and Voliva became the first evangelical preacher in the world to own a radio station. Over the years that followed, WCBD, or 'God's Station', as it was called, became a source of much pride to Voliva and he told journalists that 'It has done much to counteract the evil the newspapers and their atheistic writers have done us.'[58] The station certainly served to draw in converts from around the world, some of whom moved to Zion to begin a new life. Through the power of radio, Voliva's sermons about the shape of the earth also found a new audience on a worldwide scale, including popular author Irving Wallace, then a sixteen-year-old high-school student living across the state line in Kenosha, Wisconsin.[59] Amused by what he heard, Wallace thought that the topic would be well received by his high-school debating team and wrote to Voliva requesting an interview.

Much to his surprise, Voliva agreed. On the appointed day, Wallace arrived at Voliva's office and, with some trepidation, presented a list of proofs of a globular earth. He recalled that Voliva listened 'with great forbearance' while he reeled off a list of straightforward questions: why does a vessel disappear in the distance when it steams away? Why is the earth's shadow on the moon round? How was Magellan able to circumnavigate the earth? Then, Wallace recalled, Voliva recited his answers as if by rote:

Ships don't disappear in the distance at all. You can see a ship twenty-five miles out at sea if you look through field glasses. According to scientists, the curvature of the earth for those twenty-five miles, allowing for refraction, should be three hundred and fifty-eight feet. If the earth is round, how can you see your ship over a hump of water three hundred and fifty-eight feet high? ... As for that round shadow on the moon, the flat earth would still cast a round shadow. A saucer is round, isn't it? ... Of course Magellan sailed round the world and came back to where he started. He went round the flat earth exactly as a needle goes round a gramophone record. Millions of men have sailed round the world from east to west, and west to east. It can be done on a saucer, too. But do you know of anyone who has ever sailed round the world from north to south? Of course not. Those who tried fell off. That's why so many explorers have disappeared.[60]

Voliva seemed to have an answer for every refutation, although not all members of his church felt able to agree. Theodore Forby, city attorney and Voliva's close adviser for many years, subsequently left the Christian Catholic Apostolic Church due to differences over Christian doctrine and his inability to accept that the earth was flat.[61]

Despite frequent controversies with detractors, accusations of corruption and legal action, Voliva prospered through the 1920s and by 1927 had amassed a $5.2 million personal fortune from Zion Industries, alongside a fortune for the twenty-six institutions and industries under his control. However, his outlook for the future remained somewhat gloomy, a result of his interpretation of the prophetic books of the Bible, particularly Revelation and Daniel, and its teachings relating to Christ's Second Coming and the eventual end of the world. Voliva was a dispensational premillennialist, part of a radical Christian tradition that divided history into seven 'dispensations' or eras. As such, he believed that the world was on the brink of the final seven-year period, an age of tribulation that would be followed by the Second Coming, a thousand-year reign of universal peace and harmony (the millennium), the battle of Armageddon and the final judgement, as foretold in Revelation 20. Part of a broader tradition of prophecy and apocalyptic belief, as old as Christianity itself, millennial movements had proliferated in America through the course of the late-nineteenth century, one strand of the fourfold gospel of personal salvation, divine healing and Holy Ghost baptism that character-ized the radical evangelical revival.[62] While political visionaries, such as Owenite socialists, had looked forward to a New Age of equality and justice, various Christian sects were wedded to more specific Bible-based predictions, and date-setting for the end of the world was relatively common in such groups, often with nega-tive consequences when the predictions failed.[63] Premillennialism remained an important stream in fundamentalism throughout the 1920s, though, and Voliva naturally interpreted Revelation as lit-erally as Genesis.

Convinced that the end of the world was nigh, in September 1927 he decided to tour the earth 'before it meets its doom' and set sail for Europe, Scandinavia, North Africa and the Holy Land, planning to save souls and gather supporting proofs for flat-earth theory while he was away. He returned in June 1928, even more convinced that the world was a circular plane, and repeated an earlier $5000 offer for evidence to the contrary. As before, his challenge lapsed without takers, and little more than a year later, he set off again to preach his flat-earth gospel. On this occasion he visited nineteen countries with his entourage, including India and China, and on his return in March 1931, he boasted to the press that his knowledge of the earth's flatness had been 'reaffirmed on every hand'. Bemused reporters were further informed that the earth's shape could be compared to 'a steak on a circular plate surrounded by a rim of mashed potato'. It was easy for explorers to reach the North Pole in the centre of the steak, Voliva argued, but impossible for them to traverse the mashed-potato barrier surrounding the edge of the world. 'For one thing,' he allegedly explained, 'the weather is too cold and would stop any foolish travellers who want to go too far.'[64]

Unfortunately for Voliva and the residents of his city, trouble was much closer at hand. From 1930, the Great Depression hit Zion hard, and its industries lurched towards bankruptcy once more. Voliva tried every available tactic: cost-cutting measures, mass prayer meetings, even relaxing the lifestyle laws to improve the city's image and sales of its products. But times were hard in Zion, as they were everywhere, and his popularity continued to plummet. With each year that passed, the Independents made incursions into his political and spiritual domain until their candidate, Onias Farley, triumphed over the Theocratic Party's representative in the School Board presidential elections in 1934. Apparently Voliva was overwrought with fury: in a sermon he resolved to 'crack the town wide open' by closing the parochial schools and proclaimed that he could put Farley's sixty-three-year-old wife 'out of commission' for good, if he so chose. According to the papers, the two-thousand-strong congregation at the Shiloh Tabernacle were further cautioned: 'You people better watch out. I was brought up as a gunman. I am heavily armed now, and I'd kill a man at the drop of a hat in self-defence. Countless guards, armed to the teeth, surround me. They have orders from me to shoot at a second's warning.'[65]

Onias Farley was not intimidated and decided to reopen the public schools, which Voliva had temporarily (and, one assumes, illegally) closed. It seemed that threats and heated statements were no longer enough for Voliva to retain his already weakening hold on power in a climate of increasing diversity and change.

As Voliva's prospects grew gloomier in Zion, so did his outlook for the world in general, and his end-time predictions took on a new urgency in reaction to international events. Dispensational premillennialists had a propensity to see calamity round every corner, and for those given to looking for warning signs, world affairs in the interwar period provided ample scope. Sign-watching for the apocalypse had long been a familiar feature of Voliva's pronouncements, and he had previously forecast the end of the world for various dates in 1923, 1927 and 1930. This lay far beyond the boundaries of acceptable public discourse, and when the years in question passed without note, Voliva's authority was further undermined.[66] Nonetheless, he continued to promulgate his Bible-based predictions and, convinced he was living in the world's last days, in 1931 he officially announced 'The Time of the End'. To mark the occasion, the name of the church journal was changed from *Leaves of Healing* to *The Final Warning*, and for those who did not accept Christ as their personal saviour, all hope of salvation and renewal was duly crushed.

From that point on, the forecast worsened. In April 1934, the month of the disastrous School Board election, Voliva claimed he had divine information that Satan 'plans to take over the world – probably in September', although he revealed to journalists that he was organizing a tour of revival meetings with a sixty-three-piece brass band and a hundred of his best workers as a preventive measure. The tour, Voliva continued, would begin fifty miles south in the satanic stronghold of Chicago and, although the end of the world loomed, he expected it to 'net enough money' for him to regain control of Zion industries, currently in receivership. 'All that riff-raff,' he concluded, referring to his detractors, 'will come crawling back to me when I get the colony out of receivership. Voliva is still on top and will be when prosperity returns to Zion.'[67]

Yet neither economic recovery nor the end of the world was forthcoming. On 10 September 1935, the *Ironwood Daily Globe* reported that 'The world didn't end today, even though it was a perfectly good day for it', and went on to describe how Voliva had dispatched an understudy to the Tabernacle to inform the waiting media throng that the apocalypse would now occur on 10 September 1942.[68] In that year the world would see five eclipses of the sun and two of the moon, portents of the End, signalling irreversible doom, followed by cataclysmic cyclones, earthquakes, droughts, floods, tidal waves and volcanic eruptions. The Italian dictator Benito Mussolini would start a tumultuous battle, Voliva cautioned, and 'death will stare us in the face at every corner'.

It was, of course, another false alarm, and as the political situation shifted in Europe, Voliva rescheduled these events for 1943. On this occasion, he contended, Italy, France, England and America would unite in a revived Roman Empire ruled by Mussolini on Fascist lines, while the nations of middle and north-eastern Europe would form a 'great north-eastern Communist confederacy' that would invade the Holy Land. According to Voliva, this action would lead to a battle so bloody it would take 'seven months to bury the slain', followed by the Second Coming of Christ and the thousand-year kingdom of God on earth.[69]

Apocalyptic predictions based on interpretations of Revelation involving natural disasters and international events had long been common among premillennialists, who anticipated the appearance of a 'Beast' or Antichrist who would form a new Roman Empire as a precursor to Christ's Second Coming, the thousand-year golden age and the eventual end of the world. Although Voliva's catastrophic predictions for 1935 had been inaccurate, his personal empire met its end that year. Worn down by economic depression, failed promises and a repressive regime, the people of Zion deserted him and his Theocratic Party in the city elections and his twenty-nine-year rule was finally over. He was reduced to honorary president of Zion's bankrupt industries and became one religious leader among many in the city. The most positive development was the first presentation of Zion's famous passion play in the Shiloh Tabernacle, fully supported by Voliva as an inspiring and effective means of presenting the gospel to the public. By the end of the 1936 season, 60,000 people had seen the story of Jesus of Nazareth, although the following year's production could not claim similar success.[70] While the run commenced with good attendances, the Tabernacle and WCBD were burned down in early April by a teenage arsonist with a grudge against Voliva. The fire and the general lack of funds in the troubled city led to the closure of the briefly reopened parochial schools and an end to the flat-earth syllabus.

Later that year, Voliva was declared bankrupt with debts of more than a million dollars, and in April 1938 the new authorities selected the image of a globe as the design for Zion's compulsory vehicle-tax sticker. The *Oshkosh Northwestern* revelled in the fact that Voliva would be forced to display the picture of a spherical earth on the windshield of his car and, moreover, that it had been created by a former pupil of his church schools, which were closed permanently in 1939.[71]

From this period, Voliva avoided all such reminders of his declining fortunes by spending much of his time in Florida, where he planned to establish another fundamentalist colony. The scheme never materialized. Diabetic and half blind, Voliva died of heart and kidney disease at the age of seventy-two in October 1942, despite another prediction that he would live to the age of a hundred or more on his diet of buttermilk and Brazil nuts. Yet although he met an ignoble end, passing away in a hospital in the city of divine healing that he once owned, he left a lasting legacy to Zion besides its famous passion play. These days, it is an ordinary town, with a peaceful atmosphere and plenty of parkland, but it is said that a few old-timers educated within Voliva's fundamentalist school system still believe that the earth is flat.[72]

Chapter Seven

MAN ON THE MOON?

The Flat Earthist, working along forgotten paths, finds
many orthodox theories and speculations to be false,
which over the years have been taught as facts. This
surely is where 'assertion outstripping evidence' becomes
dangerous and if forced into young minds, a crime.

SAMUEL SHENTON,
'Blunder or Crime?', *Channel* (1962)

'To observe – think freely – rediscover forgotten facts
– oppose theoretical dogmatic assumptions'

Motto of the International Flat Earth
Research Society (1956–71)

IT SEEMS UNLIKELY THAT Wilbur Glenn Voliva made many
converts to flat-earth belief outside the city limits of Zion, despite
the growing strength of the fundamentalist and Pentecostal movements in the wake of the First World War. In Zion itself, the
Christian Catholic Apostolic Church abandoned his flat-earth
teachings, arguing that they were his personal belief rather than
official church doctrine, and even though converts remained in the
city, fourteen years and a world war passed before another believer
launched a campaign to promote his unorthodox world-view.

From the outset it was clear that Samuel Shenton's devotion to the cause equalled that of his predecessor, although any obvious similarity between the two men was limited to this one trait. While Voliva and Shenton were unquestionably in agreement about the shape of the earth, the worlds in which they lived were far removed: in terms of location and lifestyle, demeanour and temperament, they were miles and years apart. Voliva had served as overseer of a Utopian settlement on the plains of Illinois and was a formidable figure in his own right; Shenton was an average working man, a signwriter by trade, who lived simply with his wife Lillian in a ginger-brick terrace in suburban Dover. Earnest and balding, with a habit of wearing woolly pullovers and red carpet slippers, Shenton's sheer 'ordinariness' was a revelation to reporters hoping to encounter an eccentric flat-earth advocate, whose personality was as peculiar as his views. In his inadvertent subversion of social expectations, Shenton could be characterized as all the more radical; meanwhile, journalistic assessments of his background were fair. The son of an army sergeant major, Shenton was born at the Royal Artillery Barracks in Great Yarmouth in March 1903 and had remained true to his practical roots over the course of his professional life. Yet appearances can be deceptive, as many flat-earth believers argue, and the air of conventionality said to surround Shenton on first impression belied his adherence to a number of highly unorthodox ideas.

Despite his heated anti-scientific rhetoric, Shenton was mild-mannered and was never much given to looking backwards, except when considering the zetetics and the historical roots of his flat-earth campaign. Unassailably focused on the here and now, he revealed little about his early life, and his public statements about the past were restricted to the basic facts: astronomy and geography were his teenage hobbies, and he left school at sixteen a confirmed globularist. However, he often outlined the development of his interest in science and technology, and frequently dated his life story back to the First World War when he observed zeppelins and other aircraft designed on 'bird-shaped lines' for manoeuvrability and speed. Later Shenton remembered his over-whelming scepticism about the supposed innovations, and his resolution to invent a truly useful commercial cargo-carrying vehicle. Artistic yet technically minded, by the 1920s he claimed to have devised a type of airship that combined gas and engine power to rise into the atmosphere and remain stationary until the earth spun westward at 1000 k.p.h. to the desired destination on the same latitude. In effect, according to Shenton, a load could be lifted into the air above London and delivered when New York rotated to the corresponding point five hours later.

The plan clearly disregarded the atmosphere's connection to the earth in its rotation, yet having overlooked this crucial fact, Shenton wondered why no other individual had hit on his simple but ground-breaking idea. 'Think of the possibilities of such a method of world transport,' he later enthused. 'It was staggering!'[1] In the belief that he had invented a vehicle which would 'let Earth make the effort of air travel', Shenton dispatched his design to universities and government departments, anticipating patron-age, plaudits and fame. None was forthcoming and, unable to find an alternative explanation for this lack of interest, Shenton began to suspect the establishment had something to hide. 'Unknown to anybody,' he reminisced, 'I started to sift for the truth.'

From the beginning, Shenton's account of his discovery had mythic dimensions, paralleling great stories of scientific break-through for, much like Parallax, he frequently styled himself as a heroic seeker on a lone quest. Yet Shenton remained an ordinary member of the public, a stranger to the world of professional science, so he framed his story as a conspiracy narrative about his efforts to safeguard his invention and establish his individual authority in an atmosphere of establishment collusion and deceit. Underdog and self-styled hero, Shenton remembered that his truth-seeking con-tinued at the imposing reading rooms of the British Library in Bloomsbury, where he discovered to his delight that Archbishop

Stevens, a friend and zetetic sympathizer of Lady Blount's, had suggested an aircraft design similar to his own. Excited by the discovery, Shenton recalled ordering every zetetic item available to readers, and spoke of his astonishment at what he received. At this point, he confessed later, he adhered to the overriding social consensus that flat-earthers were nothing but 'cranks and fools', and in common with the majority of the population he was content to accept scientific authority and the 'fact' that the earth was a rotating globe. Now, driven by a desire to establish his own invention, and thus his expert status, Shenton questioned even this basic 'truth'. Increasingly suspicious, he continued to consult zetetic pamphlets, books and journals, until he discovered Parallax's *Zetetic Astronomy: Earth not a Globe!* and with it the solution he sought. What the authorities were concealing, Shenton decided, was the 'fact' that the earth was flat. In a matter of days, his conversion to zeteticism was complete: from that point forward Shenton was certain he had 'crawled, toddled and walked on a flat Earth since the day [he] was born'.

One consequence of Shenton's epiphany was the realization that if the earth was a stationary plane, his airship, which relied on rotation, would not have been the world-changing innovation he had originally assumed. Yet he was willing to sacrifice his invention on the altar of a discovery far greater – the ultimate conspiracy theory, perhaps. He remembered his suspicions had been multiplied by the detection of 'inconsistencies and opposing theories', together with some 'downright "chance-your-arm" assumptions', interwoven with the globular theory he had been led to accept as fact.[2] Angered by the supposed sham, Shenton commenced the task of re-examining the cosmology he had believed without question and devising an alternative model that would form an original contribution to opinions on the universe and zetetic thought.

Shenton's cosmology was based partly on his interpretation of Genesis and was founded on the idea that the universe was an endless flat plane, Mother Earth, marked by a series of deep pits. Shenton believed the disc-shaped world was lodged at the bottom of one of these holes and floated on water that had seeped through its surface, producing rivers, springs and oceans. Further to this, and in common with the Mother Earth universe as a whole, the disc-world was motionless; on numerous occasions Shenton contended that the idea of it spinning on an axis and 'rushing through space at 20 miles a second' was 'utter nonsense'. For 'how,' he asked one journalist, 'was a chap to land an airplane on a runway in high wind' while contending with such complicating factors? In reality, Shenton countered, if the earth was truly shooting through space at such phenomenal speed 'you blasted-well wouldn't dare put your nose out of the door'. He evidently dismissed the law of gravity as 'hocus pocus', claiming that it was just a theory that 'Newton himself denied'. Through the years that followed, numerous journalists were treated to a practical demonstration of this view, as Shenton, like Voliva before him, attempted to provide convincing direct proof by dangling objects such as ashtrays from his fingertips during interviews while contending, 'There's no such thing as gravity . . . It's only the weight of the article which makes it fall.'

Taking his inspiration from an interpretation of Genesis, Shenton continued that in the time of a man called Peleg, whose name means division, the earth was riven asunder and a great continent was locked beneath the North Pole. Shenton believed this lost landmass was the mythological island of Atlantis and that flying saucers often visited our planet from this underground base. His views about the South Pole were similarly unusual, for in common with other zetetics he believed it to be an impenetrable ice barrier that encircled the disc-shaped world. Also in keeping with Parallax, he argued that the sun, moon and stars circled within a few thousand miles of the earth; more specifically, the sun was thirty-two miles in diameter, 3,000 (rather than 93 million) miles distant, and neither rose nor set above the flat earth plane.

As an alternative, Shenton suggested that the sun cast a narrow beam 'like a flashlight moving over a table' as it traced flat circles that diminished and expanded above the earth on a 365-day cycle.

This movement explained why the southern (or, to his mind, outer) hemisphere enjoyed summertime while Dover suffered the depths of winter. The moon, meanwhile, was also thirty-two miles in diameter, rather than the official estimate of 2160 miles, and it floated in the sky a mere 2550 miles distant from the earth. Instead of reflecting the sun's light, as is generally believed, Shenton concurred with Parallax that the moon was a self-illuminating body whose different phases were explained by its movement across the disc-shaped earth from east to west, setting twenty-eight minutes later each day – a phenomenon described by astronomers as the phases of the moon.

With his radical reconstruction of the universe in place, in the mid-1950s Shenton resolved to overturn conventional science and establish his interpretation as the authoritative account. The obvious tactic was to set up an organization to promote his vision, so in 1956 he founded the International Flat Earth Research Society (IFERS) as a direct descendant of the Universal Zetetic Society of old. There were notable contrasts between the two ventures, however, for Shenton's interest in alternative science and technology, alongside the deepening hegemony of scientific thinking, led him to place an even greater emphasis on factual and research-related reasoning, with religious arguments relegated to second place.

Eager to begin, he appointed himself secretary and treasurer of the IFERS and commenced operations from his terraced house in London Road, Dover. Like William Carpenter, he used experience from his signwriting business, Shenton's Signs, to produce numerous flyers and pamphlets, all of which were emblazoned with the lengthy slogan 'To observe – think freely – rediscover forgotten facts – oppose theoretical dogmatic assumptions' to describe the society's broad-based aims. Shenton soon found the new organization costly and time-consuming, but he was content to subsidize it from his own pocket, pending the anticipated influx of members willing to pay five shillings to join his crusade. Moreover, he was not alone in his faith. A willing president for the group was found in William Mills, a relative of one of Lady Blount's followers, Frederick Cook. The IFERS was set to launch its public campaign.

In November 1956, its inaugural meeting was held at Mills's Finsbury Park home amid extensive publicity. Advance advertising, along with the apparent eccentricity of the enterprise, drew quite a crowd, and among the believers and sceptics crammed into Mills's living room was the eminent astronomer, and presenter of *The Sky at Night*, Patrick Moore. Attending out of curiosity and expecting a few surprises, he was not disappointed. In 1972, he recalled that the meeting consisted of 'various abstruse papers', including a well-received presentation on aerodynamical issues by the society's secretary, Samuel Shenton. With gentlemanly diplomacy, Moore listened intently while Shenton explained that 'If gravity ended at a height of nine miles', as scientists allegedly maintained, 'then a parachutist coming down from a great height would miss the Earth altogether.' 'Where he would go then,' Moore later said, 'remained something of a mystery.' He left the meeting 'in a mood of deep thought'.[3]

A month later Moore received a flyer reminding him of the IFERS manifesto:

> The International Flat Earth Society has been established to prove *by sound reasoning and factual evidence* that the present accepted theory, that the Earth is a globe spinning on its axis every 24 hours and at the same time describing an orbit round the Sun at a speed of 66,000 k.p.h.[,] is contrary to all experience and to sound common sense. In ancient times the Earth was regarded as a plane, and this is expressed in all literature up to a few hundreds of years ago. The theory has fallen into disfavour, owing mainly to the dogmatism of modern science and popular education in schools, which leads to prejudice in favour of the globular theory from the start. It is always a pity to allow false theories to pass unchallenged, and it is hoped that the Flat Earth Society will do much to undo the harm that has been caused. Remember that the truth of the plane figure of the Earth can be shown by irrefutable evidence . . .[4]

In fact, Shenton's questionable claim to 'irrefutable evidence' was soon to face its greatest challenge. On 4 October 1957, the Soviet Union launched the world's first artificial satellite, Sputnik, to map the earth's surface, and although the satellite itself was small, no bigger than a beach-ball, the ramifications of its successful orbit of the earth were vast. Across the world pundits greeted Sputnik's robotic 'beep-beep-beep' signals as the herald of a new era, and this was undeniably the case. Disturbed by the apparent technological and scientific capabilities of the Communist superpower in a Cold War climate, the US was galvanized into action and the space race began in earnest. Yet, back in Dover, Shenton remained dedicated to his world-view. He believed that satellites moving above the earth did not prove sphericity: they were merely circling over the flat disc-world in common with the sun and moon. 'Would sailing round the Isle of Wight prove that it were spherical?' Shenton demanded. 'It is just the same for those sputniks,' he contended, for to his mind they were 'just like marbles spinning round a saucer'. With this, he focused on convincing the world in the face of the pioneering age of space flight.

Through the late 1950s and early 1960s, Shenton persevered with his flat-earth campaign. Bold and single-minded, with a practical rather than an academic bent, he much preferred public speaking to writing, and prided himself on almost never declining an invitation to lecture or debate. His enthusiasm for giving talks was heightened when the proposed audience were children, for his primary aim above all else was to 'reach the young folk' before they became irrevocably convinced of the earth's rotundity. From his perspective, countering spherical-earth indoctrination was the *raison d'être* of the IFERS, and because young people were less likely to advance well-informed critiques of his ideas, he perceived events at schools and youth clubs as useful opportunities to practise new material. Consequently, he arrived at countless venues loaded with homemade charts and banners bearing slogans in bright-coloured capitals, eager to outline his flat-earth views. Pubs, clubs, student unions, coffee mornings and church halls, no venue – or audience

– was too small for Shenton to make the mammoth effort that his unorthodox campaign entailed.

Whatever the audience's true reaction, his endeavours were eye-catching, and as his name became known, invitations poured into the society's headquarters in London Road. Youth clubs, political groups, reading circles, church associations, university astronomical societies, natural-history organizations and Round Table branches all proved keen to play host to his talks. As time went by, competition between rival branches of the Young Conservatives and Young Socialists to secure a booking became a notable feature of the IFERS correspondence, while students at Oxford and Cambridge and Mensa's Young Scientists' Club also requested to hear his society's manifesto at first hand. Furthermore, as the society's president, William Mills, had died suddenly in December 1957, the onus fell on Shenton to satisfy public demand.

Throughout this period, the Russian and American space programmes were progressing at a remarkable rate. High-speed aeronautical trials, satellites, rocket development, capsule design, astronaut training: by the spring of 1961 the superpowers were prepared for manned space flight. As with the launch of Sputnik, the Russians were the front-runners and on 12 April, the capsule Vostok 1 was launched, earning cosmonaut Yuri Gagarin the historic accolade of 'first human in space'. The race immediately intensified for the American space administration and twenty-five days later, on 5 May, the first manned flight of their Mercury programme was launched, with astronaut Alan Shepard in the capsule Freedom 7. His sub-orbital mission lasted for just fifteen minutes, but America was back in the running. Three weeks later, President John F. Kennedy announced to Congress, and thus to the world, that his country aimed to land a man on the moon by the end of the decade, and this goal became a matter of national pride.

Shenton's personal pride was likewise on the line as he began to attract international attention from journalists seeking to contrast their coverage of the Gagarin and Shepard missions with an

unusual story about hostile reaction to manned space flight. Ever
eager for publicity, Shenton was willing to court the media, and he
informed journalists that Gagarin had circled above the saucer-
shaped earth, 'like a needle on a record', rather than orbiting a
globe, as unanimously claimed. On Freedom 7 and Alan Shepard,
Shenton informed Wisconsin's *Coshocton Tribune*, on 10 May, that
the astronaut had never travelled into orbit because there was no
such thing. 'Humanity has been brainwashed into accepting round-
Earth theory,' Shenton continued, 'but [IFERS] members will
continue to proclaim the truth – that Earth is shaped like a disc.'
If Shepard had observed a curved earth, then Shenton was certain
it was 'just an optical illusion quite common in these high-flying
days'. 'Science cannot shout us down,' he declared; orbiting was a
delusion and landing on the moon a fantasy because 'The moon's
transparent, you know.' Shenton refused to believe such a feat
was possible, but even if it was he thought astronauts 'had better
be ready to come right back' as 'stars have been seen through the
moon' and there 'isn't anything much to land on'.[5]

Shenton succeeded in attracting public attention with his
remarks, and letters from America started to arrive at his Dover
home. Initially they were general enquiries, seeking further elu-
cidation of IFERS views, although a Denver correspondent
requested permission to establish a chapter of the society in his
hometown – the first of many such offers that flooded in from the
States over subsequent years. The stream of lecture invitations
continued in step with coverage of the ground-breaking developments in space. In November 1961, Shenton was especially satisfied with an appearance at Oxford University's Scientific Club,
based at Lincoln College. Back at home, he scribbled in his
notebook that the lecture to three hundred people was 'very good';
he had used a blackboard to illustrate that details about the last
total solar eclipse in England (on 30 June 1954) were entirely false,
and his 'evidence re: shadows and impossible angles of the sun and
moon' was reportedly well received. All in all the event appeared
to have been successful, and to crown the experience, Shenton

noted that 'the society paid me the honour of [an] invite over to Keble for coffee [and] further talk', a courtesy he assumed had never been paid to a previous speaker. Later that month, he gave a 'first-rate' lecture at a Folkestone and District Round Table meeting where, despite the threatening presence of several teachers, whom Shenton regarded as his 'worst enemies', he received a 'very good hearing' and 'a very fine vote of thanks by the ex-mayor', who said he 'prided himself on being a non-conformist' and 'had never heard such resounding talk!'[6]

It was early days for the society but, despite the space race, Shenton believed he was achieving excellent progress, and this rising optimism led him to compose a piece for Manchester University student magazine, *Parsec*. Generally he deemed such publications unwelcome vehicles for exposure due to his reluctance to place his 'hard found material' at the mercy of 'young bloods looking for sensation and fun-and-games', but having recently received indications that young people seemed to be 'thinking for themselves', he had changed his mind. The resulting article was one of the most in-depth pieces Shenton ever prepared, although he admitted that he found presenting 'the most heterodox, derided and controversial subject [with] proofs and logical arguments' in a limit of 1000 words an arduous task. 'It would take much more than that,' he protested, 'to sweep out the indoctrination ingrained in young people since they entered the educational mill.' 'You think freely?' Shenton challenged his readers. 'Never in your life!' However, he subsequently admitted to *Parsec*'s editor:

> I have no hope whatsoever of 'converting the masses' or breaking through the mighty barriers which surround modern accepted science. My personal efforts are to foster and keep alive the valuable spirit of DOUBT both within myself and the few I may live to contact.[7]

Shenton had learned his Parallax well; his dedication to the scepticism and objectivity supposedly fundamental to zetetic philosophy lay at the core of his campaign. He undoubtedly viewed his work in historic terms, both as a tribute to past zetetics and a gift to the next generation, and a deep-seated conviction that he was working for the good of mankind enabled him to continue notwithstanding a somewhat fatalistic world-view. There were thousands upon thousands of hefty academic volumes about the 'still unproven, fantastic "globite theories",' he complained, and he sought to re-establish some balance by stating his case through the medium of *Parsec*. Water levels, the fallacy of space flight, the 'great absurdity' of gravity, the see-through moon – Shenton whipped through the A–Z of IFERS ideas at a rapid pace in his article, meeting the word-limit in the process.[8] Although it seems unlikely that the piece secured new converts to his cause, his apparently wacky agenda seemed to appeal to undergraduate humour and branches of the IFERS were subsequently established at higher-education institutions the length and breadth of Britain, Manchester University among them.

Shenton's outlook and his mood were to grow gloomier, however, when astronaut John Glenn piloted the Mercury programme's Friendship 7 spacecraft on America's first manned orbital mission on 20 February 1962. He successfully completed three orbits around the earth, achieving some awe-inspiring views of our planet. On his return he was hailed an American hero, and a note from Shenton was among the reams of congratulatory messages awaiting him. Attached to an IFERS membership card made out in Glenn's name, the message read simply, 'OK Wise guy.'[9] Shenton had reached a verdict about 'space' flight and no evidence could convince him otherwise. During interviews, he blithely informed the world's press that orbital space flight was an illusion and 'Gagarin, Glenn and Co. had circled over a flat surface like a toy airplane on the end of a piece of string'.[10]

The Vostok and Mercury programmes continued to test the boundaries of space travel through 1962 and 1963 and Shenton continued to lecture to astronomy and geography societies at the universities of London, Leicester and Liverpool, more often than not at his own expense – financially, emotionally and at a cost to his health. During this period he admitted that publicly defending 'the heterodox subject of "Flat Earth" takes a lot of me', and he suspected that such activities contributed to the two strokes he suffered in 1963; the second left him with distorted vision, a disaster for a sign writer.[11] Despite the warning signs, Shenton continued to court the media. In January 1964, the *New York Times* carried a story about the IFERS and there was further press coverage in June. During a parliamentary debate, Conservative backbencher and free-market advocate Enoch Powell had likened his opponents, the economic planners, to flat-earthers, and the Labour Prime Minister, Harold Wilson, had reportedly slung back the insult in turn. Shenton was outraged and immediately wrote letters of complaint to the politicians involved and to the national press.

The story had a natural appeal for the tabloids: under headlines like 'Enoch Sends Flat-earth Men Round the Bend', they gleefully reported Shenton's claims that schools, colleges and universities were 'media of mass brain conditioning' where the 'unproven, godless, globular theory' persisted. 'What hope was there,' Shenton had commented in despair, when even the Secretary of State for Science and Education Quintin Hogg 'fondly imagines that he lives on a whirling planet!?' For Shenton, the issue was unjust and depressing: flat-earthers were 'honest searchers who will admit to gross error', rather than the fanatical lunatics that people supposed.[12] Despite his protests, the semantics were beyond his control: 'flat-earther' had long been utilized as a multi-functional term of abuse for those who deviated from commonly held opinion or clung to dogma in the face of seemingly incontrovertible proof. Enoch Powell was later to discover this at first hand when he was labelled a flat-earther following his controversial 'rivers of blood' immigration speech on the anniversary of Hitler's birth in April 1968. He was sacked from Edward Heath's shadow cabinet, and widely discredited; his political career and reputation did not recover from the scandal.

By contrast, the negative associations of being branded a 'flat-earther' did not act as a deterrent to Shenton: a true believer and a tenacious soul, he wore the label with pride as he continued to dispute the facts of space flight throughout 1965. In April that year Soviet cosmonaut Aleksei Leonov became the first man to complete a space walk, during the Voskhod 2 mission. Predictably, Shenton advanced cutting criticism of the footage of this remarkable event. While journalists eulogized the cosmonaut's seventeen orbits, Shenton claimed that Leonov, like Gagarin before him, had merely traced an 'egg-shaped' ellipse above the disc-shaped plane. 'We are not on a globe cavorting through "space" at some 20 miles a second,' Shenton again insisted, for how could 'space' capsules, travelling at four to five miles a second 'ever compete with the Earth's greater speed'? From his perspective, the latest developments merely emphasized the pressing need to encourage the young to reconsider the nature of the earth they inhabited before propaganda and indoctrination took hold.

Meanwhile, teenagers who read Shenton's statements were amused and perplexed, as a letter from a young amateur astronomer addressed to Shenton subsequently showed:

> I have seen ships disappearing from view over the horizon, seen the circular shadow of the Earth on the moon, proved to myself by my own astronomical observations that our sister planets Jupiter and Saturn are globular and rotate about their peculiar axis, and have no reason to believe that scientists are out to hoodwink us about the shape of our planet. Will you please tell me your own ideas for I am dreading that otherwise I have been living under false impressions for the first 19 years of my life?[13]

Two months after the Soviet mission, stories about America's Gemini 4 mission hit the newsstands. Launched on 3 June, the four-day expedition was intended to evaluate the effect of prolonged space flight on the spacecraft and its systems, and demonstrate and evaluate extravehicular activity – the first American space-walk. Yet despite the successful sixty-two-orbit mission, featuring a space walk by capsule pilot Ed White, Shenton claimed the flight had proved, rather than demolished, his flat-earth case. Having scrutinized newspaper pictures of White walking in space with the curved outline of our planet as a back-drop, Shenton informed the near-constant stream of reporters who telephoned his home, 'They prove that Earth is shaped like a plate', and the curvature of the horizon was an illusion, the result of 'natural distortion caused by a wide-angle camera lens'. He even claimed that White was not walking in space but had actually remained 'earth-bounded'. The supposed pre-programmed orbits of the astronauts could never pass the impen-etrable ice-barrier surrounding the edge of the world, while the firmament of heaven would further obstruct attempts at space exploration. In conclusion, Shenton averred that 'things' kept dropping from the firmament, and the fantasy of space flight was mere 'power politics', a clever deception by governments and technical experts to milk money from the public to finance their existing schemes.

From 10 June Shenton's comments were published in papers across Britain and America as an intriguing aside to the main event, and he was deluged with letters from readers, the curious, confused and convinced. Among the correspondents were three television producers, based close to the Manned Space Center in Houston, Texas, who said they wanted to organize an IFERS chapter to provide research assistance: 'Perhaps with the home of the astronauts so geographically close to our studios,' they sug-gested, 'we can even determine the type of plate the Earth resembles.'[14] On 14 June, a Californian building contractor joined in the fun, with allegations that he had been 'battered' with the 'ridiculous notion' that the earth was a globe since childhood. He had been 'heart-warmed', he wrote, when he heard of the society's existence, for it had been a relief to realize other people had 'the intelligence to preclude the ridiculous'. It was obvious to anyone, the man continued, that if the earth was a rotating sphere, 'all of the objects upon it would long since have been catapulted into space, making tell-tale punctures in the sky'. Like the television producers, he sought to join the IFERS, claiming his commitment was genuine.[15]

The Gemini 4 mission marked a notable change of pace for Shenton's campaign, and over the next few years he was to receive letters of all descriptions from a vast range of people across the world. Academics, housewives, teachers, journalists, dentists, lawyers, students, engineers, doctors, schoolchildren, vicars and scientists from Poland to Puerto Rico, the Lebanon to Japan, wrote to him with their views and disputes, their knowledge and ideas, about the world and the universe. In the meantime, through the latter half of 1965, Shenton continued to focus on the international media, providing comments and interviews whenever he was asked. That winter he appeared on the popular Canadian programme *The Pierre Berton Show*, and gave a thirty-minute radio interview for the Australian Broadcasting Commission, which was subsequently aired across the Antipodes and the Far East.

Yet whatever his determination in the face of convincing and widely publicized evidence of the rotundity of the earth, Shenton felt increasingly isolated as the years, and the space missions, passed. IFERS president William Mills had not bequeathed any money to the society to further its work, and a local schoolmaster who had taught his pupils 'the truth' had dashed any hope of providing assistance by relocating to Devon on his retirement. Dogged by correspondence, Shenton complained that dealing with a fraction of the questions posed by correspondents had nearly killed him; indeed, the majority of incoming letters were demands for a summary of his theory with supporting proofs attached. Swamped by paperwork, he recognized that he required staff to issue timely replies, and an income sufficient to print the flyers and leaflets essential to satisfy the enormous demand. Unfortunately, he had neither the assistance nor the finances to keep pace with developments in the wake of the space race, and because he was running a non-profit-making organization single-handedly, he believed it would be unreasonable to demand a regular subscription from members.

Despite these constraints, in February 1966 Shenton managed to produce a mimeographed pamphlet, *The Plane Truth*, which included a circular letter for IFERS members. It opened with a catalogue of complaints: the constant ridicule and gibes from the easily led majority, the serious heart condition that had put him in hospital, the strain of nine years' lecturing on a 'suspect' subject, and so the inventory flowed. Shenton complained that at sixty-three he felt weary and discouraged, but would continue the crusade as long as interested members contributed occasionally towards costs. In the midst of sneers and jeers, he was consumed by the profundity and seriousness of his cause – to defend his interpretation of the Biblical account of creation and safeguard young minds from corruption by science. Intense and passionate, he informed his members that modern astronomy and space flight were insults to God, and divine punishment for humankind's arrogance was a mere matter of time. Yet he understood that many IFERS members did not link their 'interest in cosmological matters' to the 'ancient script', and Shenton wrote that he was prepared to welcome flat-earthers from all schools of thought into IFERS ranks if they undertook zetetic investigations and continued to 'think for themselves'.[16] The point emphasized that for Shenton, the self-taught inventor, zeteticism was a practical science as much as a belief; commitment rested as much on fact-collection as it did on an interpretation of Genesis and a desire to defend Biblical truth.

When the American Lunar Orbiter flight took place in August 1966, the outlook for IFERS membership had become less promising. The purpose of the unmanned capsule's mission, the first of a series of five, was to take pictures of the surface of the moon in preparation for the projected landing later that decade. The planned photographic survey was duly completed and the world was presented with the first close-up images of the moon, along with further pictures of a spherical earth from space. Journalists

were curious to discover Shenton's reaction to the latest developments, and among those looking for a story was Eric Clark from
the British *Observer*. He interviewed Shenton over the telephone
on 26 August, and two days later his article, 'The Day the World
(Nearly) Fell In', appeared in the paper. It opened with an
uncharacteristic confession from 'the world's number one flat-
earther': the latest picture of the earth from space had 'really
knocked' him and had come as 'a terrible shock'. Shenton's dismay
had been short-lived, though, and spurred by the need to legitimize
his theory, he soon reconciled the real-life picture with his mental
image of a flat earth. 'What the photograph shows is not really the
Earth at all,' he told Clark, 'but one of the non-luminous bodies
between us and the moon.'

However, Shenton admitted such 'proof' was decimating IFERS
membership and numbers had recently plummeted from over a
hundred to twenty-four. He was well aware, he said, that the
public branded the flat-earth minority 'silly old cranks', and at
one time he had felt so marginalized he was even 'afraid to walk
down the street'. 'What people forget,' he lamented, 'is that we
are only trying to help the children find some values.' According
to him, the round-earth fraud was instigated as an 'anti-God
move', but now children were being indoctrinated for less sinister
motives by a lazy society content to take 'expert' knowledge for
granted.[17]

When Shenton read the story he was content with its 'fair
coverage', and because it was reprinted in the *Washington Post* and
the *New York Times*, a good number of letters resulted. One
Durham correspondent confided that, considering the Cold War
stand-off between the two space powers, he had always been
suspicious of so-called 'space feats', while a Presbyterian minister,
based in St Andrews, eagerly requested further information. His
interest was echoed by a man from Syracuse, who said he had
believed in a flat earth since childhood and the recent photos had
done nothing to undermine either his conviction or his intention
to join the society. Meanwhile, Shenton was contending with a

deluge of written requests and telephone calls from reporters around the world.

Month by month, the space race and attendant news stories gathered momentum and each week that passed brought in a new round of media enquiries regarding the latest achievements. The pace was such that Lunar Orbiter 1 was still in space when, on 21 August, another manned mission, Gemini V, was launched, with Gordon 'Gordo' Cooper and Charles 'Pete' Conrad on board. Like their predecessors, they reported amazing views of the earth, and the media hustle for Shenton's side of the story started all over again. As a result of coverage, Shenton received numerous messages, many reiterating the sentiments of a New York man, who wrote on 20 September:

> Dear Mr. Shenton,
>   You are a nut. NUT – N-U-T – NUT. If the world is flat, why hasn't anyone fallen off? Maybe your organization is a farce, but if you seriously believe what you state, you, dear sir, are crazy.[18]

As an afterthought, undoubtedly futile, the correspondent added, 'When you get this letter, it will have traveled 3,000 miles over a curved surface.'

A North Carolina man joined the chorus of condemnation on 4 October:

> All I can say is this is the most imbecilic misconception of the Dark Ages that has existed to this modern day. You, a mature man (physically, if not mentally) rejecting all scientific proofs to support a belief that has not been accepted for many centuries. I am including . . . a photograph of the Earth taken from the moon by the American Lunar Orbiter. Try to explain that into a flat plane without sounding foolish.[19]

Shenton's well-publicized views were evidently evoking frustration, alongside widespread amusement, and it was not merely adults who took the trouble to write. Children and teenagers were likewise full of reproach. Fresh from the playground, some jeered, 'Everybody knows the Earth is round', while others sent carefully drawn diagrams in an attempt to teach Shenton the true astronomical and geographical facts. Countless people concurred with a New Jersey fifteen-year-old who said he could not grasp Shenton's blatant refusal to accept clear evidence in the days of 'mathematicians and computers and things', while a little boy from Brooklyn spoke for many when he wrote with childlike simplicity:

> Why do you think the world is flat? I think the world is round. Your people think the world is flat. But it is not flat. I can prove it . . . The people in Brooklyn knows [sic] that the world is round. They can prove it to [sic]. Because we have round globes. Why do you think the world is flat? If you come out here everybody will know that your people think the world is flat. Someone should show you that the world is round. Did you see the pictures that the spacemen took of the Earth? Didn't you see the pictures on TV?[20]

Visual images, whether they were globes, photographs or television pictures, were clearly critical to how people perceived the earth's shape, and from exposure to such cultural products pre-school children could know that it was round even if they had no grasp of the words 'mathematics', 'geography', 'astronomy' and 'science'. Despite the point that, in all likelihood, many of Shenton's correspondents could not identify how they initially 'knew' the earth was a globe, or could not build a convincing case that it was, except by reference to cultural representations of nature, they could not accept, and were sometimes disturbed or offended by, Shenton's refusal to abide by commonly held standards of authoritative proof. In this sense Shenton, who was widely portrayed as an unusual specimen of suburban busybody, was running a radical and highly subversive campaign. In challenging human cognition and accepted standards of how to measure fact and decide what was true, his campaign delved deeper than tacit knowledge, the respective merits of science and the Bible and their claims to be arbiters of truth: it also raised fundamental questions about the psychological processes by which natural knowledge is made.

Psychology and paranoia were issues frequently raised by Shenton's correspondents yet, paradoxically for one so beset by thoughts of conspiracy, he frequently overlooked the issue of trust in his daily operations on behalf of the cause. Although he was highly sensitive about public corruption, indoctrination, scientific scheming and the hidden plans of the establishment, he did not apply the same level of suspicion to the private letters he received. In particular, he was quick to trust correspondence from the technological experts he purported to loathe, and frequently interpreted their letters as sympathetic and supportive when in reality they were often the reverse. Undoubtedly such misconceptions were born of blind hope that converts would be won from scientific ranks, and so contribute credibility and authority to his heterodox campaign, along with a propensity, common to flat-earth believers and extreme polemicists of any stripe, to use information to suit his agenda and aims.

Yet besides argument-building, there was a more personal factor at play: Shenton was evidently dazzled by technical and academic accomplishments and worldly success – even though his campaign against scientific authority rendered this something of a contradiction. This tendency to place faith in those least likely to provide genuine support for the IFERS is illustrated in his response to a letter from an official at NASA's Goddard Space Flight Center in Greenbelt, Maryland, who was intrigued by Shenton's 'somewhat conservative' views as reported by a Baltimore newspaper. The correspondent commenced in a friendly tone: belief in a flat earth was undoubtedly more widespread than most people imagined, and it was a comforting concept, if to his mind undeniably flawed. Nevertheless, he continued optimistically, he was sure Shenton's conviction was 'nothing that couldn't be ironed out over a few warm beers in a Dover pub', although Shenton's comments to the press about the earth's height appeared to be slightly more problematic. Apparently, Shenton had declared that humankind had been 'misguided from childhood' because 'the Earth is so much higher than we were ever taught . . . We're only on the crust of the Earth.' No doubt in common with many other readers, the NASA employee was unable to grasp Shenton's meaning:

> I should truly like to sympathize with you in your indignation and to take arms against the accursed deceivers who have withheld from us such information as they have, but I quite frankly fail to see what you are driving at. The question which must occur to all of us, Flat Earthers and Round Earthers alike, is what do you mean by 'high'? If you are referring to the distance between one side of the Earth and the other, it is of the order of 250,000 stadia divided by *pi*. If, on the other hand, you mean its height above, say, the sun, then that is equal to 93 million miles (where a mile is an astronomical unit equal to approximately 10 stadia).[21]

Having provided a clear explanation of the basis for his confusion, the correspondent awaited further clarification or a straightforward denial. When Shenton read the letter, he over-looked the gentle criticism of the flat-earth view and noted in his diary that the message was 'splendid'. Proud to have received a letter from a NASA employee, he plundered it for quotations, which were duly incorporated into IFERS promotional material, and posted a pamphlet and a membership card, which the letter-writer had not requested.[22]

The depth of Shenton's belief in his enterprise evidently rendered him somewhat gullible, while the cause made him a prime target for jokes. His incoming correspondence testifies to the fact that numerous people failed to resist a laugh at his expense, which was made all the more effortless following the publication of his address by the American media. Among those looking for fun in 1966 was a communications specialist at General Electric's Missile and Space Division who wrote to Shenton for information to 'help to subdue the panic' in the event that IFERS theories

'should be revealed as fact'.[23] His teasing was echoed by a geologist
and hydrologist from Wisconsin, who sent a long, complicated
letter suggesting various 'corrections' to Einstein's equations, adding that this would undoubtedly form one of Shenton's 'greatest
contributions to science and an absolute proof of a flat Earth'.
He waited until the postscript to reveal the entire essay as a 'pull
of the leg', although by way of compensation he restated his desire
to enrol in Shenton's 'fun-loving group'.[24]

Throughout 1966, alongside the jokes, gibes and duplicity,
Shenton received many legitimate offers, as authors, journalists,
universities, societies and television producers competed for a
lecture booking or a scoop on the latest from IFERS. In October
alone, as well as giving various newspaper interviews, Shenton
appeared with *Daily Mirror* agony aunt Marje Proops on the
programme *Late Look*, gave an interview to a University of London
student magazine and was invited to speak at venues across
England. During the same period, a group of American engineering students requested permission to establish an IFERS chapter
at Boston University, and Shenton was invited to enter a display in
the Science Fair at Whitinsville, Massachusetts. His project 'might
not be a winner', he was warned by the organizer, but it would
certainly 'stir up a bit of controversy' in the town. Thus encouraged,
Shenton embarked on the preparation of posters, pamphlets and
photos for the exhibition. From beginning to end, 1966 had been
an eventful year, and Shenton ended it with a dramatic public
statement, writing to his local paper, the *Dover Express*, on
Christmas Eve about the supposed first picture of the earth from
the vicinity of the moon,

Sir,
    Astronaut Captain Lee Scherer on TV said, re: supposed
Lunar Orbiter Moon–Earth photograph, '*We put the Earth in
it also to make a more general interest photograph and I think it
has done that.*' This may be taken as proof [of] 'doctored' presentation. The Zetetics (flat-earthers) are still unsatisfied because

the Earth, pictured as globe, perpetuates man's theories; conflicts with many scriptural statements and is basically anti-God.
Nationally, especially at Christmas, we call ourselves Christians, but allow our children to fall into the hands of . . . evolutionists . . . and expounders of men's fallible theories.[25]

Through Samuel Shenton, the double-edged zetetic assault on
scientific authority introduced by Parallax in the 1840s – the attack
on 'man-made theoretical science' together with a defence of a
certain interpretation of the Bible – remained on the peripheries of
the public domain.

Meanwhile, on 27 January, the American space programme
had been hit by its most significant and disturbing disaster to date.
The command module of Apollo 1, the first manned Apollo–
Saturn rocket flight in the programme to put a man on the moon,
caught fire during a ground test at Cape Kennedy with three
astronauts trapped inside. The ferocious blaze, the result of an
electrical fault, sped through the capsule and the astronauts, 'Gus'
Grissom, Ed White and Roger Chaffee, were almost inevitably
condemned. It was later discovered that the module's heavy,
inward-opening hatch could not have been unbolted in such
pressurized conditions, even if the men had been able to undertake
such an escape attempt. As it was, they were locked into a highly
flammable, pure-oxygen atmosphere infused with toxic gases and
temperatures topping 2500°F, and were dead (due to asphyxiation)
within seconds. People all over the world were stunned by the
tragedy, which dispelled for a time the aura of optimism and
adventure surrounding early space exploration. By contrast, Shenton believed the disaster to be the outcome of a divine curse, as he
explained in a letter to the editor of the *Daily Express*:

Are we so fascinated by 'science' that we fail to realise the true
character of the contending forces around us? Our American
brothers emphasise their motto 'In God We Trust.' But in
their modern efforts to climb into the heavens they affront
God by naming their thrusting ventures 'APOLLO' the pagan

mythical god who in ancient times was associated with sudden death! The tragic sudden deaths of the three American astronauts . . . may give watching Christians time to realise that . . . God is not mocked.[26]

Unsurprisingly, the letter was never published.

Following his questionable public statements, Shenton suffered another breakdown in health and the final collapse of his signwriting business, but through the spring of 1967 the ranks of his society swelled. Shenton was particularly pleased to enlist all fourteen students and the teacher of the physics class at Eldora School, Iowa, after they had alleged concern about the 'unscientific explanations of the Earth's spherical shape'. The class continued that, from their earliest years, they had been told to believe the earth was round, and to accept this fact on the grounds that Magellan and other explorers 'had sailed in circles'. Yet, they protested, such supposed 'proofs' seemed to 'lack completeness', and they suspected the 'scientific community' was 'deliberately distorting the reality of the situation'.[27]

In tandem with the sudden influx of members, in March 1967 Shenton received yet more attention from the world's media. Eddy Gilmore, an Associated Press journalist, had happened on a piece about him published the year before and, realizing the potential for another news story, telephoned to arrange an interview. The resulting article was reprinted in various forms in papers across the States, some accompanied by a picture of Shenton with a globe. Naturally this was an ironic statement: like Hampden and Carpenter before him, Shenton suspected globe manufacturers of conspiring in the dissemination of the round-earth fallacy as a ploy to boost sales. In the meantime, the *Buffalo Courier Express* had paid heed to Shenton's private doubts: he had recently confessed to wondering sometimes whether it would be better to 'live and die like a sheep' than 'pay the price of great individual effort to advance a little nearer to the truth – the truth of life and being'.[28]

Although the statement was dramatic, from Shenton's perspective the flat-earth concept was a profound metaphysical issue: as a 'fact' backed by the word of God, the issue lay at the heart of a 'war' between science and the Bible. This was a flawed oversimplification of the complex and context-specific relationship between science and religion in their myriad forms, yet Shenton continued to offer the general public a direct choice between two irreconcilable systems of explanation set at polar extremes.[29] While the warfare myth had long been the province of partisan interests, both religious and secular, he contended that an individual's loyalties hinged on his beliefs about the shape of the earth.[30] For this reason he perceived the flat-earth idea as intrinsic to the truth of existence – the meaning of life itself. Besides overlooking non-Christian religious traditions with diverse cosmologies and foundation texts, Shenton diverged from flat-earth advocates like Voliva on one significant point: although he made rash public statements about divine curses and often utilized Biblical imagery and apocalyptic rhetoric, he never truly focused on the idea that souls were at stake for believing in a spherical earth. Although he was a Christian, and utilized the Bible to defend his truth claims, his fixation was always with minds rather than souls, a standpoint that engendered his constant refrain about misinformation, indoctrination, brainwashing and mind control, matched naturally by an emphasis on the education of the young.[31]

As the space race continued, Shenton laboured under a renewed influx of correspondence; by now he was receiving between twenty and forty letters a day and could not afford to comply with every demand. He was also generating some unwelcome media requests: he turned down an offer to write a piece for the far-out underground magazine, *Oz*, although his anti-establishment stance and refusal to comply with 'authority' gave him something in common with its radical values. Soon afterwards he declined actor Robert Morley's request to interview him for *Playboy* on the grounds that the 'outfit' had 'the smell of Gomorrah'. From Shenton's perspective, the flat-earth issue should not be tainted by debauchery, although he wrote that he meant no offence with his 'forward wording' because he wanted to discuss collaborating with Morley on a television series.[32]

Television or, more rightly, its globe-earth bias was a constant source of irritation to Shenton, and one BBC programme in particular roused his ire during June 1967. Entitled *Our World*, it was the first worldwide live satellite programme, broadcast simultaneously in twenty-four countries to a potential audience of 400 million people. An historic event in the development of television, each nation had a short slot in which to make a contribution to the international phenomenon by displaying aspects of its culture and country to people all over the world. Underpinned by ideals of unity and co-operation, the ambitious endeavour was undermined only by the fact that, near to transmission, Russia and allied Communist states withdrew. Britain's six-minute slot was dedicated in part to the Beatles performing a new song written especially for the event, 'All You Need Is Love'. It was later to become the peace-promoting anthem of the hippie generation and that year's 'Summer of Love', but the programme that first aired it seemed to have the reverse effect on Shenton.

When he tuned into the international broadcast on Sunday, 25 June, he was sufficiently annoyed to send a letter of complaint to the BBC's viewer-feedback programme *Points of View*. He started at the top, condemning the corporation's liberal-minded director-general, Sir Hugh Greene, closely followed by everyone connected with the 'God-insulting TV programme ignorantly styled *Our World*'. He then turned his attention to the programme's anchorman, Cliff Michelmore, protesting that he had alluded to the 'fact' that the earth was round, while 'models' (globes) had been used throughout the programme 'to deceive and indoctrinate millions'.[33] Unlike the majority of Shenton's letters, this one was not disregarded and it was duly read out on *Points of View* by the presenter, Robert Robinson. As a result, Shenton generated more of what he claimed to dread: publicity, enquiries and letters.

By this time, due to Eddy Gilmore's Associated Press article in March, Shenton was receiving postcards, pictures and letters from countries worldwide; from the Arctic Circle to the Americas, Africa to Australia, correspondents were amused and amazed, angered and delighted, by his heterodox campaign. Even an American soldier with time on his hands in a Vietnam bunker wrote from the 'combat theater' eager to discover more. Faced with this next onslaught, Shenton again felt unable to manage, and he confided in letters that the sheer scope of his international audience, from scientists to space-centre workers to hundreds of youngsters, who could 'scarcely manage even a pencil', had left him stumped about how to 'hit the mark' in disseminating his message between such varied groups. Yet promoting his ideas was of paramount importance in the light of his need to gain converts, and although the 'strain of speaking on so heterodox a subject' had affected his health (he suffered from heart trouble), he hit the lecture circuit once more with a presentation at the Dover Rotary Club and an appearance on the ITV show *Tonight*, hosted by comedian Dave Allen.

By August 1967, the fifth Lunar Orbiter mission was ready for launch with the aim of completing the ongoing photographic survey of the surface of the moon. Pictures of the far side and various specified sites were duly completed, while back on earth the media followed the latest developments with avid interest. The net result for Shenton was a renewed burst of enquiries, including a telephone interview with Phillip Howard of *The Times*. Shenton was pleased by the attention, noting excitedly in his diary that it was the first time the flat-earth story had reached the pages of the prestigious London daily. He was also impressed by Howard's efficiency: one fifteen-minute late-night interview on 16 August, and the paper featuring the story was delivered to his home early the next morning. Slap bang on the front page, the article, 'Flatly, They Call Earth Pictures a Fake', commenced with Shenton's criticism of the Lunar Orbiter's picture of the world from 214,806 miles distant in space. He admitted that it had been like 'a blow to the belly', but he had soon recovered from the shock. With shades of Tommy Cooper, he denounced the image as a 'fraud, fake, trickery, or deceit, just like that'. Calling attention to suspicious lines and ragged edges, he maintained that the photograph was just 'a composite picture'.[34]

Although the story was an unusual one for *The Times*, it made an impact in certain quarters. The editor of the American magazine *Aviation Week and Space Technology* was so amused by the piece that he forwarded it to Oran W. Nicks, aerospace pioneer and director of NASA's Voyager Space Program, with a jokey note attached: 'This should convince you deep thinkers that you are on the wrong track ... I always thought you practiced "fraud, fake, trickery or deceit," as the man said. And you gotta believe it, since this is from the front page of the London Times.'[35]

The Times was not the only newspaper to cover the Shenton story on 17 August: the *Sun* followed suit, with a sensational piece under the screaming headline 'Faked! Shenton Exposes the Great Global Earth Conspiracy'. Focusing first on Shenton's conviction that the Lunar Orbiter picture was a mock-up, which he believed any of the *Sun*'s 'photographer chaps' could have managed with ease, the reporter questioned why anyone would want to perpetrate such a deception. Shenton replied that it was all part of the 'great Global Earth Conspiracy'. Motivated 'by the desire to deny God as the Creator of all things', the plot had been perpetuated by Shenton's 'worst enemies', 'parsons and schoolmasters', the 'idle shepherds' mentioned in the Bible. The journalist then proceeded to ask whether Shenton would like to fly into space and observe the shape of the earth for himself. Shenton, who, incidentally, had never travelled in an aeroplane, brushed aside the suggestion, contending the world should wrench its focus from the 'so-called upper stratosphere and concentrate on what Earth, [its] strata and rock formations ... can teach us'. Besides, he added, 'I doubt if I would survive the experience.'[36]

Sidestepping space flight, correspondents continued to comment on Shenton's self-proclaimed expertise. Soon afterwards he received a note from a Norwegian couple inspired to write after hearing that the IFERS 'was the only society in Europe to find out how our world is *really* created'. Their interest, they explained, stemmed from their home environment for 'in Norway we are all hidden in the bottom of deep valleys with big mountains on all sides, and therefore it is extremely difficult to understand how the Earth can be round as a ball'. More focused than this was a letter from Tom Pierson, a former member of the 1966–7 physics class at Eldora High School. Despite his alleged flat-earth beliefs, he had graduated successfully and was attending Iowa State University of Science and Technology where, as member number eighty, he hoped to carry on the work of the IFERS among the 17,000 other students. Shenton was glad to hear this news, but in his own life he was struggling. His heart trouble had flared up again, the condition aggravated by a laborious move to a new home in Dover. He was loath to leave the old place for the address was now known all over the world. Yet he recognized that his home in London Road had become a burden and he felt his health left him with little option but to move. Above all else, he hoped that relocation would herald a fresh start for 1968 and better times ahead.

Shenton's intention to begin the new year with a fresh start did not materialize and he was still settling in at Lewisham Road when he had a bad fall while reaching for the telephone. Although it was a momentary mishap, he was confined to his bed for several weeks. With too much time to contemplate, his problems seemed overwhelming. On 6 March, he wrote in a letter that a primary objective of the house move had been to elude the incessant pressure of the International Flat Earth Research Society, and for this reason he had named the new place 'Sharuhen', from the Hebrew, meaning 'refuge of grace' (Joshua 19:6). But 'alas', he wrote, the strategy had proved futile for 'the pressure has followed me here'. His health had deteriorated further, the staggering influx of letters continued, his signwriting business had collapsed and he was under investigation by the Inland Revenue for not having resolved business matters correctly. In the light of his troubles, he was grateful to anyone who donated money; in the past year, the society's income had amounted to less than ten shillings and he was weary of miserly enquirers who expected to receive information about his organization without even enclosing payment for postage. Americans and students were the chief culprits, Shenton complained, for in his opinion the former thought only in terms of 'big business' while the latter were self-centred and 'used to having everything far too easy'. Despite his mounting problems, though, he had decided to resume public speaking as soon as his health would allow. 'The wife', he confided, was infuriated by this prospect: having seen 'the faints, the falls and the dose of "shingles" I had just recently', her favourite saying had become '"Go on kill yourself!"'[37]

Nevertheless, Shenton remained resolute and, notwithstanding the health risks and the hen-pecking, he travelled in March to Bedford College, University of London, to debate the question 'The World – Round or Flat?', and followed up his visit with a letter of complaint to the *Daily Express* about its round-earth coverage.[38] Meanwhile, he received more mail from Scandinavia: struck by recent media coverage of the zetetic campaign, an export manager from Copenhagen had requested permission to set up a flat-earth society, adding that it would be an undoubted success for 'the Danes are very realistic people who just don't accept right away all romantic nonsens [*sic*] about a round Earth'. Shenton received a similar offer from two Norwegians, written on 20 May:

> We live on an island . . . in the Western part of Norway. Seeing only sea all the day we have a lot of time thinking. Sometimes we can see pictures in magazines showing 'The Earth'. The differences made by [camera] lenses . . . give quite another picture to the true one.[39]

As one might expect, indoctrination rhetoric was rife among Shenton's supporters, and in early July he came into contact with a zealous convert who agreed wholeheartedly with his views.

Westminster resident Gerald St. John-Culdwart had first been persuaded that the earth was flat by his late friend and IFERS member Septimus Herbage, and over the years had remained certain that:

> In this modern technological age, people are often brain-washed into believing that these photographs supposedly taken by satellites, show a globular shaped Earth; but, if a little calculation is done, it can be seen at a glance that this is just what would be expected if the Earth were flat, due, of course, to the error of refraction!![40]

St. John-Culdwart often wrote to his local paper in the hope they would publish letters supporting this view but noted that 'they tend to regard me as somewhat eccentric in my beliefs'. Neverthe-less, he remained determined that 'something must be done' to remedy the 'absurd misconception' that the earth was a globe, for until mankind was convinced of its true shape, science would be hindered by Newton's laws of gravitation. On reading this, Shen-ton believed he had finally discovered a kindred spirit and replied immediately with a copy of his pamphlet *The Plane Truth*. In response, St. John-Culdwart enclosed a large donation (by IFERS standards) of ten shillings, with further reflections on the issues at stake. England was, he believed, a 'worldwide laughing-stock' due to the population's spherical-earth conviction, but flat-earthers must continue to work to right the wrong, for 'if one believes 100% in what he is doing, then he cannot fail'.[41] He was eager to convene with Shenton to discuss these 'urgent' matters but announced that there would be a delay: he was departing to Siberia to investigate a total eclipse of the sun on 22 September. St. John-Culdwart had no doubt that the phenomenon would occur as predicted, but because the calculations issued by the *Nautical Almanac* were based on the 'false theory that Earth was a globe', he sought to verify them in person.

So, armed with sextants, chronometers, a spectrograph, and the assistance of a Russian scientist, Dr Popov, St. John-Culdwart intended, in true zetetic spirit, to collect the facts himself. If the earth was flat and motionless, as he suspected, there would be an error in the track and duration of totality (when the main disc of the sun is entirely obscured) running through Siberia's Sverdlovsk region, and he promised to compile a report on his findings for the IFERS when he returned to England.[42]

In the interim, Patrick Moore had invited Shenton to write an article for the quarterly popular astronomy magazine the *Planetarium*. True to form, Moore adopted a candid approach: he confessed that in common with the overriding majority of *Planetarium* readers he disagreed profoundly with Shenton's opinion, but echoed Voltaire's statement 'I defend to the death your right to say it.' Moore's request denoted his conviction that anti-globularist ideas would be of interest to astronomical enthusiasts and his judgement was undoubtedly accurate: scientifically minded people all over the world were curious to know how ardent flat-earth believers upheld their extraordinary world-view.

At Cape Kennedy, Florida, final preparations were being made for the seventh Apollo mission. By 11 October, all arrangements were in place, and at 11.02 a.m., the capsule was launched, with Wally Schirra, Donn Eisele and Walt Cunningham ensconced on board. The primary objective of their engineering test flight was to assess the efficiency of the newly improved command module in earth orbit and so clear the way for the proposed lunar orbit mission, Apollo 8. While the three astronauts completed their targets, Shenton was occupied with telephone interviews: first a 'pleasant talk' with Robert Mussel of United Press International (UPI) followed by a live discussion with Michael Jackson of the Los Angeles radio station KABC. Shenton was pleased with the results, noting in his diary that the flat-earth view 'must have gone over well because volumes of letters followed'. Among those in approval were a group of flat-earth believers from Sewanee, Tennessee, who claimed that Shenton's insights 'made a great deal more sense' than the statements of the 'spokesmen of our so-called "space-programme"'. Unhappy that billions of dollars of public

money were being 'thrown away' on space flight, and tired of
funding 'trick-photography' and fraud with their hard-earned sala-
ries and pensions, the group said they were keen to affiliate with
the British society to add some weight to the crusade.[43] Genuine
or not, their comments reflected festering mistrust of the alliance
between the US government and science, alongside public disquiet
about the exorbitant cost of the Apollo flights.

Matching their alleged commitment, Shenton also received
offers to organize IFERS chapters from a fifty-six-year-old associate professor of economics and ex-Second World War pilot
employed at Providence College, Rhode Island, and an NCO from
the US Air Force base in Oscada, Michigan (who thought he had
'a better chance of setting the Air Force straight from the inside').[44]
Similarly associated with the US Air Force, two sergeants stationed
in Suffolk County, New York, informed Shenton that, as much as
they 'admired the courageous astronauts', they were 'saddened by
the fact that the men were continuing to mislead the general public
as well as themselves'.[45] The sergeants were undoubtedly teasing,
but on a more serious note, the librarian at California's Edwards
Air Force base, celebrated scene of the pre-space programme test
flights, requested more information about the IFERS to supplement his files.

Despite the letters, Apollo 7 threw Shenton into another
slump. On 12 November 1968 he wrote that his 'troublesome
heart complaint . . . nearly laid me out for good' two days ago,
but as he did all of the 'donkey work' for the society it was little
surprise that he was feeling 'a bit worn out'.[46] Handling the
taxation tangle and financing the society's work from his own
'slender purse' was also proving a strain; the literature fund contained just £170, and it cost more than £1000 to publish a book.
More than ever, Shenton wanted to reprint some of the old zetetic
works; he had invested much effort in upholding the 'facts' presented by his predecessors, and had amassed a plethora of arguments against space-age 'evidence', which he wanted to incorporate

into the originals. Besides attempting in vain to save money, he had been seeking patronage, particularly sponsorship for his publication plans, for years, but was hopeful that the tide would turn now that Patrick Moore had asked him to contribute the article to the *Planetarium*. If it did not, Shenton explained to an American correspondent, he had only intended to influence 'a few of the expanding millions on Earth'. As things went, he was content to 'plod along in the hope of reaching the one or two who begin to see the breadth of the subject', in the knowledge that championing the flat-earth idea had always been a case of 'the little individual against massed (evil) orthodoxy'.

In the face of élite oppression and inevitable defeat, Shenton further reflected that he had achieved much of which he could be proud, for as a registered member 'Old Sam Shenton is the only "flat Earthist" who has entered the sacred portals . . . of the Royal Astronomical and Royal Geographical Societies.'[47]

In the wider world, the media frenzy that accompanied Apollo 7 had hardly dissipated and already the 21 December launch date for Apollo 8 loomed. The mission was bold: an enormous step from the eleven-day earth orbit tests of Apollo 7 just two months earlier, but it would cement a clear lead for America in the space race and make a massive advance towards the goal of landing a man on the moon by the end of the decade. Crucially, the three astronauts, Frank Borman, Jim Lovell and Bill Anders, would fly further than any man previously. The plan, for what was only the second manned Apollo mission, was to fly a figure of eight, orbiting the moon ten times and relying on the clockwork laws of celestial mechanics to guide the command module safely on its course and back to earth. All in all, the groundbreaking journey was a series of firsts: the first manned test flight of the Saturn V rocket, the first human voyage to the vicinity of the moon and the first direct visual close-ups of the lunar surface. While many were excited by these possibilities, Shenton elected to take a proactive stance by writing to the prestigious British newspaper, the *Guard-*

ian, on 17 December, denouncing the upcoming circumlunar flight. The letter was eye-catching: potentially libellous, it disparaged the mastermind behind the Saturn V rocket and director of NASA's Marshall Space Flight Center, Wernher von Braun, for allegedly knowing that the earth did not rotate, yet permitting the Americans to believe 'all these Apollo Lunar ventures' and 'the unproven theory that we live on a planet . . . for good Germanic reasons'.[48]

Shenton's comments remained unpublished, but the events surrounding the Apollo 8 mission were to secure him and his cause widespread publicity across Britain and the States. One objective of the mission was to test communications from the 'vast loneliness' of space, and for the first time the world could watch the weightless Apollo astronauts in a series of live television broadcasts, beamed back to earth in fuzzy black and white. It is well known that, during the 23 December programme, the astronauts turned the television camera to a window to show earth an image of itself from space, and in another broadcast on the ninth of, ten lunar orbits, on Christmas Eve, each member of the crew outlined his impressions of space travel and described the view from the capsule, before concluding with a joint statement to mark the momentous occasion. Watched by an estimated audience of half a billion people, the astronauts read the first few verses from the Book of Genesis: 'In the beginning God created the heaven and the Earth. And the Earth was without form, and void; and darkness was upon the face of the deep. And the Spirit of God moved upon the face of the waters. And God said, Let there be light: and there was light . . .'

Meanwhile, Shenton interpreted various passages from Genesis as proof of his flat-earth view. From his perspective it was unthinkable that the Bible should be utilized to add weight to the activities of 'infidel science' and he had a further personal reason for disapproving of the astronauts' statements. During one television broadcast, the plain-speaking commander, Frank Borman, had mentioned Shenton by name while commenting on the appearance of the earth from space, and the insult was compounded when the astronauts returned with further high-resolution pictures as a supplement to those taken by the Lunar Orbiter in August 1967. Their remit was to focus on the far side of the moon, on various specified targets and possible lunar landing sites, and, despite difficulties, they captured hundreds of images, which included beautiful views of our blue, cloud-cloaked planet from lunar orbit. Cast partly in shadow, due to its position in relation to the sun, for most there could be no doubt: the earth was clearly a globe.

To say that Shenton was perturbed by these developments would be a further understatement of fact. On 27 December he dashed out a second complaint to the *Guardian* arguing that the Apollo 8 'space' photographs had not proven him and fellow flat-earthers mistaken. Those who studied the Bible knew the earth was flat and were aware that Christ himself had warned of a 'great deception which might shake frail Christian faith'. 'We should have known where to watch,' Shenton continued darkly. 'The American Space Administration have, since the early days of rocketry . . . circulated photographs and maps which instill the conception that Earth is a planet.' On this occasion, however, the technological villainy had been worsened by the astronauts' exploitation of 'the opening verses of Genesis . . . as a deceptive cloak'.

Those who understood the scriptures knew the truth, Shenton declared, that science was a lie, as described in the Book of Revelation, and those very same astronauts would soon splash down on a flat, unmoving earth. 'That is why,' Shenton concluded, 'there will always be a Christian opposition to Godless men who parade man's thoughts as science.'[49] In the face of this alleged deception, one positive aspect of the Apollo missions remained: for the first time in the zetetic movement's history he had a clear external scapegoat to hold responsible for unbelief and the schemes of a hidden cabal. Previously fixated on the adversary within, identified by Hampden and Carpenter as an intellectual élite of professional scientists and their dupes, Shenton had an explicit foreign target on which to pin evil and signs of the end of the world.[50] The apocalyptic argument was critical in highlighting his bi-polar world-view: the world appeared to him in an antagonistic series of dialectical opposites – scientific knowledge or the Bible, outsiders or insiders, critics or comrades, evil or good.[51] The argument mirrored fundamentalist insistence that evolution is incompatible with Christian faith and the division of creation-evolution into two sharply polarized views, with eschewal of biological evolution elevated to a near test of faith within some schools of thought.[52]

In the midst of Shenton's apocalyptic battle, a dynamic clash of irreconcilable forces, another more mundane aspect of the Apollo 8 mission operated in his favour: as it occurred over Christmas, he had an opportunity to gather his thoughts before facing the world's media in the next round of questions and interviews. As always, the reprieve was short-lived and as soon as 1969 arrived he was confronted with a barrage of enquiries from journalists focused on a single topic: the effect of the astronauts' photographs on the flat-earth cause. In response, Shenton was frank and several stories appeared in the American papers, under headlines such as 'Apollo 8 Floors Flat Earthmen', reporting the disastrous impact of the mission on IFERS membership. On 6 January, Wisconsin's *Appleton Post Crescent* revealed that the society's ranks now numbered less than a hundred but, even so, Shenton remained optimistic: if the opinions of his members could be swayed by 'such flimsy evidence' as the Apollo 8 pictures, he resolved, 'they're of no use to me or the society'. Although he confessed that he had never converted a sceptic to flat-earth belief, he added that his wife, Lillian, 'seems to be coming round'. Nevertheless, he accepted that the possibility of attracting new members was increasingly slim in the face of much-publicized evidence seeming to support a globular earth. Predictably Shenton smelt conspiracy: 'That's where those Americans and Russians are so damned cunning,' he speculated. 'For some reason they obviously want us to think the world is round.' As with Darwinian critics, the globe idea formed the public face of something more sinister for Shenton: a satanic plot to undercut belief in God, which would hasten the end of the world.[53] According to Shenton, the superpowers were so committed to perpetuating the deception that they had 'blatantly doctored' the pictures of earth from space.[54]

Shenton's perception was certainly a minority view, and the Apollo 8 pictures were widely admired across the world. In common with numerous other publications, on 13 January, the international current-affairs magazine *Newsweek* carried a glossy eight-page portfolio of moon close-ups and 'breath-taking shots of the cloud-shrouded Earth from lunar orbit', and as a human-interest sideline, an article on Shenton entitled 'The Flat-earthers: Where Are They Now?' Despite the recent pictures, Shenton informed *Newsweek* that there was no evidence whatsoever supporting 'the theory that the Earth is hurtling through space', but 'the world is run on this belief, and so it has to be maintained'. For this reason, he branded teachers, scientists and ministers his worst enemies, but again admitted that even his wife, two children and two grandchildren were sceptical about the flat-earth concept. 'Often a man's main opposition is on his doorstep,' he lamented, although he understood that 'I am not the first prophet without honour at home.'[55]

As 1969 progressed it became increasingly clear that Shenton was bent on defending his viewpoint, even in the face of 'breath-taking' evidence generally accepted as unquestionable verification of a scientific fact. In this sense, Shenton was living proof that personal perceptions are not necessarily ordered according to external reality, in the most essential of ways, and he remained determined to make his own meaning, notwithstanding criticism from sarcastic detractors.[56] He was not present when the photographs of the earth had been taken, after all, and it was a zetetic commandment that one should never accept information in the absence of personal fact-collection.

In practice, of course, his egalitarian ideal of open investigation was more problematic because in an ever more specialized world it was impossible for everyone to gain access to the full range of information, much less understand it. As science, technology and medicine became more complex, necessity fostered a passive reliance on experts to a certain degree – for example, not everybody could fly into space to make observations and take pictures. Yet, despite this, Shenton stood by his zetetic principles. When it came to proof of the earth's rotundity, he was the world's most prominent sceptic, although he contradicted his own philosophy in his failure to apply similar scepticism to 'proof' that the earth was flat. In this way, and in common with other modern flat-earthers, Shenton's application of zetetic methodology was fatally flawed: it employed doubt on one hand and belief on the other, while the inherent contradiction went deeper still. Shenton's desire to defend the end point of Parallax's 'zetetic astronomy', the 'fact' that the earth was flat, implicitly involved an avoidance of the methodological component of the *modus operandi* – or else his dogmatic conclusion would have been falsified.

The central paradox of Parallax's counterfeit system was that zetetic research, so-called 'free enquiry', while claiming to be the ultimate in objective truth-seeking (thus co-opting the mythological image of heroic discovery), was not open at all. Every zetetic who adhered to his teaching was certainly being duped, but not in the manner they supposed: what was posted as a pure, rational investigation of all available evidence to uncover the truth was tainted by bias from the outset. In practice, while staking a claim to true science, the zetetics turned the scientific method upside-down; from Hampden to Shenton they appeared to gather facts to support a fixed conclusion, rather than the other way round. Moreover, the extremity of their assault on 'man-made science' is denoted by the fact that although zetetics argued that the earth was no more than 6000 years old (so-called young-earth creationism) and frequently criticized the concept of evolution and its most famous proponent, Charles Darwin, these largely remained side issues in their protests. From Parallax onwards, the flat-earth idea, a less prevalent Bible-based truth claim than the divine origins of man, formed the constant focal point of their campaigns.[57]

Although in terms of arguments and tactics the zetetic movement in later incarnations remained remarkably similar to Parallax's original scheme, Shenton was clearly operating in a context much different from that of his Victorian predecessors. At grass-roots level, broad societal shifts necessitated the development of new forms of defence and attack in response to facets of an increasingly complex, technological and industrialized world that Parallax could not have imagined in his wildest fantasies. The changes ran much deeper than decades of discoveries and a multitude of other advances, for in the wake of the Second World War public attitudes to science had also metamorphosed. Gone was the heady optimism of the Enlightenment *philosophes*, who sought to convince the world that their secular social science was the solution to social ills, and the visions of nineteenth-century industrialists that science and its offspring technology would harness nature and spearhead a march of progress into a glorious future. In the aftermath of war and economic depression, such sweeping assumptions had been undercut by sneaking suspicions that science might be a cause of decline and devastation, a creeping realization that the knowledge and power to create and improve could likewise be used to destroy. In many ways, in the mid-twentieth century, reality was replacing myth: technological innovation was not quite the golden egg propagandists had presented. For some it seemed more readily comparable to Pandora's box.

The shift in attitudes towards science and technology in the Western world, and in America in particular, was largely the result of war. The military potential of science and technology had been underscored by the tanks and poison gases utilized during the First World War, leading to an unprecedented upsurge of government funding towards research and development through the interwar period and the Second World War. Most notable were the further advancement of sonar and radar systems and the Manhattan Project, under the scientific direction of theoretical physicist,

J. Robert Oppenheimer, to develop the atomic bomb. The critical role of such cutting-edge technologies, especially in the final stages of the conflict, led to the continuation and increase of funding for scientific research from the US government and military following the Allied victory, boosted by the Korean War (1950–53), the launch of Sputnik (1957), the Cuban missile crisis (1962) and the continuing Cold War stand-off with the USSR.[58] The development of 'big science', whereby billion-dollar budgets and projects, research centres and laboratories, staffing levels and equipment reached unparalleled levels through the Second World War and beyond, solidified the mutually dependent relationship between the government, the military and science.

As a consequence science, critics claimed, had become tainted. Once idealized as an objective quest for pure knowledge, it now appeared to be buried deep in the pocket of the government and related defence interests. National security seemed to be setting research agendas: what projects were initiated, what questions were asked, what direction research should take. It was argued that the intellectual freedom of scientists was being constricted as their jobs and attention were focused on state-funded projects – it seemed as if they had been commandeered to direct their research towards destructive rather than constructive ends. This heavily militarized science was becoming socially and politically irresponsible, it was alleged, dedicated to weapons development and destruction rather than curing social and environmental ills and improving the lot of humankind.

Such concerns were compounded by revelations regarding the deleterious effects of pesticides, the burgeoning environmental movement and television coverage from war-torn Vietnam. Audiences around the world, and in America particularly, were shocked by scenes of defoliation, deforestation and crop-destruction, as well as graphic images of starvation, deformity and death, the consequences of herbicides, napalm and Agent Orange – so-called technological advance. In 1968, amid talk of moral blight and the student draft, the anti-war movement reached fever pitch in America. Violent demonstrations and protest marches multiplied, while the role of science, with the government and the military, was a critical focal point. Above all else, the atom bomb loomed large, the ultimate symbol of technological risk and apocalyptic anxiety. As CND went from strength to strength in Britain, science was increasingly associated with danger, risk and questionable political demands in the public mind. Even its remarkable mission, to put a man on the moon, appeared to some to be motivated by international power-play, by a desire to boost national prestige, assert technological superiority and fend off the challenge from America's Communist nemesis, the USSR.

Meanwhile, critiques of science cut deeper than its problematic practical applications: it was simultaneously placed under the microscope on a philosophical level. Notably, there was a new mood of scepticism about scientific method: the objective testing of hypotheses by experimentation and observation had traditionally been viewed as characteristic of the discipline, in keeping with Enlightenment ideals; now, by contrast, critics questioned to what extent the emphasis on cold logic had spawned destructive effects. Such ethical concerns were paralleled by scholarly work which emphasized that science was not a pure and objective search for truth, as commonly supposed, but was, in fact, a complex, ideological and bias-laden activity constrained by time-specific assumptions regarding the natural world.[59]

In the light of such shifts, Shenton's conspiracy theories did not seem so marginal, and allusions to scientific corruption, technological risk and the activities of experts in cahoots with the government and the military were frequently mentioned in the letters he received. In America especially there was heightened public criticism of the role of experts: radicals on the New Left had gone so far as to portray scientists and engineers as the secret rulers of the industrialized Western world. Based on the premise that knowledge is power, society, they alleged, had become a 'technocracy': the balance of power had been overturned. This was a reversal of the idea that science had been co-opted by the government and the military; indeed, it was suggested that the powers-that-be were dependent on science and technology to achieve the dominance they craved. Their ideology was aggressive, ruthless and hidden, their operations based on excessive rationality and cold efficiency, the assumption that biggest is best. Moreover, their hold on society seemed unbreakable: their knowledge was so sophisticated that it was impenetrable to the majority, and science had become compartmentalized to such a degree that even experts in related fields could not grasp the nuances of their colleagues' work. Society was thus ensnared: a monopoly of highly trained specialists had assumed authority in a silent revolution based on their technical know-how.

Meanwhile, Shenton had long voiced fears about a corrupt power-base and the rise of a morally irresponsible technocratic élite. In the wake of the Second World War and the atom bomb, and in the midst of the Cold War, the arms race and Vietnam, this lack of openness was particularly worrying. Technology, with its antiballistic missiles, nuclear, chemical and biological weaponry, seemed to be running out of control, and concerns were raised: what were the experts up to, and was their work for moral ends? In these circumstances, critics who had perceived negative shifts in the relationship between science and society since the war were not liable to concede, while sectors of the scientific community were themselves calling for a transformation in the direction and emphasis of their research.[60]

It is ironic that such criticisms were burgeoning when science was on the brink of what was arguably its greatest moment to date. Across the world the public were agog over the achievements and new frontiers promised by landing a man on the moon; families and friends gathered round the television sets that had beamed terrible images from Vietnam into their homes, eager to see the programmes transmitted from space. In an atmosphere of patriotism and praise, hope and anxiety, the political overtones of a space race between powers locked in competition for military supremacy were evident to those who were concerned. Thus, condemnation and glorification were parallel strands in public perceptions of science and its applications during the period and, at its most extreme, the questioning of science's place in society fed into the proliferation of conspiracy theories and paranoia. From environmental hazards to ufology, writers sought to tackle uncertainty and risk with overarching explanations identifying sources of evil and corruption, and apocalyptic rhetoric, a type of discourse once deemed extreme or marginal, became more evident in the public domain. Through the late 1960s, anxiety about technological destruction mixed with talk of a new age of peace and harmony, with scientists, once cast as heroes, now villains in the cosmic drama.[61]

In the midst of this, Shenton and some of his correspondents implied that the space programme was the culmination of a dark technocratic deception, deliberately designed for truth-destroying ends. One letter, written on 20 January 1969 by a US Peace Corps volunteer in Manila and addressed to the 'Chief FlatEarthman, Dover, England', was packed with such conspiracies:

> With poverty rampant 'from sea to shining sea' I have often asked myself why it is America was spending billions of dollars a year to send a man to the moon. I now feel it is part of an international plot which spans the centuries, perhaps beginning with Galileo, to discredit your society. In Vietnam, our mandate to rule the world was nominally opposed; in similar fashion, your organization's persistent efforts presented a challenge to our mandate to rule outer-space.[62]

If the earth was in fact flat, the writer continued, how could the American government justify the space programme's vast expense? The answer was, it could not – so the Apollo 8 Christmas flight had been undertaken to provide firm 'proof' that the earth was a globe. More disturbing still, the Peace Corps volunteer alleged that Filipinos were being indoctrinated with 'round-earth theory' by US government agents, bankrolled by the CIA. Alleging concern about the situation, the man expressed his eagerness to contribute to the 'fine work' of the IFERS and, with many others, awaited further instructions. In Dover, Shenton faced the difficult task of prioritizing his replies: he existed in a twilight world of a pen-ultimate struggle between truth and lies, and this very issue was reflected in the correspondence he received. Whom to trust? It was a key question – particularly with hundreds of correspondents demanding a response.

One correspondent provided a constant source of inspiration and support, however: Shenton's Westminster associate Gerald St. John-Culdwart. His latest letter lay among the stacks of post from strangers all over the world: he had been seriously ill with pleurisy since catching a chill in Siberia, and had only just recovered sufficiently to report on his trip to observe the solar eclipse of 22 September 1968. All in all, he believed the expedition had provided some 'rather interesting results', which would add still further to the 'overwhelming evidence that the true form of the Earth is flat'.

But his results had been gathered at a price, St. John-Culdwart confessed. Following a difficult journey across the bleak Siberian plains in a biting north-easterly wind, he and his Russian scientist friend, Popov, had finally selected a position near the eclipse site at Yurgamysh to arrange their equipment and observe the event. At the moment of totality, St. John-Culdwart recalled that he was 'shaking all over with excitement': the moment he had imagined for over a year was at hand. He and Popov started their stop-watches, and clicked them off when the sun reappeared. St. John-Culdwart checked the timings: by his calculations the total eclipse had endured for the length of time anticipated if the earth was a globe. He admitted to Shenton

> I was utterly heartbroken; although I had adopted a truly Zetetic attitude, the thought that we would be proved wrong had never really entered my mind. I thought of all the other evidence pointing towards the Earth's being flat, and won-dered if all the time we had been wrong.[63]

However, the next day St. John-Culdwart had happened to

meet a group of astronomers who had been stationed at a different spot, checked their findings and discovered that the duration of totality was just thirty-seven seconds, instead of the forty-three seconds predicted on the basis of a spherical earth. He decided that the error, approximately ten per cent, was surely due to the fact that the earth was flat and motionless. Although he admitted that Popov's results relating to the parallax of the sun were 'rather unsatisfactory' for his purposes, he had already concluded that the true value of the trip lay in the timings of totality he had made. Armed with this new evidence, he told Shenton that the time had arrived to convey the true facts to the public, yet warned there was also a need for caution because

> . . . in this day and age, anyone with ideas differing from those of the majority, [is] liable to ridicule. So many people I speak to regard me as a crack-pot; and now that Apollo VIII has supposedly sent back photographs showing a round Earth, the situation is even worse. They laugh in my face when I try to explain that the curvature is probably produced by distortion, resulting from the use of a wide angle lens. Anyway, we must continue to ignore this, and push on with presenting our facts to those who are willing to listen. I feel sure that one day we shall succeed, and until then we must be prepared to accept the fact that the majority of the population are like sheep being led by a few, who purposely want to conceal the truth.[64]

With this final expression of faith, St. John-Culdwart closed with characteristic optimism, sharing his hope that 1969 would bring some success for the flat-earth cause. He was sure that it would.

By February, Shenton was still contending with the unprecedented publicity generated by Apollo 8, and the subsequent deluge of letters and enquiries from the States, when Patrick Moore arrived at his Dover home with a film crew to interview him for his most important television appearance to date. The premise for the programme was ingenious: Moore intended to explore the kaleidoscope world of 'independent thinkers', various champions of unusual notions, in a series of serious interviews for a prime-time BBC2 documentary. Despite his dislike of the corporation's programming and its on-screen identifier, the famous whirling globe, Shenton was happy to oblige. The interview would be a rare opportunity to explain his views in detail without 'ribbing', the exposure would provide welcome publicity, and he had developed quite a respect for the kindly if, to his mind, misinformed Moore. Yet not all potential participants shared Shenton's enthusiasm for the BBC project. Moore had also been angling to interview Gabrielle Henriet, a correspondent of Shenton's who had invented a similar 'loitering' aircraft and believed that the sky was solid. In his guide to independent thinkers, *Can You Speak Venusian?* (1972), Moore recalled that, despite repeated invitations to take part in the programme, Madame Henriet had declined on the grounds that if she faced a television camera her false teeth would fall out. Moore remarked that he 'would be the last to deny this possibility' (in fact, to him, 'nothing appeared to be more probable'), but he said he remained 'very sorry about it'.[65]

In the wake of Apollo 8, Shenton was feeling the pressure of his ongoing struggle to publicize the flat-earth view. Increasingly weary of hostile and mocking media attention, he expressed his annoyance in letters to the sharp-witted journalist Bernard Levin, and an old Luton friend, Basil Roberts. He began with the subscription issue, for Shenton remained concerned that the IFERS was attracting curious bystanders and sensation-seekers rather than committed flat-earthers willing to pay for the literature they received. As for the workload, he again complained about the deluge of interviews and correspondence, which had resulted in 'distorted, slanted and lying reporting . . . in many newspapers all over the world'. The outlook was not promising: during a lengthy period of dealing with interviewers, by telephone and in person, Shenton moaned that not one member of the IFERS had provided him with 'the slightest encouragement', despite 'knowing that the N.A.S.A. people and all scientists were rejoicing at our seeming defeat by the moon orbit'. In early 1969, running the society seemed a thankless task. Yet Shenton concluded his unappealing account of his position by asking if Roberts would visit to discuss taking over the society.[66] Ten days later, he informed Roberts that he had suffered another serious setback in health and had been terrified that he would not 'pull through'. Following urgent treatment, and with Lillian's constant care, he could now sleep at night, but did not feel sufficiently strong to meet Roberts in person.

While Shenton was undoubtedly a fatalistic character, his outlook appeared a little brighter during March. The month opened with a rambling letter from Gerald St. John-Culdwart tearing into a pamphlet issued by the Admiralty on the last total solar eclipse in England, on 30 June 1954, as a devious work of official misinformation:

> I find it very distressing that documents like this, giving information, which I am sure the persons who compiled it knew to be false, can be widely distributed without anyone questioning it. Surely some of the scientists realise the falsity of it all? I realise that in official circles they are probably withholding the facts, for fear of public panic; but would there be any panic? In my opinion people would prefer to learn the truth, and would feel more secure knowing that the Earth is at rest, rather than spinning through space at 18 miles per second.[67]

St. John-Culdwart was correct in one respect: Shenton's correspondents often mentioned finding the notion of a motionless planet more palatable than that of a flat earth. But despite the prevalence of doubt, St. John-Culdwart wrote, 'Our goal may not be so remote as we feared', although there remained an evident need to 'convey our knowledge to *more* people'. He suggested a host of tactics to tackle the matter: appointing IFERS members to investigate specific aspects of the flat-earth issue – eclipses, rotation and so forth – to muster arguments for the campaign, a committee to collate and report on this research, and a regular half-guinea subscription, half of which could be set aside to cover publication costs. St. John-Culdwart was sure fellow members would welcome his proposals and, together, they 'might be able to convince this indoctrinated earth of the truth'. There was power in numbers, he reminded Shenton, and provided members took an active interest in the society's work, the truth would eventually prevail. Admittedly, overcoming widespread opposition to flat-earth theory would need 'a great deal of determination and strength', but St. John-Culdwart was willing 'to take the ridicule and insults' in the belief that victory was merely a matter of time. Fired with passion for the cause, he ended by asking Shenton if he could pay him a visit during his Easter holiday in nearby Deal. Shenton greeted the proposal with enthusiasm and, never having met previously, the two men set a date: Wednesday, 9 April. On the day, St. John-Culdwart confided, he would have much to reveal, and with the upcoming Patrick Moore programme and regular newspaper coverage, he remained convinced that 1969 would be 'a great year for the Society'.

Indeed, the initial wave of criticism that accompanied the Apollo 8 mission seemed to be subsiding, and as Shenton recovered from his health scare, he received a number of sympathetic letters from New Zealand, New Jersey, Pennsylvania and beyond. Among them were communications from a Chicago lawyer, a Boeing Aircraft Company employee working on the Apollo space programme, doctors from Pittsburgh and Edmonton, and an airport worker from Florida. Even a group of physicians working at the US Naval Submarine Base in New London, Connecticut, wrote expressing their 'alarm' at the 'terrifying hoax being perpetrated by governments and press alike'.[68]

But Shenton remained depressed by his quixotic quest. Over recent months his heart trouble had impeded his lecturing activities but he did not want the public to suspect that, after 'twelve years' hard slog', he was dodging the difficult position regarding Apollo 8 and the photographs from space. It was just an 'awful job', he wrote, encouraging people to 're-think'. Unfortunately, at this point, he was stunned by more unwelcome news: Basil Roberts, his long-standing Luton correspondent and fellow flat-earth believer, had committed suicide with an overdose of sleeping tablets in mid-March. There was more tragic news to follow. On Wednesday, 9 April, Shenton was eagerly anticipating the long-awaited visit from Gerald St. John-Culdwart when he opened an ordinary-looking letter that had arrived in the morning post. In it St. John-Culdwart's nephew reported that his uncle had been killed in a car accident on the journey to begin his holiday in Deal. Shenton was devastated and wrote back immediately, expressing his deep sadness at the loss of one with similar ideas concerning the 'real structure of the Earth'. It was a 'terrible affair', he said, which would concern him for a long time to come.

In the wake of the death of two of his closest associates, Shenton continued to plough on in his usual way, considering invitations to lecture at Oxford University Scientific Society and Maidstone Grammar School, and a request to establish an IFERS branch in San Luis Obispo, California. Meanwhile, questionnaires and letters from schoolchildren in Florida, Alabama, Chicago and London flooded in, alongside a lengthy questionnaire from a class of Swedish schoolchildren posing familiar questions. What is on the other side of your supposedly flat earth? Which countries lie at the edge of the disc? Has anybody looked over the edge?[69]

Shenton found contending with such repetitive questioning tedious and time-consuming; increasingly, he seemed to spend his life metaphorically 'on the edge', constantly justifying heterodox views and coping with the impact of public ridicule on a worldwide scale. In mid-April, he started to have blackouts and was compelled to cancel another round of interviews, informing one journalist that he was 'glad to have any pulse left'. The missed media opportunity was a pity, he continued, for he was planning to stress the ease with which the human mind can be deceived. However, he had little option but to cancel. 'Imprisoned in his bedroom', he could not reach his typewriter, let alone cope with intensive cross-questioning from incredulous journalists looking for a scoop.[70] He told one *Birmingham Evening Echo* reporter that he had suffered a couple of strokes, been in and out of hospital and was increasingly 'world-weary'. Ground down by the sniggers, he admitted, 'You walk down the street, you pop into the pub ... and you get odd looks ... The pals you used to have shy off you. But you've got to go on speaking about it ... It's getting more than I can manage. I'm looking for someone younger to take over.'[71]

Indeed, during April and May all Shenton could do was sit in his bed and read the stacks of mail that arrived at his home. Some correspondents were more supportive than others, yet even a detractor reflected politely that 'in a world where everything is in a set pattern, it is nice to know that there are still people who doubt'.[72] Among them was a seventy-eight-year-old Dutch immigrant in Ontario, who informed Shenton that he had believed the earth to be flat for forty years. Despite having always been ridiculed, he said that 'No-one can tell me any different because a round ball is nuts.' In spidery scrawl, the man insisted he 'had all [his] faculties together', notwithstanding advancing age, and now that he had found the IFERS he was keen to assist in the campaign.[73]

Another Canadian citizen, a retired air-force pilot, chimed in with a similar offer of support and the suggestion that 'A number of astronauts have obviously been brainwashed and could probably be persuaded to become honorary members of [your] society.'[74] While Shenton was undoubtedly pleased to receive such letters and the inspiration they evoked, he still lacked the strength to reply to the vast majority. Frequently he received puzzled follow-up enquiries, especially from those who had enclosed money, and some correspondents, such as the chairman of the Crouch End Young Conservatives, lost their patience and fired off stinging criticisms of Shenton's alleged bad manners. Shenton did not generally reply to these letters, either: resigned to criticism and settled in martyrdom, he said all he wanted was a quiet life. This in itself was a tragedy: a peaceful existence was inconsistent with being the leading public proponent of the world's most infamous erroneous idea. Something had to give.

By summer 1969, another more pressing factor was working against Shenton – that of time. A quiet life was an unattainable prospect in the run-up to the Apollo 11 moon landing, scheduled for 20 July. Well within John F. Kennedy's end-of-the-decade deadline, the three-man crew of Neil Armstrong, Buzz Aldrin and Mike Collins had won the task of flying to the moon and landing the spidery lunar module on its surface. The moment was of the utmost historical, technological and political significance, and Armstrong's eventual 'giant leap for mankind' fed a media frenzy of vast proportions long in advance of the mission's official launch date.

All over the world editors rummaged for original and entertaining Apollo news stories, and the human-interest angle of the International Flat Earth Research Society offered the perfect lightweight accompaniment to meaty, hard-hitting coverage of NASA-related events. Consequently, the pressure on Shenton was intense, and sometimes became too much for him to bear. On 28 June, he felt compelled to write a private letter of apology to journalists at three leading papers, the *Los Angeles Times*, the London *Times* and the *Minneapolis Tribune*, after losing his temper and hanging up on them during an interview. Filled with remorse after this embarrassing *faux pas*, Shenton admitted he 'certainly paid the penalty of stress afterwards'. His most critical error, he believed, had been his attempt to reply to the journalists' 'very insistent questions' regarding the size, shape and composition of the moon, when in fact, questions about the sun and moon 'do not apply to the "Flat Earthist" [as they] do not relate [to] or fall within his field of investigation or Zetetic searching'. From Shenton's perspective, his mistake only proved he had 'not learnt his oft-repeated lesson . . . not to be sidetracked or diverted by questions related . . . to heavenly bodies'. Circumventing the critical point that evidence for the rotundity and rotation of the earth (or otherwise) is based in part on the appearance, position and movement of these other heavenly bodies, he concluded his argument with the contention, 'I should have maintained that our field of research lies in the tangible evidences of Plane, Stable, Unmoving Earth.'

Despite the controversy, Shenton remained keen to deal with the media, and through July he continued to field calls from reporters around the world. Now he was careful to foster an image of knowing calm, and a run of stories appeared in the American papers reporting that the moon landing was causing little disturbance at Shenton's home in the quiet Dover suburb of River. But that is not to imply Shenton deceived reporters. He was perfectly candid about the terrible impact of recent Apollo missions on the IFERS, whose ranks now numbered less than eighty. 'I just can't seem to get through to the young people these days,' he griped. 'Science teachers have deluded them from the cradle.' Nevertheless, according to the *San Jose Mercury*, Shenton remained unshaken in his belief that 'The United States is practicing a great deception [using fake earth photographs from space] to simulate the Earth as round.' This, and the supposed moon landing, remained a 'great delusion for the rest of the world'.[75]

Such anxieties about education and misinformation were pressing topics for Shenton sympathizers in the days of the first moon landing. On 14 July, a group of friends from West Lafayette, Indiana, expressed their concern for IFERS prospects 'as a result of this country's space program' and requested further information on the grounds that 'We do not like to have our children taught in school from an early age that the Earth is round without anyone being able to present the other side of the argument.'[76] Their concerns echoed creationist fears about the teaching of evolution, and were backed by an ex-army commander, who wrote to Shenton from the office of the Registrar General of Georgia on 10 July. He likewise branded science teachers the principal culprits in 'destroying our fundamental beliefs' and reminisced that where he was born and raised, in the Southern Appalachian Mountains, belief in a flat earth was commonly held. The correspondent was sure that, as long as membership dues were within their reach, many mountain people would be keen to join the society to safeguard their folk-belief, and as a mark of his faith he was happy to offer his services in recruiting new members from his native district. He believed the prospects looked promising. Even though, on Sunday, 20 July, an estimated 600 million people, a fifth of the world's

population, had watched the grainy images of Armstrong's landmark steps, the man described in a subsequent letter how his
brother, who resided in their hometown, had found only five
people who believed that Apollo 11 had 'actually reached the moon
and the crew had actually walked on it'. For many the event
seemed too amazing to be true and the ex-army officer was sure he
could exploit such scepticism in his home state.[77]

Through the late summer of 1969 the stream of seemingly
supportive mail to Shenton continued unabated and, in keeping
with previous letters, much of it marked a backlash against the
Apollo 11 moon landing. The idea that the event was staged, or
that the photographs were faked, was not uncommon, but to
accept this as proof of a flat earth was quite another. However,
many of Shenton's correspondents made, or claimed to have made,
this 'giant leap' in logic, and flat-earth believers became some of
the most vociferous proponents of the moon-landing hoax. Yet
despite the genuine and pseudo-genuine letters of support, the
moon landing and subsequent publicity affected Shenton's health.
He became more and more fragile and felt increasingly persecuted.
He told one correspondent that he had finally collapsed in late
August and this latest setback had 'curtailed the small opposition
I maintain against the seeming all-powerfulness of accepted science'. Although Shenton believed the 'Apollo 11 moon venture'
showed 'an astounding, elementary fault in concept', he was not
well enough to attempt to rectify the momentous mistake.

Naturally, correspondents across the world were unaware of
his condition, and letters – the good, the bad, and the ugly –
continued to arrive at his Dover home. The Halifax branch of the
British Association of Young Scientists and Manchester University's environmental-design department both invited him to debate
the flat-earth question, a Utah man congratulated him on his
'imagination and good sense of humor' for allegedly offering a
special-effects award to the Apollo astronauts, and several incredulous correspondents wrote asking, 'Does your society really exist?'

Through all of this, ideas about the shape, size and movement of the earth were an enduring source of fascination for children, and teachers were keen to feed their curiosity. Many saw the educational potential of the unorthodox idea and wrote for pamphlets to encourage classroom discussion, while a minority claimed to be teaching their classes that the earth was flat and that they required material with which to illustrate the point. One Bromsgrove schoolboy even wrote:

> Until recently I believed the Earth was round, but my teacher at school, himself a profound believer that the world is flat [who] wrote to you some time ago, gave the class a talk on the subject. I am now, I am pleased to say, a believer that the world is flat.[78]

By whatever means, it appears that Shenton had the occasional success in attracting interested parties, and in 1969 he finally found the successor he had been searching for since William Mill's untimely death in 1959. During the past few years Ellis Hillman, geologist, principal lecturer in environmental studies at the North East London Polytechnic, and left-wing Greater London Council (GLC) member, had displayed an avid interest in the IFERS. Shenton was delighted: as a scientist, academic and politician with a base at County Hall, Hillman, a well-known and well-connected London character with posts on numerous high-level public committees, councils and boards, would lend a reputable and authoritative edge to the society and its beliefs. But, Shenton's wife, Lillian, was not so sure. Although she was, at best, only semi-convinced of the earth's flatness, and this might have been a diplomatic manoeuvre for the good of her marriage, she was naturally protective of her husband's efforts, and doubtful of Hillman's professed beliefs. Yet blinded by Hillman's status and renewed hope, Shenton overlooked her suspicions and subsequently invited Hillman to become president of the IFERS. Initially Hillman, who was frank about never having believed the earth to be flat, was reluctant to accept Shenton's offer and recalled contacting Patrick Moore to ask his advice. According to Hillman,

Moore was encouraging: 'For God's sake, keep it going,' he allegedly exclaimed, 'we must have heretical people in the world of astronomy.'[79] Besides Moore's enthusiasm, there was a second persuasive factor: at the time Hillman was planning a postgraduate course on the development of ideas connected with the shape of the earth and he believed it would assist his academic research to accept Shenton's offer.[80]

Regardless of his new recruit, Shenton continued to run the society single-handedly from his spare room. Throughout the winter, correspondence of all descriptions thudded through his letterbox, and among the more interesting was an irate letter from a Chester man written on 3 December 1969. The correspondent was straightforward: he sought to join the IFERS because he was tired of 'all this American "Apollo" rubbish going on'. Evidently disturbed, he continued,

> These vulgar buffoons are evidently trying to fool us just for the prestige they can gain for it. The film they produced of their first Moon-shot was obviously produced in a studio. I find it damned nauseating that people, particularly British people, should value so highly the worthless pebbles these foul liars claim to have brought back from the moon. For the latest 'voyage' to the Moon [Apollo 12], these Americans have committed themselves to a one mile walk on it. Not having a studio big enough, they pretended that their television cameras had broken down. This virtually proves that the Americans are lying. It is too much of a coincidence to be true, especially backed up by the fact that they also claim that their film cameras were out of focus to explain the lack of films to make their exploits more believable.[81]

The letter was suffused with xenophobia, British cultural supremacy, circumstantial evidence and conspiracy theories, yet it continued with praise for the British, for at least

> ... some of us are not fooled by this childish trash. We are still firm in our belief, in absence of evidence to the contrary,

that the Earth is flat and that all photographs, supposed to prove that it is spherical are patently counterfeit.[82]

For this correspondent, the existence of the IFERS proved that in reality England, rather than America, was the land of the free. This was just the type of rhetoric Shenton frequently articulated, although not all such unorthodox opinions met with his approval. As the world's most famous proponent of the world's most infamous alternative idea, he was an established target for a diversity of theories: people wrote to him about topics from LSD to astrology, alien abductions to UFOs, and on rare occasions a combination of the above. Among such colourful letters he received were several from a self-confessed Ottawa schizophrenic, who asked about the likelihood of being hypnotized by an unknown force when alone in a room, whether humans could shape-shift through the use of chemicals, and if it was possible to communicate through jewellery, such as chain bracelets.

If this was not sufficiently unconventional, there were several letters from an Illinois sci-fi writer, who always signed off with threats to sue if Shenton dared steal his novel ideas. His latest letter, months after Apollo 11, was similar in tone, packed with questionable statements such as 'our solar system is a huge Magnesium atom', and he had also concocted a startling new theory about the moon landing. Allegedly the 'biggest put-on since Gengiss Kahn's [sic] funeral', the writer believed that five seconds before lift-off the astronauts had been dropped in a capsule to a mile inside the earth. As the empty Apollo rocket shot into space, the astronauts plunged down, landed on and explored a fake moon, from which point they were transported underground to the Pacific Ocean, taken to the splashdown area by submarine and fastened to the ocean floor. Meanwhile, Apollo 11 – which was filled with lead, incidentally – came falling through the sky from space. When it hit the water and sank, the capsule containing the astronauts was brought simultaneously to the ocean surface. If anything, the hypothesis was certainly extraordinary; indeed, it seemed that for this sci-fi writer, life had begun to imitate art and even Shenton, exceptionally broad-minded about alternative theories, did not grace his letter with a reply.

By Christmas 1969, Shenton had been in his sick room for three months, but still managed to 'straddle a typewriter across his body', as he informed one correspondent, and compose the odd letter. The outlook was rather dismal: unlike other zetetics he had no close confidant, society or community for shared discussion, moral support or the reinforcement of his beliefs. It seems that he had little face-to-face interaction with other flat-earth believers, and the inaugural meeting of the IFERS in November 1959 remained the only gathering on record of the organization's members. Although Shenton's apocalyptic rhetoric never extended to descriptions of an eventual state of perfection or hopes of transformative social change, during 1969 and 1970 his tendency towards negativity spiralled, leading him to characterize himself as a 'watcher' for End-times.[83] Isolated, depressed and cooped up with thoughts of conspiracy, IFERS letters became both a lifeline and an outlet, and Shenton recounted his woes at great length to his correspondents – enemies, critics and complete strangers alike. Shenton had once been the epitome of gentle politeness, but years of trials and tribulations, self-inflicted and otherwise, meant that now even the simplest requests could elicit a strained response. In one particularly graphic letter to a Canadian clergyman, he claimed that he had 'not been to bed for about a year'. A life 'spent propped up between chairs' was his lot, while 'whacking great ulcers down each shinbone' had prevented him 'sleeping, thinking or reading'. Even worse, he wrote, their 'vile stench and the weeping makes me wish to hide away; had it not been for my wife's constant attention, I don't think I could have stuck it!'[84]

There can be little doubt that these were sour times for Shenton and his flagging campaign, and his correspondence seeped negativity as a mark of his shift in mood. He increasingly deplored the whole 'self-poisoned world', which was 'liable to kill itself' with 'science advancement in technology'. The atom bomb loomed large in his Armageddon, and in the face of suicide through man-made science, he urged his supporters to make a 'thrusting stand to really shake the theoretical world'.[85]

For Shenton the ultimate seriousness of the issue justified and shaped his rhetoric, which, like much End-time and conspiracy theorizing, was designed to persuade the public through drama, fear and shock. Although his success was limited – due, one supposes, to the high falsifiability of the flat-earth idea – he still believed he had support scattered across the world.[86] Whether or not such correspondents were simply masquerading as flat-earth sympathizers, no one except the writers themselves will ever know for certain. Yet for every supportive letter-writer – bogus or real – there were numerous willing to jab Shenton with insults, side-swipes and snubs. Such merciless remarks ranged from 'My mother feels that you are a ridiculous bunch of holdouts' to 'I fully realize that many people have called you quacks and attention-getters', and among this, there was pity in abundance. Many people wrote to Shenton seeking to explain aspects of astronomy, geometry and gravity, enclosing photographs of the earth from space or diagrams painstakingly constructed with compasses and set-squares to add emphasis and clarity to their well-proven points. Caring correspondents frequently wrote of how they were only attempting to be of assistance and Shenton should not take offence, while one particularly polite British housewife ventured, 'I am rather doubtful about your "pit" but perhaps the details of the universe are allowed to be personal.' Echoing her confusion, one Surrey teenager complained, in late 1970, that in the days of 'computers and astronauts and things it seems difficult to know what to believe'.[87]

While the letters addressed to Shenton were a motley assortment, so too were the lecture invitations – from the Guide and Scout Group at Oxford University to the philosophy department at the University of Leeds: flat-earth beliefs continued to fascinate a remarkable range of potential audiences. But however intriguing the letters or tempting the invitations, by the autumn of 1970 IFERS post was piling up unanswered at the society's headquarters in Lewisham Road. Shenton's health was failing fast, and he was not well enough to field the endless stream of enquiries. His condition deteriorated through the winter of 1970–71, and he died in a Folkestone hospital on 2 March 1971 at the age of sixty-eight. His death certificate listed congestive cardiac failure and hypertensive heart disease as the causes of his demise, conditions doubtless exacerbated by the strain of persistent campaigning through the ground-breaking era of space flight.

Despite his financial troubles, a constant constraint on his campaign, Shenton left his wife reasonably well provided for, and she continued to open IFERS mail. One of the letters she received that March was particularly poignant. From the self-styled 'only flat-earther in Montgomery, Alabama', it expressed his feelings of isolation and despair due to the lack of like-minded people in his hometown. He had attempted to tackle the issue head-on by giving a public lecture about his beliefs but the audience had jeered and mocked. The situation had intensified when he found it impossible to deflect their persistent questioning, and matters had been further complicated when he tried to research the subject and discovered that his local library held no information about the earth being flat. Shenton seemed to be the one person in the world who could provide advice and understanding, alongside the zetetic books and pamphlets the writer said he so 'urgently required'.[88]

As Shenton had died, this literature was never sent to Montgomery, Alabama, but assistance was on the horizon. In addition to Ellis Hillman, a second potential successor was keen to inherit Shenton's IFERS, and he was based much closer to the letter-writer's Alabama home.

Chapter Eight

THE VIEW FROM THE EDGE

In the Middle Ages people believed that the earth was flat, for which they had at least the evidence of their senses:[1] we believe it to be round, not because as many as one per cent of us could give the physical reasons for so quaint a belief, but because modern science has convinced us that nothing that is obvious is true, and that everything that is magical, improbable, extraordinary, gigantic, microscopic, heartless, or outrageous is scientific.

GEORGE BERNARD SHAW,
Saint Joan (1923)

Now, Shaw is exaggerating, but there is something in what he says, and the question is worth following up, for the sake of the light it throws on modern knowledge. Just why do we believe that the earth is round? I am not speaking of the few thousand astronomers, geographers and so forth who could give ocular proof . . . but of the ordinary newspaper-reading citizen, such as you or me.

GEORGE ORWELL, 'As I Please,'
Tribune, 27 December 1946

'We're on the level'
Motto of the Flat Earth Society of Canada

AMONG THE STACKS OF correspondence that lay unanswered in Shenton's study at the time of his death was a standard airmail letter from Canada. Dated November 1970 it expressed forty-three-year-old philosophy professor Leo Ferrari's interest in Shenton's IFERS and offered his fervent support for the campaign:

> People are placing far too much blind faith in science these days and not enough in their own senses. They have made a veritable god out of the new sciences. I am most interested in valiant efforts to resist this flood of blind faith aimed at idolizing science. In addition, the flatness of the Earth is an evident truth derived immediately from basic experience.[2]

Ferrari hoped his enthusiastic comments would elicit a reply from Shenton, but he did not receive a word in response. As he had been expecting a letter packed with gratitude and strongly worded opinions, he wondered whether the English society was still in existence, but he decided to plough on with his plan nevertheless. Over the past few weeks he and his friends had been laying foundations for a Canadian flat-earth society and they were not prepared to allow a rival organization to prevent them realizing their goal.

At first glance, establishing a flat-earth society seemed an unusual pastime for a middle-aged professor of philosophy with a BSc in science, but Ferrari was no ordinary academic. His curriculum vitae testifies that even his basic career path was unorthodox for a man of his profession. Born in Australia in 1927, he had worked as an industrial chemist for seven years after completing his degree. He soon discovered, however, that a technical career did not fulfil his artistic and literary inclinations and, in search of an outlet, he started to teach himself philosophy at night. Captivated by the subject, he decided to move to Canada to study for an MPhil, followed by a Ph.D. The qualifications led to various teaching posts, and in 1961 he was finally appointed professor of philosophy at St Thomas's University, New Brunswick. The position proved to be a job for life, and as the years passed, Ferrari made quite an impression on those he met through his role. Wild-haired and wacky, he had an innate love of the extraordinary, the unconventional and the ridiculous, and was renowned among his friends for his ingenuity, his sense of humour and his talent for making beer. Blessed with natural exuberance, a novel idea or a good joke could inspire him to leap around, plunge into handstands or roll on the floor with laughter, all of which provided light relief from the pressures of university teaching and his scholarly study of the life and work of St Augustine. Indeed, more than anything, Ferrari feared taking life too seriously and he viewed the dry pretensions of academia with disdain. In letters he grumbled that he could not be constantly respectable, and that even living in a college town was sometimes restrictive and dull. By way of diver-sion, he spent his free time with a like-minded coterie of Bohemian friends and it was from this group that, in November 1970, the Flat Earth Society of Canada (FESC) was born.

Boisterous and talented, the circle revolved round one of Canada's most renowned literary figures, the hard-drinking Nova Scotian poet, Alden Nowlan. Best known for his introspective poetry collections, chronicling the effects of regional environment and the experiences of ordinary people, Nowlan wrote about what he knew. Born in rural poverty in the backwoods town of Stanley in 1933, he had left school around the age of eleven, and worked as a labourer in a sawmill among a succession of such jobs. Largely self-educated, he was later employed as a reporter on various provincial newspapers, and it was during this period that he clawed his way to acclaim as a writer. In 1967, he received the Canadian literary establishment's ultimate accolade, winning the Governor General's Award for his poetry collection *Bread, Wine and Salt*. But prizes and praise were not enough for Nowlan: he longed for sufficient money to write full-time. His chance came in 1968, when at the age of thirty-five he was appointed the Univer-sity of New Brunswick's first 'writer-in-residence', a semi-academic position that provided a regular income, time to write and a little house half-way up a hill on the edge of the campus. Delighted with the opportunity, he moved with his wife, Claudine, to the neat university town of Fredericton to take up the post.

The literary and social scene there was soon enlivened by his presence. Throughout 1969 and 1970, Nowlan's house at 676 Windsor Street, known as 'Windsor Castle', became *the* place to be for a colourful stream of aspiring young writers and poets, while Nowlan was only too happy to hold court in his den, fortified by gin. In a sedate town, characterized by its elm-lined avenues and elegant government buildings, Windsor Castle represented another world, where wild parties and deep discussions endured long into the night.[3]

During 1970, Leo Ferrari, professor at nearby St Thomas's University, was a welcome face at these bacchanalian gatherings, along with young author and poet, Raymond Fraser. An enthusiastic, gifted twenty-year-old, Fraser was well suited to the vibrant, creative scene. Dedicated to his craft and fond of a good time, he had first met Nowlan in 1961 after writing him a fan letter; despite the age difference, the men soon discovered they had much in common. Most notably, they both hailed from rural working-class backgrounds and had devoted their lives to writing rather than passing judgement on writing, the latter branded a pastime fit for phoneys and academics.[4]

In the 1960s a close bond developed between these 'drinking and thinking' buddies, and it was during one of their spirited *tête-à-tête* on the night of 8 November 1970 that the conversation turned to philosophical questions of science and religion. What intrigued them was the question of authority: in past centuries people had been willing to believe wholeheartedly in religion, yet now they seemed blindly to accept the truth as taught by science. Why weren't people willing to think for themselves? How come they weren't capable of carving out their own world-view based on individual experience? Surely knowledge should spring first and foremost from the evidence of one's own senses? Wasn't the shape of the earth a perfect example of such a phenomenon?[5] 'Leo,' Nowlan called to Ferrari, who was talking in the corner, 'do you believe the earth is flat?' With characteristic quick wit, Ferrari hit back, 'Sure the Earth is flat – any fool can see that.'

The phrase had appeal and the idea took hold – Ferrari had been rehashing with Fraser's wife, Sharon, the recent collapse of his marriage and was especially glad of the distraction from his woes.[6] Thus inspired, the friends started to plot: they could form a society, even publish their own journal if they wanted, while making a worthwhile point and irritating the Establishment by questioning one of the cornerstones of human knowledge. The flat-earth idea proved irresistible: it was a non-conformist cause, packed with potential for controversy and jokes, and their enthusiasm for an organized group did not falter in the cold light of day. Over the next few weeks Ferrari, Fraser and Nowlan set about establishing the FESC, a tongue-in-cheek idea with some serious philosophical implications.

Their first task was to allocate positions of responsibility and decide the society's structure, tactics and aims. Leo Ferrari was the obvious choice for president: his profession, his doctorate and his knowledge of ancient and medieval philosophy would lend the society an authoritative edge, essential if the group was to claim credibility and succeed in making its point. Meanwhile Nowlan and Fraser tussled over their respective roles. Nowlan plumped for 'symposiarch', meaning the master of a public debate or discussion, while Fraser, claiming envy of his friend's impressive title, settled for the more commonplace chairman of the executive.

Their attention then turned to organizing the executive, with a constitution, manifesto and plan of attack. During the winter of 1970–71 the writers canvassed their friends. The Ontario poet Gwendolyn MacEwen was duly appointed an executive vice president over a few drinks in a Toronto bar, much to the amusement of Nowlan, who speculated about what the eavesdroppers at the next table said when they departed.[7] Closer to home, Claudine Nowlan was appointed the society's executive assistant while Fraser's wife, Sharon, was named executive vice president. Old drinking buddies from Windsor Castle proved similarly keen to become involved, and two promising young poets, Jim Stewart and Al Pittman, were assigned pivotal roles. 'I have come to believe that the concept of a round earth was formulated for no other reason than to give a few so-called "thinkers" something to do,' Stewart blustered, in a letter to the society. There could be little doubt that the round-earth idea was 'based on myth and ignorance', he continued, as the fallacies of globularist 'reasoning' were striking:

1. They cite Columbus as having proved the roundness of the earth – they forget, gentlemen, that he sought to reach India by sailing west; they also seem to forget that he failed miserably.

2. If, as they contend, the earth is round there can be no up or down. Who gave them the right to place North America above South America? Why don't they leave cartography to those who know what they're doing?

3. They contend, gentlemen, that the earth rotates about an *IMAGINARY* axis between the POLES. Their absurdity is self-evident.

4. Their original theory has been modified so many times that I submit to you, that for all THEY know, the earth changes shape with the seasons! They went from a round earth to a pear-shaped earth to one SLIGHTLY FLATTENED at the POLES. I suspect the latter stems from guilty conscience.

5. The whole issue has given them licence to speculate ad nauseam as to what lies at the CENTRE of the earth. As children they probably dissected a golf ball and now expect the earth to contain a small rubber ball or a bag of white paint![8]

As a mark of Stewart's loyalty, he was appointed chancellor of the FESC.

Subsequently, 'educator, poet, short story writer, gourmet, folk singer and raconteur', Al Pittman, was awarded an official title of his own. As he lived on Fogo Island, an isolated rocky outcrop twelve miles from the north-eastern coast of Newfoundland, he turned this into a selling point for the campaign. Wasn't it near the edge of the earth, which lay somewhere between Fogo and Greenland? he queried. Surely, since there was nothing but an icy stretch of the Atlantic between him and the edge, he should qualify for an executive position in the Flat Earth Society? He could even keep a lookout and take tallies of the unfortunate few who disappeared over the rim.[9] Back in Fredericton, the suggestion was appealing and Nowlan announced to members that the FESC had taken the 'very practical step' of appointing Pittman 'official representative at the edge of the earth'. 'Mr. Pittman has stationed himself on a rockbound island between Newfoundland and Greenland where he can observe the edge from his kitchen window,' he reported in letters, while it was rumoured that when the representative at the edge was fortified by a potent brand of Newfoundland rum, called Screech, he could muster the courage to peer over the edge into the 'Abysmal Chasm'. On such occasions, Nowlan went on, Pittman could 'no doubt . . . hear the piteous cries of those poor souls who – martyred by an inflexible science – have become imprisoned in those abysmal depths'. The United Nations had a responsibility here, Nowlan declared: it should fence the edge and lower ropes into the chasm to enable any survivors to clamber back to earth.[10]

With their sights set supposedly on the world stage, Ferrari, Fraser and Nowlan were keen to enlist foreign members in the society and add a truly international edge to the campaign. In late 1970, Keath Fraser, a Canadian acquaintance based in London, was appointed plenipotentiary to the United Kingdom, while Nicholas Catanoy, a Romanian radiologist and writer living in Paris, became plenipotentiary in France. The next stage, Nowlan informed Raymond Fraser, was to locate a flat-earth believer in New York. That should be a simple task, he joked, and with men in London, Paris and New York, he thought that the society would sound as 'impressive as hell'.[11] While they contemplated the matter, Nowlan and Fraser considered the tone the campaign should take.

In mid-November, Nowlan wrote to Fraser, who was living in Montréal, that the public persona adopted by the society was vital and he wanted to guard against it appearing too slapstick. 'It would be best,' he told Fraser, 'if some people took us literally – as they do the British group.' Better still, he added, 'if they're not quite sure whether they should take us literally or not'. Under these circumstances, he thought the society's journal, which they had elected to call the *Official Organ*, should strike a careful balance between the serious and the absurd. Perhaps it should even sound a bit pompous, with 'footnotes and all that academic jazz'? Nowlan suggested. 'Leo would be good at doing something like that,' he continued, adding, 'Sometimes I suspect that he really and truly does believe the world is flat.'

Whatever the beliefs held by their president, Nowlan had high hopes for the crusade. Conferences with pretentious papers, appearances on national television, coverage in newspapers, even a mention in *Time* magazine: anything seemed possible for a group that propounded the world's most infamous alternative idea. But none of this would be possible without the essential veneer of solemnity, Nowlan realized, although he was quick to add that the society's members should not actually *be* serious about the flatness of the earth.[12] Mindful of this approach, Nowlan informed Fraser that he had drafted some aims for the group:

1. To restore man's confidence in the validity of his own perceptions. For more than fifteen hundred years man has been blinded by metaphysics and coerced into denying the evidence of his senses. The Flat Earth Society stands for renewed faith in the veracity of sense experience.
2. To combat the fallacious deification of the sphere which, ever since Galileo dramatized the heresies of Copernicus, has thwarted Western thought.
3. To spearhead man's escape from his metaphysical and geometrical prison by asserting unequivocally that all science, like all philosophy and all religion, is essentially

> sacramental and, therefore, all reality, as man verbalizes it,
> is ultimately metaphorical.[13]

'But maybe you'll think that's _too_ solemn?' he asked.

Writing from Montréal, Fraser was in full agreement with his approach. 'The only way the FES[C] can realize its full potential,' he replied, was for it to be a serious organization. 'You can ruin your own joke by laughing at it,' he continued, so the only option was to 'put on a solemn and sincere face to the world'. As for the _Official Organ_, Fraser thought that, in keeping with the earnest façade, it had to be a 'grave and even pedantic vehicle', while grand-sounding titles, such as 'plenipotentiary', would add yet more weight to their cause. He believed they were off to a storming start with Nowlan's aims, however. 'A few more like that,' Fraser commented drily, 'and we all will believe the Earth is flat.'[14]

The FESC continued to keep the collective spirits of its founders afloat during New Year 1971, and provided a much-needed boost in certain quarters. Nowlan, in particular, was feeling overworked and he complained to Fraser that he was 'as busy as a grasshopper in a frying pan' with 'horseshit stuff', like writing a speech for a conference in Toronto and compiling newspaper columns. Amid all of this, he was missing his friends and desperate to find time to compile a new book of poems. In the meantime, Fraser offered solace.[15] 'We're good men,' he reminded Nowlan, 'and there's nothing better than when we're having a drink together.' 'I wish we could do it often,' he went on, 'and some day we will when one of us gets rich and we can all hang out in a magnificent villa of 40 rooms on a hill by the sea.' Until then, he had decided to plough on with his novel-in-progress, and advised Nowlan to concentrate on his next, long-awaited book. 'Everyone is high on your poems,' he told Nowlan, 'and rightly so.'[16]

While he attempted to bridge the distance from his friends in Fredericton, Fraser sought to disseminate the flat-earth message among his contacts in Montréal. In late February 1971, his proselytizing led to his first public appearance for the FESC on a local Canadian Broadcasting Corporation (CBC) radio show, *Duncan's Dregs*. There, he insisted that the footage of the moon from the 1969 landing bore more resemblance to the rugged province of Newfoundland than the faraway lunar surface. He subsequently reported to Nowlan that the interview 'came off OK'. He had kept the all-important tone right – 'very serious, very rational and quietly earnest' – and must have seemed relatively convincing because the radio station received several serious enquiries from listeners after the programme was aired.[17]

In the interim, 350 miles away in Fredericton, Nowlan and Ferrari had been compiling the first issue of the *Official Organ*, along with a manifesto and application forms. These specified that the organization was open to 'persons of integrity' with a serious commitment to the issues at stake. To ensure that this remained the case a procedure was established whereby prospective members were required to write an essay explaining why they wanted to join the group and sign an official pledge: 'I loathe and detest the vicious delusion claiming the Earth's rotundity, and do hereby solemnly promise to combat the Globularist Heresy in all its disguises.' Nowlan and Ferrari hoped that these regulations would facilitate the rigorous screening of applicants before they were accepted into the society's ranks. 'We could not afford to accept [just] anyone who applied,' Ferrari later wrote, 'we had to be on our guard against the lunatic fringe', and to a greater extent than Nowlan and Fraser he remained dedicated to removing applicants 'deficient in soundness of mind and of moral character' and safeguarding the society's unique tone.[18]

In addition, Nowlan and Ferrari also designed a motto, 'We're on the level', to encapsulate the society's 'common-sense' message. In a pamphlet, they asked,

Can as much as one person in a hundred give a rational proof for that supposedly global shape which is so deeply embedded in the popular imagination? The answer ... must be in the negative ... Imagine it if you dare, half the world living upside down, millions of people being whirled around at 1,000 mph – flashing through solar space at 67,000 mph and through intragalactic space at some 600,000 mph, not to mention all the spinning . . . No wonder that modern man is so disorientated and confused! . . . If one cannot be certain that the immense Earth beneath one's feet is essentially flat, absolutely immobile and at the very bottom of the universe, what then can one be certain about?[19]

In the midst of preparing pamphlets and flyers, Nowlan had further contemplated the character of the flat-earth campaign. 'We're going to have to decide if the FES[C] is to be a private joke or a public conspiracy,' he told Fraser, in March 1971.[20] They decided on the latter and he began to pester Ferrari to write the first in a series of FESC tracts, 'The Global Fallacy as a Cause of Racial Prejudice'. An original idea, Ferrari's thesis was that the inhabitants of countries at the top of the globe felt superior to those at the bottom, when in fact there was no top or bottom to the world. Even if one assumed that the world was spherical, he argued, the 'top' and 'bottom' had been arbitrarily selected, result-ing in racial discrimination against those in the south. 'How can the globularists, their hirelings and dupes, seriously claim to believe that all men are created equal when they teach that some men are eternally fated to hang like bats from the bottom of a globe on which other men stand upright?' The only solution was a flat earth. It would assist in the elimination of prejudice by placing all continents on the same footing and, in the meantime, to redress the balance, he urged FESC members to hang their maps of the world upside-down. Nowlan approved of the suggestion and the tract. 'This would be just the right kind of leaflet,' he told Fraser, 'funny in a way, but also sensible and even practical.' Inspired by the scheme, he thought they should follow Ferrari's lead by drafting tracts two and three. It would be good if they could have them printed as individual pamphlets, Nowlan added, but the problem, as usual, was the cost: 'I wish some financial angel would give us a hundred bucks,' he griped, for a publication fund would enable them to send letters to 'likely looking celebrities', inviting them to become honorary members, and he had an idea that 'something like that might appeal, say, to Bill Cosby . . . or even, God knows, Norman Mailer'.[21]

So the decision was made: the society developed from a conversation piece at parties to a full-blown public organization that people petitioned to join. The reasons applicants gave were varied, ranging from the hard-hitting 'dread of the alleged advances of science and technology and mistrust of "expert" wisdom' to the absurd 'I've always been suspicious of round things.' While strangers concocted inventive reasons for seeking to join the FESC, the founders continued to recruit from among their acquaintances and friends. Most notably, in June 1971, Fraser enrolled the best-selling Canadian author Farley Mowatt, famous for numerous books on nature and native life, at a time when his own writing had fallen into a slump. Kicking around in Montréal, wondering whether to start something new, Fraser had already sent two novels to publishers and heard nothing back. As one would expect, the situation was grinding him down: 'It's hard to keep writing and writing [when] all you do is fill a trunk with manuscripts,' he complained to Nowlan.[22]

Meanwhile Nowlan had received a disturbing letter from the former schoolmarm in Stanley, the village where he was raised. A 'semi-official communiqué' on behalf of the folks back home, the letter informed Nowlan that he had been branded a pariah for publishing unflattering portraits of life in 'Desolation Creek', one of his pseudonyms for his impoverished and allegedly unwelcoming hometown. Reading between the lines, the letter implied that, as a wealthy author, Nowlan should send the struggling locals some money to soothe hurt pride. This was little more than blackmail, he decided, and after drafting fantasy replies for fun, he decided to overlook the demand.[23]

As Nowlan and Fraser exchanged their troubles by post, Ferrari's exploits on behalf of the society continued in notable style. Still recovering from his divorce, Ferrari was experiencing a wild and elusive phase 'walking an emotional tightrope, with madness underneath', as Nowlan later described it. Unhappy with the political power-play at St Thomas's, where competitive young academics were vying for pole position, he decided to throw his abundant energies into the Flat Earth Society of Canada's public campaign as a distraction from the egotism, back-biting and stress. By June 1971 he was particularly focused on an opportunity presented through the academic network that he had grown increasingly to loathe. The occasion was the annual meeting of the Learned Societies of Canada, held at Memorial University in St John's, Newfoundland, where he was due to present a philosophical paper on Augustine of Hippo, which had no connection with the flatness of the earth. But on seeing numerous learned professors gathered together, the opportunity to have some fun, and possibly 'deflate a few stuffed shirts', was too tempting for him to overlook. Hoping to provoke a reaction, Ferrari erected a placard about the society outside the hall where the delegates were meeting, and curious enquirers were invited to leave their details in an envelope attached to the stand. The outcome was gratifying, Ferrari later reported: there were sixty responses, including a request to call a certain telephone number. When he did so, Ferrari found himself talking to a representative of the CBC who interviewed him later that day.[24]

Nowlan reported subsequently to Raymond Fraser:

Now hear this: a fellow from the CBC saw the placard and to make a long story short Leo was interviewed on both radio and television in St. John's. He said the radio wallah had the right spirit but the television man was a bit asinine. I gather, however, that Leo bested him. At one point the interviewer asked (with a look of pitying scorn) 'But, Dr. Ferrari, how do you explain the fact that the Earth appears round in the pictures taken from space by the astronauts?' And our Leo answered, 'Simple. No doubt you're familiar with Einstein's

20. *Right.* Lady Elizabeth Anne Mould Blount (1850–1935), founder of the Universal Zetetic Society.

21. *Below.* Front cover of the Universal Zetetic Society's official magazine, the *Earth not a Globe Review.*

22. *Below right.* Joseph Holden (1816–1900), ardent flat-earth lecturer from Otisfield, Maine.

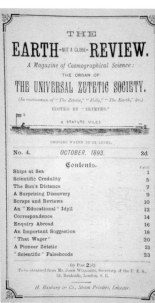

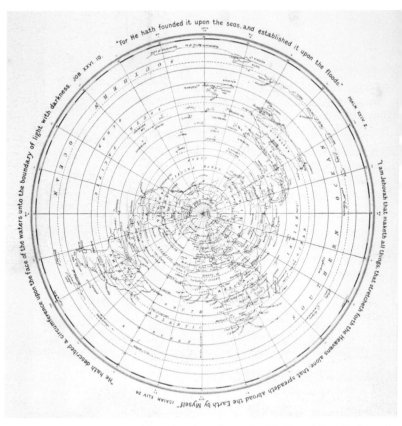

23. An Edwardian flat-earth map, taken from the work of dedicated East End zetetic, David Wardlaw Scott. He refused to accept that the earth was a planet because to his mind it could not be in orbit around the sun.

24. As this cartoon in the *Earth not a Globe Review* shows, zetetics were highly critical of churchmen who turned their backs on bible-based flat-earth belief.

25. *Right.* Zion overseer and major flat-earth proponent, Wilbur Glenn Voliva (1870–1942) as pictured in the *Chicago Daily News* on 9 April 1911.

26. *Below.* Proofs and refutations: Aristotle's classic evidence for the rotundity of earth (pictured left) with Voliva's counter arguments (pictured right).

LINE OF SIGHT

SHIP DISAPPEARING

VANISHING POINT

Aristotle's proof: The disappearance of a ship sailing over the horizon.

"Merely an optical illusion caused by perspective," claims Voliva.

CURVED SHADOW OF EARTH

MOON

MOON

ECLIPSE OF SUN

The curved shadow of the earth during an eclipse of the sun or moon.

"The sun and moon have both been visible during an eclipse."—Voliva.

DIPPER

SOUTHERN CROSS

NORTH STAR

STARS SET IN LOW DOME

The changing aspect of the heavens in different latitudes, some stars appearing and others disappearing, proves the earth is round. Or does it?

"The stars are set in a hemispherical dome so close to the earth that all cannot be seen at the same time," is Voliva's explanation of this point.

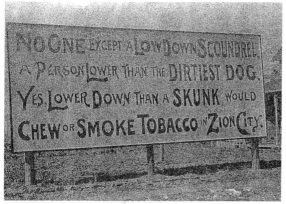

27. One of the notorious warning signs displaying Zion lifestyle rules.

28. Teaching tool: Wilbur Glenn Voliva (inset) and his flat-earth map, which was used in Zion's church-run schools.
It appeared in print under a headline repeating his infamous $5,000 challenge for proof that the earth was not a globe.

29. *Above.* Shenton lecturing on his alternative bible-based cosmological system at an unknown venue.

30. *Right.* The *Sun* newspaper's coverage of Shenton's society frequently adopted a mocking tone, as here from 17 August 1967.

FAKED!

Mr Shenton exposes the Great Global Earth Conspiracy

EARTH .. SEEN FROM A SATELLITE

I MUST say this for the Flat-Earthers: they don't scare easily. You might have thought that the latest pictures of a satellite's-eye view of the Earth, taken from 214,000 miles up, would have shaken them. Not at all.

Mr. Samuel Shenton, secretary and guiding light of the Flat Earth Society, had the answer. Mr. Shenton always has the answer: *Faked.*

"You see those lines running across the photograph, especially at the bottom? That shows it is a composite. Mocked up, probably, from a static model.

"One of your photographer chaps could do it easily."

Well, perhaps he could. But why should anyone want to perpetrate such a deception?

Mr. Shenton has the answer to that one, too. It's all part of the great Global Eart' Conspiracy.

"For centuries we have all been indoctrinated to believe that the earth is a tiny pill, bounding through space at 20 miles a second, and all the things that go with that. Your children are being indoctrinated at this moment.

"The whole world is run on this belief and it has to be maintained."

by RICHARD LAST

gravitational theories later on."

They are also convinced that the Earth extends more or less indefinitely beyond the ice barrier (the Antarctic) and, that,

31. *Left*. Shenton's sign-writing skills on display at the Whitinsville Science Fair, USA. He ploughed much time and energy into the exhibition but the eventual gains were few.

32. *Below*. Media interest in Shenton flourished in the run up to the Apollo moon landing in July 1969. This piece appeared in the *Birmingham Evening Echo* on 17 April 1969.

It's a flat, flat world

SAYS THIS MAN...

LOOK HOW WE'VE BEEN CONNED ALL ALONG ...

33. *Right*. Leo Ferrari, President
of the Canadian Flat Earth Society

34. *Below*. Flamboyant founders
of the Canadian Flat Earth Society
dressed for the Stuart Monarchy skit.
Left to right: Raymond Fraser,
James Stewart, Alden Nowlan
and Leo Ferrari.

35. *Above*. Charles K. Johnson of the
IFERS with long-time correspondent,
science writer, Robert J. Schadewald,
and their wives. Left to right: Charles
Johnson, Wendy Schadewald, Marjory
Johnson and Robert Schadewald.

36. *Left*. Schadewald's membership card.

37. What lies over the edge is an enduring source of curiosity, as letters to flat-earth
believers routinely proved.

38. The flat-earth concept has a natural appeal to human cognition – explaining
the timelessness of the idea.

theory of the curvature of space. If space *is* curved – and modern physics is based on that assumption – the Earth, from space, would appear circular. It's a simple optical illusion.'[25]

With the Flat Earth Society of Canada firmly in the public spotlight, Ferrari followed his trip to the conference with a visit to Al Pittman at the edge of the earth. Tucked away on Fogo Island, Pittman resided in a rickety house by the ocean at 'Lord's Cove', named after a long-dead aristocrat who, according to local legend, had left his treasure buried thereabouts. Ferrari was greeted by home-made posters celebrating the society and, after a night of heavy drinking with Pittman, ventured to Brimstone Head to find the much-discussed edge. What Ferrari discovered, only he and Al Pittman will ever know for sure, but suffice to say he later made much of 'the elemental terror of gazing down into the Abysmal Chasm'. 'Thank God I was duly fortified with copious quantities of Newfoundland's national drink, Screech,' he added, 'otherwise I doubt that my nerves could have withstood the ordeal.'[26]

Later Nowlan told Fraser:

I wish you had been here the night Leo returned from Newfoundland; he lurched into the living room, carrying a placard about the size of a newspaper page, a roll of posters, an overstuffed envelope and a rock about as big as a pumpkin . . . The rock came from Brimstone Head, Fogo, and Leo intends to mount it under a glass and have a plaque made reading:

This rock from Brimstone Head, Fogo Island, Newfoundland, commemorates the first visit, June 13 1971, of the first president of the Flat Earth Society of Canada to the edge of the Abysmal Chasm.

He and Al evidently went out at night, after tossing back a few jars of the other, to find a suitable stone.[27]

In Montréal, Fraser was amused by the news of Ferrari's antics and hopeful that they would get a good response from the brochures dispatched to the '60 learned persons' whose curiosity had been piqued at St John's. 'For one thing,' Fraser suggested, 'as successful academicians they're probably all rich, and for another they will not wish to appear ignorant when confronted with our irrefutable research and logic.'[28] As they awaited feedback from academia, Nowlan was kept occupied processing application forms from prospective members in every Canadian province. 'Sometimes you must suspect that all I ever think or talk about is the Flat Earth Society,' Nowlan joked later to Fraser, 'when actually I only devote about one-tenth of one per cent of my time to that august organization.' Be that as it may, Nowlan thought it was time that the founders had a meeting in person to discuss policy matters. 'Leo and I don't want to appear to be running the whole show,' he told Fraser, and a 'beer-soaked session of the Governors' in Fredericton was definitely in order.[29] Up in Québec, Fraser was pleased by the suggestion. Increasingly frustrated, he still could not find a publisher and he replied that it would be good to travel to Fredericton, Poets Corner and Flat-earth Hideout, to spend some time with his friends.[30]

Back at Windsor Castle, Nowlan was also becoming depressed. His state of health had been precarious since he had suffered from cancer in 1966: he had been on heavy medication ever since undergoing surgery to remove a tumour from his throat. The drugs had unpleasant side-effects, particularly on his weight, and in early 1972 his health problems were compounded by a bout of the flu. Added to this, Nowlan told Fraser, money was tight, he was drinking too much and poison-pen letters from the flesh-and-blood ghosts of his Nova Scotian past were continuing to arrive.[31] In February, he jokingly suggested that he was suffering from 'midsummer urban madness' and said that he was reluctant to accept a radio-show invitation because he did not want to argue the flat-earth case in a live interview again. 'It's too much work keeping all my arguments in my head,' he grumbled.[32]

More positively, young poet and FESC 'chancellor' Jim Stewart had finally recruited a plenipotentiary to oversee the mission in the United States. The new appointment had occurred quite by chance,

Nowlan wrote, for Stewart had been reading the sex spoof *Ronald Rabbit Is a Dirty Old Man* by the American detective-story writer Lawrence Block, when he noticed its preface contained the statement 'The Earth is flat.' Stewart had rushed straight to his typewriter and duly invited Block into the fold.[33] In the same month, the Romanian-born French dramatist and theatre-of-the-absurd pioneer Eugene Ionesco likewise enlisted, while FESC membership cards had been the cause of some dispute. On crossing the border between Canada and the United States a member had produced his card as proof of his identity, but the American immigration officer in charge was distinctly unimpressed. On examining the offending document, he accused the member of being 'some sort of Commie nut' and denied him entry into the US on suspicion that the Flat Earth Society of Canada was a Communist front.[34]

Despite the 'subversive' character of the society, Canadian television networks had grown increasingly curious about it. In April, Ferrari and Nowlan were interviewed for *Take Thirty*, a nationwide show hosted by Paul Soles and Adrienne Clarkson (now Governor General of Canada), in which Ferrari played the nutty professor with globes, wall charts, a pointer and, as the grand finale, the stone from the edge of the earth.[35] The campaign seemed to be proceeding from strength to strength, and the only concern that haunted Nowlan, as Shenton before him, was the workload involved. More than ever he wanted to see the society adopt an independent life of its own, with members sharing the 'dogsbody work' of addressing envelopes, writing tracts and so forth. 'At the moment,' Nowlan reflected, the campaign remained at the stage that if the founders 'said to hell with it', that would be the end. Ideally, they should be able to sit back and enjoy the fun, but the only way to achieve this, he calculated, was to 'give the thing a shove' and attempt to enlist a couple of thousand members to contribute to the crusade. His plan was not as ambitious as it might initially have appeared. Throughout the summer of 1972, he and Ferrari continued to receive letters from prospective members, who included academics, astronomy students, marine biologists, teachers, doctors and a folk-singer.

Besides membership, prospects were brightening on a personal level for Ferrari and Fraser. A research trip to Germany enabled Ferrari to relax and enjoy himself in the wake of his divorce, while Fraser received a Canada Council grant to complete his first short-story collection, *The Black Horse Tavern* (1973). A welcome wind-fall, he used some of the money to buy a small boat, *Spanish Jack*, for $650, and moved to the river near his native Chatham to live and write during the summer. In the interim, back in Fredericton, Nowlan was juggling two trying projects: a history of Campobello Island for the Roosevelt Campobello International Park Commission, and his novel, *Various Persons Named Kevin O'Brien*. With his friends dispersed far and wide and deadlines to keep, he was feeling strained.[36] The nationwide airing of Ferrari's appearance on *Take Thirty* on 27 December provided him with some amusement, however, and in Montréal, Fraser was impressed. 'The show was very good,' he told Nowlan. 'You and Leo came off perfectly, and Paul Soles had the right attitude.'[37] Ferrari was similarly pleased, subsequently informing Soles that 'the society greatly appreciates your objectivity and meticulously balanced exposition of its tenets'.[38]

Soles applied to become a member of the FESC, and a steady stream of correspondence from people around the country resulted from the show. In fact, adults and children from British Columbia to Saskatchewan requested further information, basing their enquiries on a variety of motives. Among them was a community schools co-ordinator from the flat prairies of Manitoba who claimed that his part of the country needed a group to criticize the basic tenets of the scientific age. Such a group would be happy, the man continued, 'to combat subversive propaganda and learn from imaginative groups such as yours', and although there were 'a few globular heresiacs [*sic*] on the prairies', he joked that the people there were 'renowned for their common sense attitudes'. Under such circumstances, he believed it would be possible to 'mine this rich vein of common thought that has made Western Canada what it is' and stated his willingness to help with the campaign.[39] In the meantime, the FESC received a similar offer of support from an Ottawa man, relieved that a coherent society of scepticism existed in Canada. Allegedly in full accord with the FESC's arguments about visual evidence and mass brainwashing, the man berated the 'passive almost moronic acceptance of commonly held conventions' and the public's immediate aversion to any unorthodox idea.[40]

The campaign continued to make advances through 1973. In January, Ferrari received an encouraging letter from an astrophysicist, Bryan Andrew, who had seen *Take Thirty* and wanted to include an article on the society in the Algonquin Radio Observatory magazine. 'I am afraid the Observatory is based on principles contrary to your own,' Andrew added, 'but it all seems to work somehow.'[41] Similarly amused by Ferrari's television appearance, an Ontario man asserted that the earth's horizon only appeared to be curved because human eyelids are shaped that way, while a Vancouver woman griped that traditional science had been allowed to run amok, and groups promoting the reality of a flat earth, aliens, UFOs and suchlike were social necessities, much needed to poke at orthodox egos.[42] A geography professor from Queens University, Kingston, Ontario, echoed her interest, albeit in a more serious tone, with a request for a position paper outlining the society's arguments. Although he admitted 'certain associations with the global heresiarchs committed to a spheroid', he reported an enduring fascination with ideas about the earth's form and cartographic conventions that biased mankind's view.[43]

In February, Ferrari built on burgeoning public interest by appearing on another television programme, *W5*, while newspapers around the country picked up on the human-interest story. In the society's hometown, the *Daily Gleaner*'s headline 'Fredericton's at Hub of Flat Earth Circle' was followed by a front-page interview with Ferrari. In it Ferrari invoked Einstein's theory of the curvature of space to explain why the earth appeared spherical from the moon. 'People don't trust their senses anymore,' Ferrari continued, 'there's such a contradiction between what we can see through first hand observation and what we've been brainwashed to believe.' Society seemed to be suffering from a kind of cultural schizophrenia and, for him, the organization's most crucial role was to highlight this discrepancy, particularly as people did not seem consciously to realize that this was the case. Content to play devil's advocate, Ferrari mischievously claimed that 'people were strange': 'Once if you said God was dead they'd burn you at the stake. Now they say, "So what else is new?" Yet they believe without even questioning that the Earth is round. All cultures are convinced that they have the answers . . . but all they have is a precarious point of view.'[44]

What do we really know? For Ferrari, the unconventional philosophy professor, that was the central question.

Incoming correspondence illustrated public approval of the article, and a Nova Scotia woman informed Ferrari she was planning to save her pension to finance a trip to the edge in her power wheelchair.[45] Another reader, a psychiatric nurse from Montréal, told Ferrari how much she had enjoyed the piece and her subsequent debate on the subject with her four young sons. She had been especially interested to note their reactions to the flat-earth idea and reported that the three younger boys, aged between ten and thirteen, had been relatively open-minded and willing to discuss the concept at least, while her fifteen-year-old was disturbed by the mere suggestion that the world might not be as he believed. Since her profession, as 'assistant head-shrinker', entailed 'blowing people's minds', the nurse thanked Ferrari for providing her with an opportunity to do the same with her own 'little guys' and asked if she was eligible to join the FESC.[46]

By 1973 the society had fifty-four members, including a newspaper editor, an astronomer, a university president, and many of Canada's leading literary figures. Although enrolment figures might have been higher, Ferrari remained determined to uphold his policy of quality over quantity, and was perfectly open about his intention to avoid recruiting 'fundamentalists' and 'cranks'. The society was based on the belief that both science and religion spoke ultimately in metaphors, and Ferrari was an agnostic who looked to enlist individuals who did not take such matters too literally or seriously. 'People are intolerant in religious matters because they are fundamentalists, lacking a sense of metaphor,' Ferrari claimed controversially, 'and, as I have said before, they lack a sense of metaphor because they have no sense of humour.'[47] Yet despite Ferrari's discriminating membership policy, dues and donations rose in line with the society's ranks. By March, the group had the princely sum of C$23.93 in its coffers, although Ferrari claimed money was a thorny issue for FESC founders. 'We realize,' he wrote, 'that there are many things in life that money cannot buy, including a sound view of the universe and the Earth's place in it', and while the society was established as a non-profit-making organization, the founders had remained reluctant to open their first bank account. 'It takes guts and conviction to go up to a teller and say that you want to open an account for the Flat Earth Society,' Ferrari reminisced.

Ferrari was soon to cause more consternation through an interview with J. William Johnson, published in the *Saturday Review of Sciences*. Johnson began the piece by noting his bewilderment when he heard a 'slightly nasal' Australian voice emanating from his television set: 'We openly proclaim that the Earth not only is flat but may well be square,' Ferrari had reportedly declared. When they later met in person, Ferrari insisted he was quite serious about the statement and presented Johnson with a flyer outlining the society's aims. 'You think we're crazy, don't you?' Ferrari asked.

'On the contrary,' Johnson had replied, 'from what I've just read, I get the feeling your stand on the shape of the earth is not really what you're trying to say.'

Ferrari leaped on the point. 'Exactly,' he admitted. 'We're not really obsessed with the shape of the earth. We say it's flat to dramatize our desire to keep our God-given senses from being numbed by technology. To provide some resistance to the forces of conformity.' Echoing the neo-romanticism of radical sixties science critics, Ferrari insisted that the FESC had been established to promote the idea that a scientific world-view was not the only possible perspective. Myths and mystery were dead, he continued, and human world-views required adjustment before rationality ran out of control. 'Believe what you see!' he declared; the point was critical thinking. Invoking G. K. Chesterton's remark that 'A man should always question the strongest beliefs of his age, for those convictions are invariably too strong', Ferrari said that the FESC intended to present an annual award to those who 'defied the dictates of popular prejudice and made an outstanding contribution to the Cause of Common Sense'.[48]

On reading the article, Ferrari believed Johnson had successfully captured the anti-authoritarian spirit of the society, and a high-school teacher from Birmingham, Alabama, wrote to him to agree. 'I've been preaching your – our – gospel to my physics classes for some time,' he added.[49] This letter was one of a number from the United States, and as a result, the founders of the Flat Earth Society of Canada began to reflect on its name. Surely, since the world was flat, their mission was to all humankind and the name should stand without national allegiance to reflect this scope. With this point in mind, in mid-1973 the organization dropped 'Canada' and became simply the Flat Earth Society (FES).

Meanwhile, the society had been receiving attention from one of the nation's leading scientists, for Dr Bryan Andrew, the astronomer who wrote to the society in the wake of *Take Thirty*, had published an article on it in the Algonquin Radio Observatory magazine. Entitled 'In Defence of the Flat Earth Society', he had tailored his piece to suit a scientific audience. 'The first reaction of most of you on hearing of the Flat Earth Society was probably one of scorn, disdain and disbelief,' he began. 'Or to put it more simply "Cranks".' Andrew admitted that this had been his own reaction on first encountering the society through *Take Thirty*, and it was some ten minutes into the programme before he had begun to suspect that he was being duped. The experience made him question his own conditioned reflex, and after reading some FES literature he had seen that the society was more than 'a bunch of loony jokers indulging in some relatively harmless fun'. Although such a judgement certainly had some relevance, he admitted, he informed his readers that there was an undercurrent of seriousness to the society's purpose that was deserving of further consideration.

Most telling for him was the assertion that, while everyone accepted without question that the earth was a globe, not one person in twenty could have provided, before the age of Apollo, one cogent reason why this was so. Although everyday common sense told most people that the earth was flat, they continued to believe implicitly that it was a sphere – because this fact had been imparted to them repeatedly and with confidence. For Andrew, this was a critical point about the public understanding of science and the social role of expertise, and was 'surely an appalling condemnation, not only of our own failure as scientists to explain ourselves, but of the unthinking gullibility of man'. That a fact becomes truth was not due to its demonstrability or even to its implicit veracity but to repetition, lazy-minded acceptance and the authority with which it was told. Andrew noted that it was an uncomfortable fact of life that 'if you say something often enough and assertively enough, eventually what you say will be believed without question'. In a world where increasing specialization created communication difficulties and the appearance of authority could outweigh the substance of an argument, Andrew argued that the public were at risk:

> In an age of overwhelming technological and sociological change . . . experts, self-anointed or CBC-appointed, continue to offer us facile answers to complex questions. We are told that all we need is universal love, the abolition of the motor car, health food, communes, moustaches . . . We are told that all we need are price and wage controls, no strikes, law and order, everybody working, and short hair.[50]

For Andrew what people truly required was the ability to think for themselves, and by pinning their campaign on an eye-catching, heretical idea the FES was successfully drawing attention to this crucial point. Although this was undoubtedly backward logic, the society was raising issues about critical thinking, the importance of questioning 'authority' and those supposedly in the know. In the modern world, Andrew asserted, the FES stood as a laudable example because, in its unique way, it represented an appeal for belief in oneself and in one's own senses, defence of the epistemo-logical rights of the individual and a vindication of the attitude 'if some bureaucrat or theorist or psychiatrist or social planner thinks he knows better than I do what is good for me, then he'd better be prepared to prove it'. That, Andrew concluded, was an attitude worth any scientist's support.[51]

In challenging expertise and underscoring individual sense-experience the FES was not as far removed from Shenton's British society as one might initially assume. Indeed, the idea that science should be based on empiricism, experience and the evidence of one's own senses rather than the authority of published works was a key theme for early modern natural philosophers, who grappled with the issue of how to transform witness experience into trust-worthy knowledge about the natural world.[52] While Ferrari's ideas had a historical precedent, he was happy to exploit Andrew's resounding endorsement of the society in his public lectures.

During 1973 further positive developments were to transpire. Nowlan had finally finished his first novel, *Various Persons Named Kevin O'Brien*, and disappeared on a working holiday to Campo-bello Island to continue the research for his official history. Leo, meanwhile, went camping in California, where he hoped to further his medieval studies and promote the society's work. More prom-ising still, he was not alone on the trip. Earlier that year he had met a new girlfriend, and together that summer they clocked up approximately 12,000 miles in Ferrari's 1966 Chevelle. While Ferrari was enjoying 'wine, sunshine and the best of company' in California, he joked on a postcard to Nowlan that he was 'heartily

gratified to see how incontestably flat the Earth is out here'.[53]
While in California he gave several lectures on the topic, including
one at the University of California, Santa Cruz, about which he
reminisced:

> There was much laughter and clapping and I found it difficult
> to be entirely serious in front of an audience containing many
> people already known to me. I'm afraid I hammed it up a bit
> – due to the fact that my glass of 'water' contained white
> Californian wine. Anyway, I got several applications for mem-
> bership. One guy in the faculty told me I could make a good
> living out of the FES here in California – but that I would
> attract two opposite groups – those who understood the spirit
> of the society and those who saw it as another 'gimmick'.
> California apparently attracts many of the latter – those who
> want to opt out of the 'rat race' and come here for that reason.
> California seems to contain both aspects – the dream and the
> nightmare of the US.[54]

Whatever the case, the *Santa Cruz Sentinel* published a fair
report about the lecture by the 'pixyish, bushy-bearded' professor,
covering all of his main points: that the astronauts had been
brainwashed, that people prioritized theory over experience and
the average man or woman in the street had become 'so obsessed
with scientific thought' that they were, in fact, 'subconsciously
disturbed'. Inspired by wine, Ferrari had called for an end to
excessive rationality and for people to think with their hearts as
well as their heads, for he was sure they would feel more secure
if they trusted their instincts and simply accepted nature as it
appeared. 'The scientific view of reality is not the truth,' Ferrari
insisted, 'it is just an aspect of the truth.'[55]

The lecture at Santa Cruz was followed by further radio interviews and television appearances on behalf of the society, and
Ferrari returned to Fredericton with fond memories of his trip.
The vegetation and climate reminded him of his native Australia,
and his freewheeling adventure was an experience that academic

routine could not match. By the autumn, however, he was back on the treadmill of university teaching, including a course on St Augustine, which, due to New Brunswick's status as a bilingual province, had to be delivered in French. By this point, Ferrari was beginning to feel over-burdened. Academia was a distraction from his work as president of the Flat Earth Society, he joked, and he was falling behind with his correspondence on the organization's behalf. Letters to the society were indeed piling up, including some from physics professors and medical students, who had written offering encouragement, and one from a man who lived on the flat prairies of Saskatchewan, insisting that the view from his window provided irrefragable proof that the earth was flat and anybody who refused to trust his eyewitness testimony was welcome to visit and make observations at first hand.[56]

From Ferrari's perspective it was a significant point, for in the days when the ultimate proof of anything seemed to be 'I've seen it on TV', he continued to believe his group had a serious philosophical purpose to counter-balance the fun.[57] Tongue-in-cheek, he complained, 'We of the Flat Earth Society are regarded as being cranks, nuts or fools because we have the audacity to believe what we see', while the mindless majority were rewarded with the dubious titles of 'normal' and 'sane'. Thus he declared: 'If there were not a Flat Earth Society, one would have had to invent it, because the modern mind, with all its feigned scepticism, feels safer in knowing that there is [such a] society to serve as an object of derision. It's like those people who need to feel safer about their sanity by labelling others as "nuts".'[58]

In May 1974 it was an angle that Ferrari took again in a keynote address to an international symposium of geodesists. In retrospect he viewed this occasion as the most memorable and successful of all of his talks, but at the time he felt threatened by the audience's scientific attainments. Later he confessed to having taken the precaution of peppering his talk with buzz-phrases to render it palatable to even the most 'globular' of the geodesists present, but was relieved to discover the audience took his defence of planoterrestrialism in the spirit that it was intended. Contrary to expectation, the Flat Earth Society received a few more applications as a result and its membership rose to ninety-seven.

Applications to the society and media coverage proceeded hand in hand through the winter of 1974, and among all of this the undoubted highlight was Alden Nowlan's article for *Weekend* magazine, a mass-circulation supplement to over forty newspapers nationwide. His story, entitled 'Would you Buy a Used Globe from this Man?', focused on Ferrari's role as president of the FES, and was accompanied by various pictures of him stamping on a globe. The piece opened with a description of Ferrari's antics and his view that what people called 'sanity' actually denoted a narrow, rationalistic outlook and a dreary, sterile state of mind. All the 'experts' had provided were intricate calculations, unintelligible to the average person, and the pictures supposedly taken from space.

But nobody knew if the images were true, Ferrari asserted, and as for television coverage of the Apollo missions, the science-fiction series *Star Trek* appeared more authentic. In the light of Watergate, the notion that the whole stunt had been faked was not unthinkable, and Ferrari claimed there was a lot to be said for not being sure of what went on in the world. Alleging distress about humankind, he proclaimed the 'cult of certainty' had taken over, and modern society was obsessed with the desire to explain everything according to science. 'Human beings have lost touch with the Earth and the infinite mysteries,' Ferrari declared again, for 'mankind was happier when it was prepared to make its peace with the unknown'. For him, science was only one of the many windows through which mankind could look at the universe, and a very small, murky window at that; just a highly human and conventional means of negotiating the 'Great Mystery of Reality'.[59]

Nowlan's profile of the FES president was well received and, somewhat ironically, he received more feedback on the article than for any other piece he had written during his career. But not all of it was complimentary, as an Ottawa doctor complained in a letter to *Weekend*:

According to the article, among the members are computer scientists, university men, lawyers, physicians, geographers, astronomers and seamen who 'don't take themselves seriously'. I would suggest that nobody else should ... Admittedly, one often needs a large sense of humor to get along in this sickly great society ... But when supposedly grown-up men find time for and put work and energy into such a nonsensical and organized activity[,] behaving like a bunch of infants, the situation changes. If professional people exhibit such a grossly irresponsible behaviour, one must wonder who is going to teach our young people to take seriously anything at all?[60]

Nevertheless, as a result of the article, Ferrari was invited to play the mystery guest on the hit television show *Front Page Challenge*. 'Very curious how our casual acts set off a chain of reactions,' Nowlan reflected to Ferrari. 'One moment you're in the front room at Windsor Street [being asked if the earth was flat], and the next moment, or what seems like the next moment, you're being questioned by Pierre Berton [a permanent panelist on *Front Page Challenge*] before an audience of 750,000 people.'[61]

In the broad scheme of things, the comment had some poignancy, for while the society was going from strength to strength, friendships between the founder members were faltering. The major difficulty was geographical distance. Desperate for a change of scene, Ferrari had taken a year's sabbatical and was based at the University of Montréal's Medieval Institute. Big-city life was proving radically different from serene Fredericton, and in an atmosphere of political upheaval beset by protests and murders, Ferrari speculated that he might have been safer in Los Angeles. 'Many an evening, I think about how nice it would be to be in your front room,' he wrote to Nowlan, 'having one of our timeless talks.' However, he had now finished a 'funny-serious' study, *The Earth Is Flat! An Exposé of the Globularist Hoax*, and had dispatched the then 25,000-word manuscript to several publishers. 'Wouldn't it be a laugh,' he joked, 'if the only enduring thing that I did on my sabbatical was to write *The Earth Is Flat!*' That said, he believed he had at last written something that would sell to the general public, in contrast to his scholarly work on St Augustine, and 'hopefully,' he quipped to Nowlan, 'the book will also effect a much-needed change in perspective. Let us hope and pray that the Truth will triumph!'[62]

As Ferrari hatched plans for literary greatness in Québec, back in New Brunswick changes were afoot for Ray Fraser, who had left *Spanish Jack* for a tumbledown house with a leaky roof in the village of Black River Bridge. Subsequently he set to work on his first published novel, *The Struggle Outside* (1975).

A hundred miles away, Nowlan was feeling abandoned. In September 1975, he bought a car in a drunken frenzy and suggested that although he was 'the world's most gutless driver', he might make the trip to Black River Bridge to down a few drinks with Fraser and discuss the FES. His loneliness was short-lived, however, for by the autumn Ferrari was back in town, accompanied by his new girlfriend, Lorna Drew, a 'wonderfully wacky nurse', who wrote poetry, played the piano, collected paintings and had also, coincidentally, been an early member of the FES.[63] With the return of the president, Nowlan thought the society should be reactivated: there was now sufficient money in the kitty to publish a few tracts and he suggested Jim Stewart's 'Newton's Nonsense' and a 'terrific piece' by an American member, 'Advice to Plano-terrestrialist Parents on What to Tell their Children', as worthy candidates.[64]

Alongside the preparation of book manuscripts and tracts, Ferrari remained keen on public speaking in his capacity as FES president. The activity undoubtedly lent an edge to his life, and he admitted this was one of his major motives in involving himself with the campaign. 'You know, I got into all this because I found academic life as dull as ditchwater,' Ferrari confided to the astronomer Bryan Andrew, 'besides my innate certitude about the flatness of the Earth, of course.'[65] That said, lecturing on the subject required a certain humour and flamboyance, and these were traits Ferrari exploited to the full during his numerous public appearances. 'Audience reaction has always been sympathetic and concerned,' Ferrari remembered in *The Earth Is Flat!*, 'although it was not always easy to decide whether they were more concerned about me or my subject matter.' According to reports, the lectures were never humdrum and evoked all sorts of reactions, from irate arguments to tears of mirth and threats of physical violence. 'Fundamentalists of Science,' Ferrari joked, could become aggressive when faced with evidence that challenged their deeply held beliefs and he recalled that 'one such fanatic' had tried to disrupt the 'serious tone' of a talk by complaining loudly from the back of the hall that he could not hear Ferrari on account of the earth's curvature. Despite this incident, Ferrari said that the most belligerent audience he ever encountered was a group of grade-five students. 'They really gave me a hard time of it,' he reminisced. 'One young girl asked me if the Earth was really flat after all, how come that every day in the week had a different name? In order to play for time on that one, I asked where was yesterday right now. Immediately a bright young fellow in the front row shot up his hand and blurted out the evocative answer: "In China, sir."'[66] Public advocacy of flat-earth belief was an educational experience, Ferrari discovered – about human nature, if nothing else.

Through the mid-1970s, the Flat Earth Society was not the only diversion by which Ferrari, Fraser and Nowlan sought to elude the mundanity of everyday life. Together they cooked up a number of schemes at Windsor Castle, and another of these, the 'Stuart Restoration' campaign, hit the headlines in early 1976. On this occasion, the scam revolved round Jim Stewart, the young New Brunswick poet and chancellor of the FES. When drunk one night years before at Windsor Castle, Stewart had told Nowlan about his Scottish heritage and his family's distant connection to the Earl of Fife. In reaction, Nowlan's fertile imagination ran riot. He had an encyclopedic knowledge of genealogy and royal lineage, based in part on a lifelong fascination with his own Irish background, and he immediately constructed a fantastical vision that challenged the course of history. Was Stewart perhaps descended from Charles Stuart, 'Bonnie Prince Charlie', who had been defeated at the battle of Culloden in 1746? Moreover, with Stuart's death wasn't the Earl of Fife the rightful heir to the throne? If one did not accept the Hanoverian line as the legitimate rulers of Britain, which was certainly an issue at the time, surely their own Jim Stewart had a lawful claim?[67] The scene was set: the Jacobite cause was resurrected at Windsor Castle and, as with the Flat Earth Society, grand-sounding titles were shared among the group. Stewart was King James III, head of the Stuart Monarchy in Exile, Nowlan became the Duke of Wexford and Prince of Fortara, Fraser's time on *Spanish Jack* earned him the title First Lord of the Admiralty, and Ferrari was christened Archbishop of Canterbury to reflect his theological interests.[68]

With the names in place, the joke rumbled on for years, and in January 1976 Fraser decided to write up the story in a spoof article, 'Royally Wronged: Why James III is Just Plain Jim', for *Weekend*. The piece argued that Jim Stewart was King James III, true heir to the thrones of England, Ireland, Scotland and France while his predecessor, James II, had been cruelly robbed of the crown by Parliament and the Protestant Church on the grounds of his Catholic faith.[69] Accompanied by a mock-serious picture of the major players in fancy costumes, the whole affair had overtones of *Monty Python's Flying Circus* and most people saw the joke, including Queen Elizabeth II, who reportedly laughed when she was shown the article by ardent genealogist Lord Mountbatten.[70] As one might expect, the difficulty lay rather with those who did not appreciate the off-beat humour, and Nowlan received several heart-wrenching letters from people who had read the article and sought assistance with tracing family members or establishing religious or political rights.[71]

Again, in common with the FES article, criticism of the tongue-in-cheek campaign swiftly followed, and one *Weekend* reader wrote a letter denouncing Ferrari, 'the Rolf Harris of Maritimes academia', and his 'clique of third-rate writers and their pantomime'. Although the comments seemed harsh, as time passed it appeared that the joke had backfired, particularly after CBC began hounding Nowlan and Ferrari for an interview on the programme *90 Minutes Live*. Neither was keen and one evening at Windsor Castle the subject caused a rift. In his absence, a drunken Nowlan blamed Fraser for concocting the Stewart monarchy scam, provoking Lorna Drew, now married to Ferrari, to remind him that it had been his idea. It was a risky manoeuvre. Nowlan was renowned for his opinionated rants when he was drunk and this tendency was exacerbated by the fact that, among the many people he ostracized over the years, he had never liked Lorna Drew. Predictably, he flew into a rage, and Ferrari and his wife decided to leave. The argument signalled the end of the Stewart monarchy episode and further controversy was to follow.[72]

Despite the undignified collapse of the make-believe monarchy, the Flat Earth Society remained largely unaffected, although it had stagnated since Ferrari's sabbatical and stacks of mail had awaited him on his return from Montréal. Nevertheless, the FES now had a hundred members paying dues of three dollars a year and the ranks included land surveyors, attorneys, air-force pilots, geodesists, geologists and astronomers, alongside an undertaker, a judge, a few farmers and some real-estate agents. But Ferrari realized that support for the society was not self-generating, and in 1976 he undertook a series of public-speaking engagements to reinvigorate the mission and set to work devising a plan for the upcoming Learned Societies Conference to be held at the University of New Brunswick in the spring. Placing himself in yet more debt, Ferrari financed the printing of two thousand FES brochures, a new edition of the *Official Organ* and further copies of various tracts, which he planned to distribute to delegates on campus. In addition, he persuaded the university library to erect an exhibit on the flat-earth issue, 'thus getting back at them,' he joked, 'for that atrocious display they had on Copernicus in 1973'. At one point he even considered asking some of his students to put on a play: one member, an American philosopher, had composed a ten-page dialogue in which Plato and Socrates discuss the flatness of the earth and Ferrari thought it would be hilarious if students re-enacted it while the learned delegates were visiting town.

Although his plans were ambitious and costly, Ferrari calculated that there would be five thousand academics on campus for the conference and he was hoping to enrol a thousand new members in the society. To allow for this anticipated influx, and to depressurize the screening process, Ferrari and Nowlan established a new rank, the associate member, which involved partial rights for three years before promotion to full-member status. Ferrari reflected that 'Even if we do get the occasional nut in as an Associate Member, then he (or she) will not have gotten right in. We will have three years to make up our minds about such cases.'[73] In the meantime, he would continue to make good use of his 'Hope He Goes Away' file to store letters received from suspected cranks.

Over the next couple of years, relationships between the founder members continued to deteriorate. Nowlan's alcohol consumption, and the consequent tantrums, seemed to be on the increase, and this had an especially damaging impact on his relationship with Ray Fraser. Albeit close, their friendship had always been changeable, fuelled as it was by wild drinking and squabbles, and it was further undermined when Fraser decided to join Alcoholics Anonymous. He wisely elected to keep his distance from Windsor Castle for a while, and other friends followed suit after their own disagreements with Nowlan. For a time even the usually unflappable Ferrari kept a low profile at what was effectively his second home, and although Nowlan and his wife Claudine continued to undertake much of the dogsbody work of answering FES letters and posting literature, by the late 1970s the continuation of the society was mainly due to its president. Yet Ferrari's steadfast efforts could not prevent it falling into decline and even Fraser's relocation back to Fredericton with his wife in 1980 did not signal a reversal of fortune. Fraser's ongoing sobriety continued to annoy Nowlan, who felt abandoned. For his part, Fraser made a series of well-meaning attempts to interest the poet in the benefits of getting off the bottle, even visiting Windsor Castle armed with AA pamphlets, but his efforts were unappreciated and, as events were to prove, fruitless.[74]

Throughout this period Ferrari continued to be the focus of media attention in his capacity as FES president, but the workload at St Thomas's was such that he had taken to shunning publicity wherever possible. Added to this, he had developed something of a grudge against publishers, or so he jokingly wrote. Having published more than two dozen articles on St Augustine, he had failed to experience the same success with his revised and expanded 330-page manuscript, *The Earth Is Flat!*. Too risky, too off-beat, too entertaining, too scholarly, too long, too funny, not funny enough, Ferrari had been through the whole gamut of possible responses and was tired of publishers telling him that *The Earth Is Flat!* was 'not our kind of book'. Even the promise of an introduction by famous author and FES member Farley Mowatt was not sufficient to sway the many publishers in America, Canada and Britain who had considered the manuscript over the past five years, and in 1980, Ferrari finally decided to publish *The Earth Is Flat!* himself. It was a weighty task, however, and although he was not part of the administrative hierarchy at St Thomas's, his academic duties were sufficiently distracting to interfere with his plans.

By the 1980s Nowlan and Fraser were experiencing markedly more success. Nowlan was on the way to his thirteenth volume of poetry, while since 1966 Fraser had published four collections of poems, a novella and short-story collection, a biography of the boxer Yvon Durelle, and two novels, the second of which, the acclaimed *The Bannonbridge Musicians* (1978), had been runner-up for the prestigious Governor General's Award that year.

Whatever their achievements in their respective fields, though, the founder members of the Flat Earth Society all discovered that their association with it generated the most attention from the media and the general public. This was a source of irritation for Nowlan in particular, but it appeared to be an inescapable fact.

The flat-earth idea was too intriguing and controversial; the founder members had never needed to whip up publicity for it seemed to seek them out. Paperwork and expense were generated in turn, and gradually it became clear that the society had outgrown itself. The joke wasn't funny any more. In February 1981, long-standing member Tom Dolan wrote from Baltimore to ask Nowlan about the situation. 'I've heard nothing!' he exclaimed. 'Is Leo well,' he wondered, 'or has he befallen the evil scheming of the Globularists?'[75] Nowlan replied a week later, explaining the current state of affairs:

> The Flat Earth Society is in a state of suspended animation – principally because it became too successful. We reached the point where running the Society was pretty much like running the Rotary Club or Alcoholics Anonymous. There were bank accounts (very small bank accounts) to be audited, letters to be addressed and stamped and posted; and while everyone wanted to share in the fun, nobody wanted to share in the work or help with the finances. My wife and I simply got tired of doing all the clerical work (and paying the printing bills). The membership fee was just a joke. I suggested to Leo and the other locals that we raise it to the point where we could hire part-time clerical help and publish a quarterly magazine. But the feeling was that this would be too much effort to put into a joke. I say that the Society is in suspended animation rather than dead because Leo still makes speeches from time to time and gives the occasional radio or TV inter-view. Do you know that we had two Nobel Prize winners in our membership before we went into suspended animation? And also the Chief Astronomer for Canada. Yes, we really did. Not bad for an organization founded at 4 a.m. in my living-room in little old Fredericton.[76]

The period of 'suspended animation' did not last. Leo Ferrari, the member most concerned about the 'lunatic fringe', started to receive crank calls and, unnerved, he abandoned his flat-earth talks and moved on to other projects. Meanwhile, two and a half years after his letter to Tom Dolan, Nowlan had a heart-attack in the shower at Windsor Castle. He was rushed to hospital in Fredericton, but the outlook was not hopeful. An alcoholic with health problems since his early thirties, he died on 27 June 1983 at the comparatively young age of fifty. Ironically, the often absent Raymond Fraser is the only founder member of the society who lives in Fredericton now. He remains dedicated to writing – poems, short stories, biographies and novels – and when asked about the Flat Earth Society, he replies with a twinkle in his eye.

Chapter Nine

THE CALIFORNIAN CONNECTION

'When we navigate the ocean, or when we predict an
eclipse, we often find it convenient to assume that the
earth goes round the sun and that the stars are millions
upon millions of kilometres away. But what of it? Do
you suppose it is beyond us to produce a dual system of
astronomy? The stars can be near or distant, according
as we need them. Do you suppose our mathematicians
are unequal to that?'

O'Brien to Winston Smith in Room 101,
GEORGE ORWELL, *1984* (1949)

'Restoring the World to Sanity'

Byline of the *Flat Earth News* (1972–89)

DURING THE SAME PERIOD that the Canada-based Flat Earth
Society was challenging the 'fallacious deification of the circle' for
a mixture of motives, a more traditional type of zetetic campaign
was under way in America. While they were equally unorthodox
in claiming that the earth was flat, the two organizations could
not have been more dissimilar. The critical distinction lay in the
delivery of their agendas: while the Canadian society had comic

overtones, the president of the International Flat Earth Research Society of America (IFERSA), Charles Kenneth Johnson, was thoroughly in earnest about his work for the cause.

In his late forties when Samuel Shenton died, Johnson was born on his father's cattle ranch in Tennyson, Texas, on 24 July 1924, and raised in nearby San Angelo. A 'natural sceptic', so he said, Johnson explained that he was 'plagued or blessed or whatever you want to call it with having a critical mind' and was 'able to find the unreasonableness in ideas and things' from his earliest years. Instead of 'just accepting things', he would always 'detect a flaw'.[1] Later he reminisced that this tendency persisted when he started school and was central to his rejection of a spherical earth. Speaking in the early 1980s, Johnson remembered that he felt 'brainwashed' by teachers and this suspicion was magnified when he first saw a globe when he was in the second or third grade at school:

> Now they brought out this globe. It wasn't like today, they didn't have globes everywhere, and people didn't say globe every few minutes. They put out this globe, and started the propaganda on it. I didn't accept it from the start. You can see the thing is false! It's quite obvious. I can see it today the same as I saw it there.[2]

To Johnson, a globe seemed unreasonable and illogical – he said it just did not make sense. He remonstrated with the teacher: 'I say that there couldn't possibly be any truth in it, a ball with a cap of water on it, and the ships going over the edge. And the water hanging there. Why doesn't this water fall?' In reaction the teacher sent Johnson home from school and told him to fetch a bucket, fill it with water, swing it round and watch how the water was held in the bucket by centrifugal force. Johnson remembered that he went straight home and tried the test:

> I can see myself now. I got a bucket of water and I whirled it around. It didn't come out, but I saw at the same time that it

had nothing to do with the globe. This was absurd. So I knew there was a lie here. Maybe I was extra smart or extra something for my age. I see other things they told us in school weren't true. But that was the big one. I knew it and I've always known it. It's a complete world of lies, telling you things they know when they really don't know it. I always knew that the Earth was not a ball. I pondered for a while about how I could prove otherwise, but I didn't want to spend any time dwelling on it. I just knew that the Earth is flat.[3]

Although the water stayed in the bucket, as Newtonian laws of motion implied, Johnson told a *Chicago Tribune* reporter that there was a discrepancy in his observations that he found impossible to rationalize. 'The water was rigid in the bucket,' Johnson remembered, 'it wasn't moving, like people, cars, trucks, trains, and animals do', so for him the experiment did not provide sufficient proof to 'explain why people don't go flying off into space when they move about'.[4]

As with many flat-earth believers, Johnson contended that the globe concept defied common sense, and having inferred his conclusions from a straightforward test, he was proud to emphasize that he was never taught that the world is flat.[5] Although he was raised a Nazarene (a Methodist-based Protestant denomination), attended Sunday school and 'knew that the Bible was a flat-earth book', it was the result of his own experiment that constituted irrefragable proof. That the earth is flat was a 'fact' he believed he had discovered for himself through personal observation, objectivity and a simple direct test – the scientific method, no less, although in common with other flat-earth experimenters the way Johnson inferred conclusions from results was somewhat problematic. By contrast, however, he viewed himself as a free-thinker, a term implicitly bound up with flat-earth belief through the self-contradictory zetetic approach. On these grounds, he reflected, 'You might say I've been a flat-earther all my life.'[6]

With hindsight, Johnson considered the experiment to be the

catalyst that prompted a quest for knowledge that transformed the rest of his days. From when he was eight years old, while other children were playing, Johnson was roaming the libraries 'wildly in love with books and learning'. He remembered, 'My world was strange.' During this period he became a voracious reader and once he had acquired his own library card he was preoccupied with 'searching things out all the time'. He later recalled that it was hard to think of his junior high-school lessons and textbooks in the light of this personal quest. 'If I could have had the chance to go to special classes or school where one could "move on",' Johnson supposed wistfully, rather than being '. . . stalled and wasting so much time [there was] no telling what I'd have done.'[7]

Motivated by a refusal to bow to authority and a love of learning, Johnson remembered that he continued his studies without restriction through his teenage years. Such activities frequently brought him into conflict with authority, and Johnson recalled that although he argued with teachers until they became infuriated he remained unconvinced by their interpretation of facts. On a more negative note, while placing emphasis on his personal search for knowledge, and thus on his own authority rather than what his teachers said, he recognized that his education was haphazard and his mastery of grammar particularly poor. 'But basically,' Johnson concluded, about his formative experiences, 'I consider my mind as pretty logical and not warped as bad [as the majority of the population].'[8]

Although the majority might have begged to differ, Johnson said he 'just knew that the Earth was flat'. To this extent, and in common with many flat-earth advocates, he portrayed his 'finding' as a life-changing insight and a ground-breaking revelation, comparable to the 'eureka' moments ascribed to noted discoverers in traditional historical accounts. Although in this sense Johnson co-opted the mythological glamour of science in the zetetic tradition, one notable point remained: he lacked evidence to support his conclusion about the shape of the earth. That was to come in the early 1940s, when he was living in Fort Worth. Having abandoned the school system, he had continued to pursue his solitary educational path, plundering public libraries for information and imbibing works on psychology, science and literature by authors such as Freud, Einstein, Dickens and James. As a consequence, Johnson claimed, when he turned sixteen, he was 'a little better read up on everything than the average 25 year old college graduate', and it was his bookish habits that led to the flat-earth evidence he sought.[9] 'It was through Voliva that I actually learned how to prove it,' Johnson said later, for one day in his local library he happened on an article in *Harper's* magazine about the Zion overseer's beliefs:

> I knew in a split second, when I read in *Harper's* magazine, just check the water. I said, 'My God! Why didn't I think of that?' I vowed that the minute I'd get to a lake, I'd check it. I knew from that second then how to prove it.[10]

Eager as always to discover more, Johnson wrote to the ailing Voliva in order, he said, to 'get the facts', and he subsequently received a vigorous reply supporting his unorthodox world-view. With Voliva's death soon afterwards and the abandonment of his teachings by Zion's founding church, that line of enquiry was permanently closed, and Johnson found himself sidetracked from his studies by his work as a mechanic during the Second World War.

The years passed and Johnson relocated to Arizona, Los Angeles and, finally, San Francisco, where he was employed as an aircraft mechanic occasionally involved in space-shuttle work, and later as a factory supervisor for a heavy-equipment manufacturing concern. Yet despite his evident technical bent, Johnson did not forsake his flat-earth beliefs, and during the 1960s space-race publicity he was heartened to hear of Shenton's British society, courtesy of the American press. In 1965, he wrote to Shenton seeking general information and a membership form. A correspondence developed between the two men, which lasted, in fits and starts, until Shenton's death in 1971.

It was at this point that Lillian Shenton was faced with a

difficult decision: what to do with the stacks of books and papers that remained from her husband's chaotic campaign. According to Johnson, who claimed Lillian was a fellow believer, she was anxious that her husband's work would not have been in vain. Although the IFERS had a president in the Greater London Council member and North East London Polytechnic lecturer Ellis Hillman, she distrusted his devotion to the cause. While Hillman was intrigued by the topic – he had read all there was on zetetic astronomy at the British Library – the fact remained that, whatever his enthusiasm, his intellectualism, his connections and his desire to preserve Shenton's papers for posterity, he did not believe the earth to be flat. Even though Hillman enjoyed lecturing on the topic, and had continued to do so since Shenton's demise, to him it remained just an 'interesting idea' that he found he could defend effectively in public talks and debates. On paper, literally and metaphorically, Hillman was Shenton's natural successor and had a rightful claim to that role. Meanwhile, Johnson contended that Hillman's lack of conviction disqualified him from a legitimate claim to the presidency and it was Samuel Shenton's dying wish that he, Johnson, should inherit the organization.

Whatever the truth, Lillian Shenton found an awkward, yet workable, solution with Hillman. He borrowed a van to transport Shenton's papers to the archive department of the North East London Polytechnic, later claiming that he had arrived in time to prevent Lillian disposing of the collection in a dustbin outside her home.[11] Destruction averted, the papers were crammed into cardboard boxes and transported to the archive – based appropriately, one might conclude, in Barking – where they became an addition to the collection of the Science Fiction Foundation (SFF), which Hillman had helped to found.[12]

Here, to the probable amusement of academic administrators, the confusion of correspondence from Dover proved a source of interest for some unconventional enquirers over subsequent years. On repeated occasions, staff members were compelled to clarify that the SFF had no connection with the IFERS and that the polytechnic's custodianship of Shenton's papers did not imply support of his society's creed. Hillman, meanwhile, continued to give humorous lectures on the flat-earth idea as IFERS president, although he limited such engagements to two per year despite numerous requests.

By now, Hillman was president of the Lewis Carroll Society, re-established in 1969, and held a number of high-level public positions besides presidency of the IFERS and his academic post. At one time vice chairman of the Inner London Education Authority and a governor of Queen Mary College, University of London, he also served as Labour mayor of the North London borough of Barnet in the mid-1990s. Such positions restricted the time available to him for tongue-in-cheek promotion of the flat-earth cause, while his ideological differences with Johnson limited their communication − seemingly to Hillman's relief. The divide between the two men was highlighted in an interview shortly before Hillman's death in 1996 when he jokingly described Johnson as a 'nutter'.[13]

As lines were drawn between sincerity and comedy, fiction and fact, Lillian shipped part of the original collection to San Francisco so that Johnson could continue the society's work. Although he was appalled that some of the society's papers were lodged in a science-fiction collection, he remembered experiencing 'a great surge of feeling' when he received the package.[14] At that moment, Johnson said, 'I knew what all my life had been for, what all the experiences I've had are for.' It had been, he continued, 'to prepare me . . . for now!' In this way, Johnson believed he had heard God's call and felt ready, willing and able to take on the work. 'If it calls me for life,' he concluded at the time, 'so be it.'[15] Johnson evidently garnered a sense of personal identity and a framework of meaning from his flat-earth quest, and consequently transformed his life to clear the way for his new role. In 1972, he retired from his job after twenty-five years' service, incorporated Shenton's organization as the International Flat Earth Research Society of America and Covenant People's Church, and relocated for a final time to

California's Mojave Desert, where he had purchased some land, 'God's five acres', for $10,000.

From that point on, Charles Johnson became a dream subject for every journalist looking for an angle: an almost lone voice in the wilderness, he seemed to fulfil the stereotypical image of a wacky Californian conspiracy theorist to a T. Yet while California was considered a haven for eccentrics matched only by England in the popular psyche, Johnson claimed that his location was one reason why Shenton had wanted him to inherit the campaign. As the years of negative media coverage rolled by, Shenton had grown increasingly bitter about rampant ridicule in the land of his birth, and had believed, on the basis of letters he received, that Americans would be more open to the flat-earth cause.

For his part, Charles Johnson relished the challenge that lay ahead and proclaimed himself 'the last iconoclast', who had assumed leadership of Shenton's society as if it were 'the mantle of Elijah'. Supported by a small private income, he established campaign headquarters in an office at his remote home, 'The Old Chateau', deep in the desert, half a mile from the nearest neighbour and twenty miles east of the small city of Lancaster. Here, in his ramshackle home with only his wife and their 'four-legged children', a menagerie of cats and dogs, for company, Johnson lived a separatist lifestyle reminiscent of other sectarian groups. In such self-imposed quarantine, he had much space and time to contemplate the ways of the world he had left behind. According to reporters, one of Johnson's favourite pastimes was watching sunsets from his porch while smoking a cigar, and it was frequently remarked that from the Johnson yard, the earth indeed looked flat.

It was one of those places, like the Cambridgeshire fens, where a vast, bleak expanse under a sweeping skyline could play tricks on the mind if one were so disposed. Yet there was one striking difference from the oozy English fenland: the parched desert landscape – dust and sand, horizon and sky, stretched for miles, with only the odd tumbleweed or Joshua tree to punctuate the seemingly endless flatland. All in all, apart from Old Bedford

Level, Johnson could not have selected a more appropriate base for his campaign. The view from his hillside home swept over Lancaster, while Edwards Air Force base, the location of the pre-space programme test flights, was situated nearby in Palmdale.

With his new home and alternative vision in place, by 1972 Johnson was prepared to tackle the public's world-view. Crucially, he was not alone: his wife served as IFERSA secretary. He had met Marjory, a quiet, dainty woman, in a second-hand store in San Francisco where they were searching for the same record, Acker Bilk's evocative bestseller, 'Stranger On The Shore'. For Johnson it was an auspicious moment: during their first conversation he not only discovered that Marjory shared his taste in music, she likewise believed the earth to be flat. In interviews Johnson often eulogized his good fortune, while in letters he confided that until their chance encounter the couple had both been 'living in Heartbreak Hotel down the end of a lonely street in San Francisco'.[16] Isolation behind them, Charles and Marjory married in 1962 and honeymooned in Reno. 'Stranger On The Shore' remained their song.

While love and romance were evidently important to Johnson, the relationship produced equally interesting developments on the campaign front. Most notably because Marjory was Australian, Johnson believed she had direct proof of the earth's flatness. He often said that she had been shocked and offended when she emigrated to America in 1944 and discovered that her homeland was nicknamed 'Down Under'; she later swore an affidavit that she had never hung by her feet from a globe. As far as Marjory was aware, 'everybody in Australia' thought the world was flat, and she was willing to assist her husband in proving that this was the case. Johnson placed such faith in her direct sense-experience that he challenged reporters, 'Who can argue with a witness?'[17] Thus, in a new development for evidence of the earth's flatness, Marjory became living proof – a valuable human exhibit Johnson utilized on numerous occasions in the public defence of their cause. He often informed journalists that when Marjory had sailed on the ship from Australia to America she 'did not get on it upside down, and she did not sail up around a globe'. Marjory was certain that she had sailed straight across the flat ocean, and that, Johnson asserted, should be considered incontrovertible evidence that the earth was a plane.[18] To bolster his case, he encouraged Marjory to testify in person about how she had not hung by her heels from her homeland, while he frequently produced photographs of his smiling wife in Australia – standing the 'right' way up.

With Charles and Marjory prepared for the journey ahead, they did not lose sight of the historical roots of their campaign. In keeping with their zetetic predecessors, they recognized that, for credibility's sake, it was vital to place their work in a long-established tradition and consequently stressed the legacy of research and ideas left to them by 'authoritative' sources such as Voliva and Parallax. The historical narrative undoubtedly provided them with a sense of identity and community, albeit with those long since deceased, along with much-needed reinforcement of their views. In a survey of their forerunners, Johnson placed Shenton on a par-ticular pedestal as a 'Knight for Truth' and 'one of the greatest men England has ever produced', and in their absence, Samuel and Lillian were appointed respectively honorary president and honorary secretary of the IFERSA.[19] Yet Johnson's debt to the zetetics of the past extended further than quotations, references and praise, for his view of the universe was a continuation of the ideas promoted by the prominent flat-earth campaigners who had gone before. Equally he held that the earth was a vast flat disc, floating on water, with the North Pole at its centre and a wall of ice, 150 metres high, round the outside edge. This barrier formed the South Pole of popular legend and, like Parallax, Johnson was sure that what lay outside was beyond the realms of human comprehension.

As for the earth's size, Johnson was similarly vague; he pre-sumed that it was infinite and 'didn't really have a size'. To him, it was simply 'world without end'. Two points about which he was certain, however, were that gravity was non-existent and that the earth was not a planet; it was the static centre of the universe around which the thirty-two-mile-wide sun and moon circled, three thousand miles above the disc. The main difference between these bodies, Johnson explained, was that the sun radiated 'hot light' and the moon 'cold light'. Meanwhile, he argued, a thousand miles beyond the sun and moon lay the stars and the dome of heaven.[20]

As for alleged 'proofs' of the earth's rotundity, Johnson also adhered to the official line laid down by past zetetics: circumnavigation was akin to travelling round an island; the phenomenon of ships disappearing over the horizon was an optical illusion; sunrise and sunset were tricks of perspective; lunar eclipses were the result of an unidentified dark body pausing in front of the moon; and as for solar eclipses, Johnson explained that people did not need to investigate that point for 'The Bible tells us the heavens are a mystery.' 'You know,' Johnson continued, in the *Montréal Gazette*, 'a lot of things that seem to be are not . . . and if people will only examine the evidence, they'll realize the Earth is flat.'[21] On this subject, the fundamental point to which Johnson returned, on successive occasions, was Parallax's original 'proof': the flat surface of water. If the world were round, he asserted that for every six miles of water there would be a drop of twenty-four feet. Consequently, Johnson reasoned, large lakes would have humps in the middle, and the 101-mile-long Suez Canal would feature a 6600-foot-high hill of water, a concept he branded 'absurd'. Spurred on by this belief, and inspired by Parallax's experiments and Voliva's statements, Johnson ran tests on Lake Worth (near Fort Worth, Texas), Lake Tahoe (on the California–Nevada border), and southern California's Salton Sea, and alleged that all three bodies of water, along with the earth's surface, were entirely flat. Backed by convincing proof, or so he believed, for Johnson it was now a matter of disseminating his truth to the rest of the world.[22]

The tactics for his campaign were the familiar ones adopted by those with a cause to promote: membership of a society and publication of a regular magazine. But although his approach was conventional, Johnson's quarterly *Flat Earth News* was unlike the majority of subscription publications available to the public. Most noticeably, its byline was 'Restoring the World to Sanity' and the unusual slogan was matched by the extensive use of capital letters, boldface type and a frantic narrative style. Scandalous and polemical, the magazine's subject matter was similarly eye-catching: a heady mix of flat-earth theory, outrageous conspiracies, Biblical quotations, offbeat political commentary and anti-vivisectionist advertisements, which could not fail to catch the attention of those who happened to flick casually through its pages. In addition, the *Flat Earth News* carried regular features, including the William Carpenter-inspired column, 'One Hundred Proofs that Earth is not a Globe', and an oasis of calm, 'Marjory's Corner', in which Johnson's wife contributed her ideas about life along with Biblical extracts and poetry.

For his part, Johnson published copious material on scientific fraud and the supposed 'death of science', and the paper was packed with quotations assigned to authorities from Roald Amundsen to Darwin, Goethe to Freud, designed to bolster his flat-earth case. Readers could not fail to notice the militancy that seeped through every issue of the paper, yet Johnson's writing style, particularly his typographical quirks and shambolic grammar, belied his intellect. He was undoubtedly a well-read man, who sometimes complained that his literary shortcomings were due to the fact that his pen could not keep pace with his mind. Johnson was also witty, and inventive wordplay, a particularly striking rhetorical strategy, became a characteristic feature of *Flat Earth News*. Only here could one find sputniks transformed into 'spookniks' and evolution as 'evilution', while Copernicus became 'Copernicious' and 'grease ball' was utilized as a universal term for a spherical earth, because for Johnson (who, as we know, dismissed gravity) if the earth were round mankind would slide off into space.

It was an unconventional argument and, to many, Johnson must have seemed misguided, but from his perspective he was endeavouring to assist the human race. Kind and compassionate, if mercurial and paranoid, his heart, alongside his meagre income, was dedicated to the 'Flat Earth Work'. Consumed by his mission, he wrote of feeling 'like a Professor in a wild reformatory' where the children had not been raised correctly; as a consequence his 'job' was to attempt 'to reach them' rather than 'get mad'.[23] Above all, Johnson wanted to make the world a better place: on a millenarian mission, flat-earth theory was a central tenet of a more broad-based social improvement plan. 'Facts, Logic, Reason, Sanity, also known as Common Sense' were the society's watchwords, and he frequently spoke of his altruistic desire to encourage people to 'use their minds logically', or less positively, to assist them to 'come out of the herd and become sane'. From his perspective, Johnson was seeking to assist 'the human stragglers' among the eating, drinking, breeding 'herd of beasts', which constituted unthinking society as a whole.

Through the 1970s Johnson felt he was having some success, and by the end of the decade, the IFERSA had two hundred members paying an annual subscription of seven dollars, although their commitment is questionable in light of the fact that only a fraction paid their dues after the first year. Conscious of his society's reputation, Johnson was sure to emphasize that, contrary to popular perception, 'kooks' did not dominate the group: doctors and lawyers constituted the biggest percentage of members by profession, while airline pilots, Zion residents (in all likelihood taught that the earth was flat in the Voliva-run schools) and even a state governor had likewise enrolled in the society. With the exception of a poison-pen letter-writer in South Carolina, hate mail addressed to the IFERSA was rare. Moreover, Johnson claimed that it was not merely a hardcore of fee-paying members who supported his cause: he had thousands of unregistered followers who concealed their beliefs to avoid animosity in their daily lives. For people of this type, Johnson was happy to provide confidential membership, a safe haven and a spiritual sanctuary in the IFERSA, and his recruitment flyer welcomed individuals of 'goodwill who seek the truth[,] also known as the Facts'. Meanwhile 'stupid,

mindless, brute beasts with two feet whose only aim is to scoff'
were deemed ineligible for membership and every applicant was
required to sign a statement agreeing never to defame the group.

Although research suggests that Johnson had many serious
members, some inquisitive non-believers also negotiated the selection procedure and chief among them was Robert Schadewald, a
science writer who assisted in making the society known to a broad
audience with a series of fair-minded articles. Entirely frank about
his globularist stance, he was careful to adopt an ethical and factual
approach in his dealings with Johnson in his numerous publications about the society. Yet even this led to dispute. Much to
his amusement, Schadewald was expelled from the group on the
grounds of his 'spherical tendencies' and on another occasion
received a letter from Johnson's attorney forbidding him to publish
anything pertaining to the IFERSA. Yet despite this and myriad
differences over orthodox science, the men remained largely on
good terms and corresponded with one another for many years.
On one occasion, Johnson even asked if Schadewald would be
willing to take over the FES, an offer he declined due to his
globular convictions.[24]

Importantly, all who wrote letters to the society received an
answer from Johnson, for he saw it as his God-given purpose to
help mankind reach 'real knowledge' about the shape of the earth.
Employing familiar zetetic rhetoric, Johnson believed he was
shining a light in the darkness, carrying a torch for his perception
of truth. The society was a mission, a calling, and this commitment
involved a heavy workload, for the IFERSA received about a dozen
letters a day, approximately two thousand a year, from locations
as distant as Finland, India and Iran. Besides answering them,
Johnson accepted public-speaking engagements whenever the opportunity arose. Well aware that the IFERSA lacked credibility, he
admitted that audiences frequently laughed when he began to
speak. He always remained confident, however, that they would
eventually 'calm down' and give him a fair hearing, for that,
Johnson said, was all that he wanted.

Besides lecturing to societies, university and high-school students, branches chapters of the Elks, the Kiwanis and the Rotary Club, Johnson was interviewed over the years on a number of radio stations and the major American television networks. In the late 1980s, he even starred in an advertisement for Dreyer's ice-cream, and although his plans for a zetetic research centre, a national flat-earth convention and a lecture tour in a motor-home never materialized, media coverage guaranteed that the American people knew he was there. That was enough for Johnson. He did not care for making money or collecting converts and refused to proselytize. Pragmatic and self-assured, he reasoned that people would either understand what he was saying or not; the 'Flat Earth Work' was just his and Marjory's 'very Life', while in private Robert Schadewald doubted that the couple ever had a break-even year.[25] Yet they did not care for worldly riches or material goods and never resented having to live below the poverty line. Besides, having chosen their path, Johnson explained that he and his wife required little – 'a few beans, plenty of coffee and tobacco, a roof' were enough:

> Life is so fleeting, seems to me makes little difference what I eat, what I wear, just feel I must make my life count ... to have [done] something worthwhile, to be an aid in God's Creation. I can look myself in the mirror without shame, and look the world in the eye ... I know I'm doing my 'bit', and am true to my vision, so can't complain.[26]

Consistent with their eschewal of the material world, the Johnsons remained content to serve as witnesses and 'servants of the God of Truth', accepting the fellow seekers of knowledge who happened to cross their path.[27] Johnson characterized himself and Marjory as 'the avant garde'. In line with mythic and Biblical traditions of a chosen people, they were an example to society: they had cracked an age-old puzzle and stood 'way ahead of the pack'.[28] From his perspective they were innovators rather than eccentrics to be pitied: it was science that was the 'old time stuff'.

Johnson understood, however, that this was a minority view. The mission was difficult and isolating, and he accepted that, although he also admitted to Schadewald that this 'does kind of get to you sometimes'. Essentially a two-person sect, with a complex set of values and limited external support, Johnson confessed that from time to time he was 'overcome by the sad realization that Marjory and I are so alone, but unafraid, in a world we never made', not merely in relation to the flat-earth concept but on all that the idea entailed.[29] On the whole, though, he was philosophical about his efforts to subvert the status quo, commenting that opposition was 'no skin off [his] nose' because he was 'only trying to teach logic'.

Johnson's seemingly paradoxical emphasis on straightforwardness was further reflected in his egalitarian moral code, for he saw himself as a decent, straight-talking Texan, who despised duplicity and crooked dealing:

> I was born and raised in west Texas, where my father a cattle rancher of the Old West taught me to be honest and true, to tell the truth[,] treat every one fair and square, and fear nothing or anyone, confident that if your [sic] 'right' you'll win in the end ... This is the basis of my dedication to the International Flat Earth Research Society and to keeping alive this Truth or Fact, [that the] Earth [is] flat, [and] wherever the chips fall I am 'for' the Facts![30]

Obsessed with truth and honour, Johnson frequently berated the lack of honesty in the world: people seemed willing to sell their souls for money and prestige. By contrast, the Johnsons gained a sense of peace and satisfaction from living a simple life, dedicated to their beliefs. As Marjory put it, the 'Flat Earth' was just one part of 'working out the right and true way of life'.[31]

In the light of this, the activities of their Canadian rivals were always bound to cause dispute. Johnson first caught wind of the organization in June 1974 and immediately wrote to Fredericton for further information. Leo Ferrari did not reply. Two years later, in September 1976, Johnson read the *Saturday Review of Sciences* article on the Flat Earth Society (FES), and decided to make contact again. It was an animated letter. Delighted by the prospect of a like-minded campaigner, Johnson wrote that it was 'a very happy day' when he learned of the Canadian society and discovered that Ferrari was an Australian by birth. Excitedly, he reported that Marjory was also from Australia, and for many years he had felt that their birthplace had some sort of 'special destiny'. Pleasantries aside, he said he understood that the 'folks' in Canada knew of the IFERSA yet had chosen not to inform him about their parallel campaign. Johnson said that this was fine and emphasized that it was not his intention to interfere with FES work. Merely curious, he asked Ferrari for further details. 'I feel sure at the core we can't be too far apart in aims,' he enthused. 'I do try to practice what I preach, to think and seek and search out reasonable ideas and concepts.' Then he signed off, proclaiming that he could 'hardly wait' to hear from Ferrari, who he hoped would reply by return of mail.[32]

Six months passed, and still Johnson waited for a response. On 14 March 1977, his patience evaporated and he wrote a further letter. On this occasion, he requested a copy of the society's journal, the *Official Organ*, but as Ferrari had chosen to be 'somewhat secret', Johnson stressed that he had no intention of using his name in public or calling attention to St Thomas's University. His sole purpose, he explained, was the zetetic desire to enlarge his view and to practise his 'religion of getting and holding onto the Facts'. This, he believed, could only be of benefit to himself and, in time, the rest of the world. He continued that Ferrari was honour-bound to send the information because he had publicly adopted the label 'Flat Earther'. Like it or not, he was thus intimately connected to the mainstream campaign. If Ferrari proved willing to comply with his demands, Johnson offered 'thanks from the bottom of [his] heart' in advance. But if Ferrari refused to act decently and reply with a copy of the magazine, Johnson warned, 'I will then know for sure you are some kind of enemy of the Flat Earth Work.'[33]

Whether Ferrari heeded the warning is unknown, but he did reply, enclosing an FES application form and a pamphlet, rather than a copy of the *Official Organ* as initially requested. Finally the truth about the Canadian society began to dawn on Johnson. Missing the nuances, albeit well disguised, he was furious that the flat-earth idea was being used as a 'gimmick' and furthermore 'to entertain and promote' the 'atheistic society' he sought to oppose. In Johnson's opinion, Ferrari was absorbing the publicity that belonged by rights to the genuine campaign and was a false prophet guilty of 'muddying the waters of truth'.[34] In June, Johnson gathered his thoughts in a robust letter of complaint to the Canadian society. Full of recriminations, he listed a string of offences: Ferrari's campaign was causing confusion over the identity of the true flat-earth society and what it represented, and for two years Ferrari had ignored his letters, electing to conceal the character and tone of his operations. Now was the time, according to Johnson, that Ferrari needed to know that 'We here represent the direct line of the Flat Earth Society', and as the Australian-Canadian was an illegitimate interloper, 'some explaining is due from you'. Further to this, he demanded information about Ferrari's religious beliefs and more samples of his work, then highlighted errors in the pamphlet he had already received:

> Some errors: you mention you are fighting a 'lone' fight. And you 'presume' to tell what the Flat Earth Society 'stands for' and you the 'President'!!!!! Now according to [a] Boston [news]man you have muddied the waters regarding the Flat Earth Society![35]

For Johnson, the whole affair was a complicated priority dispute and, in keeping with his opinion of Ellis Hillman, whom he saw as a 'double-dealer', the Canadians were bent on taking the work from those who truly believed. In private Johnson criticized the group as 'idiot devils', and Ferrari as a 'creep', a 'jerk' and a 'rat'. That Ferrari was an 'expert' philosophy professor led Johnson to conclude that he was a 'criminal idiot', for whom there was no hope. 'If somebody went that far along an idiot system, and co-operated that much,' said Johnson, referring to Ferrari's academic background, 'his character certainly is no good.'[36] Yet despite his anti-intellectualism, Johnson informed Ferrari in the midst of the controversy that he would postpone final judgement until he had gathered the full facts in accordance with zetetic teaching. His decision was pending for a while. Ferrari ignored his irate letter and, with his tongue firmly in his cheek, later confided his 'disappointment' in the Californian group to Robert Schadewald.[37] Meanwhile, Johnson's letters had been stored for posterity in Ferrari's 'Hope He Goes Away' file, his permanent resting-place for correspondence from cranks.

Johnson's mission was certainly unorthodox, even in the light of previous zetetic campaigns. Although the marriage of scientific and religious imperatives was familiar, his tactics were more complex than those of his predecessors, particularly in terms of meanings and scope. In keeping with tradition, his stance remained a matter of scriptural authority in preference to the findings of conventional science and the advocacy of an extreme form of creationism in accordance with this approach. Importantly, like Voliva, Johnson professed to loathe fundamentalists and scientific creationists who used Biblical inerrancy and a literal interpretation to oppose evolution while simultaneously embracing the idea of a spherical earth. For Johnson, it was 'common sense' that the Bible was a 'flat-earth book', it was 'impossible to believe in God and believe in the lunacy of a spinning ball Earth!'[38]

In private letters he reserved special venom for the heart of the creation-science movement, the Creation Research Institute (CRI), established in 1962, and he denounced its leadership as 'demented devils', a 'criminal gang', 'phonies' and 'the worst enemies of truth'.[39] Johnson believed that the CRI was actively undermining the Bible while publicly claiming to defend it and was undoubtedly in league with 'Satanic Grease Ball Science'.[40] Intriguingly, Robert Schadewald believed that Johnson's bitterness stemmed partly from the fact that he had approached the institute's founding member, director and co-author of creationist landmark text *The Genesis Flood* (1961), Henry Morris, with an offer to amalgamate, which was categorically declined.[41] Whatever the veracity of this claim, the CRI had no connection to the IFERSA so it was natural for Johnson, with his dualist world-view of polarized extremes – a grand dialectic of evil and good – to assume that, as outsiders, Morris *et al.* must be operating on behalf of the adversary's cause.

Johnson, therefore, placed himself in direct opposition to the CRI, and a call to divine authority was central to his campaign. The IFERSA was incorporated as the 'Covenant People's Church' and the 'fact' that 'round-Earth theory' was 'anti-God' frequently received a vociferous airing in *Flat Earth News*. Throughout his campaign, Johnson argued that a globular earth was central to science's evil plan to 'destroy all faith in God', while the high-flying space programme was specifically designed to destroy belief in a firmament or heaven. 'The moon landing . . . leaves people as little insects with no hope,' Johnson once told a reporter, '. . . there's no right, no wrong, nothing nothing nothing. This is why our youth are going to pieces.'[42] For Johnson 'the spinning ball thing just makes the whole Bible a big joke' and under *Flat Earth News* headlines linking the twin 'facts', 'Earth is Flat . . . God Exists' and 'God says Earth is Flat', he discussed the issues and claimed the ultimate authority for his cause.

Besides his strident defence of a scriptural world-view, Johnson adopted another tactic to add weight to his campaign: reversing the traditional meanings of science and religion, so that science became 'superstition', 'a form of idiot religion with no God', while the Bible was 'fact'. On numerous occasions, Johnson declared that the twentieth century was the 'Dark Ages', the most superstitious era of world history, and that it was his ultimate objective to 'replace the *science religion* with sanity'. In keeping with this approach, an IFERSA advertisement maintained that scientists were nothing but a 'gang of witch doctors, sorcerers, tellers of tales [and] "Priest-Entertainers" for the common people'. Scientists were the 'kooks', while the body of knowledge they peddled was a 'false religion, the opium of the masses'. By contrast, Johnson cast himself as a heretic to the 'religion of science' who was 'bringing on the Age of Reason'.[43]

Such reversed meanings continued with Johnson's contention that the spinning-globe idea was a '100% religious doctrine', a 'way-out occult weird theology of the old Greek superstition', and a 'Blind Dogmatic Article of Faith' for the 'illogical, unreasoning "herd"'.[44] As far as Johnson was concerned, science was a religion run by 'evil vampire type people who concoct theories and . . . tie up people's minds . . . and prevent more and more improvements'.[45] The science 'religion' was invented, founded and named in 1840 and '100% unrelated to facts'. In fact, science was allegedly rooted in dogma to such an extent that Johnson branded it as a 'sun-worship religion', implying its basis in irrational, primitive beliefs. In keeping with this redefinition, scientists were 'pseudo-scientists' who peddled a pack of unfounded assertions: that the earth was round, that the world was 500 million years old, that the universe was an emanation of 'chaos', that the sun was 93 million miles from the earth and that men took their origins from monkeys. If these were established facts, Johnson argued, he would be prepared to 'bow before the truth', but because he believed hard evidence to be lacking, the 'pseudo scientists will have to prove it if they want us to believe it'.[46]

Equally serious for Johnson was his belief that science obstructed people from 'coming to real knowledge about why we are here and how to live' with 'the Religion-Theology theory of evolution'. In common with creationists, he held that science was a satanic device and the 'official science doctrine' of evolution by natural selection was solid proof of this scheme. Johnson saw Darwin's 'discovery' as a vicious concept, by which 'the most evil . . . the most cunning, [the] hardest [and] coldest' are the natural victors in a competitive free-for-all, where creatures had no choice but to prey on one another to survive.[47] For Johnson, 'evilution' was senseless and wrong, firm evidence that scientists were determined to undermine all that was good in the world. Just as Darwinism had materialist implications, so too did round-earth theory; in fact, they were facets of the same anti-God phenomenon. In lectures Johnson contended that 'the same gang' who informed people that they had 'hatched out of hot slime in a swamp' were 'the same creeps of the greaseball religion'.[48]

While Johnson identified scientists with atheism and material-ism, a sweeping and inaccurate generalization, he argued that the practice of vivisection provided further proof of scientific malevol-ence and moral depravity.[49] In every issue of *Flat Earth News*, Johnson reminded his readers that animals were being poisoned, bludgeoned, trapped and cut by 'the same scientists who claim that Earth is a globe and they and Buck Rogers orbit it'. Such scien-tists were uncontrollable animal torturers, Johnson asserted, the 'debased, degenerate, perverted, maggot-infested brain of the devil in the flesh', and he dedicated himself to ending the 'holocaust' they had started. To persuade his readers, grotesque pictures of starving ponies, puppies in cages and kittens in clamps accom-panied strongly worded articles in *Flat Earth News*, while Johnson, an animal lover and vegetarian on ethical grounds, promoted an end to animal testing through advertisements for the American Anti-Vivisection Society (AAVS) and People for the Ethical Treatment of Animals (PETA). He was proud that he had always been an advocate for the 'dumb', and never expected credit for this work, for he believed that 'getting it done is the main thing'.[50] All in all, while combating the globe concept remained his major preoccupation, Johnson's case against science was broad: not only was it anti-God, it was also anti-Good.

In reaction, Johnson claimed, like modern creationists, to take the truly scientific approach. While zetetic predecessors, such as Lady Blount and Voliva, had emphasized the Biblical authority of their truth claims, Johnson also chose to underline facts, personal experience, objective reality and 'scientific' proof of the earth's flatness, harking back more frequently to zetetic ideals than scriptural quotations to bolster his case. An appeal to Baconian empiricism (data collection) was a focal point of his campaign and Johnson's phrase 'Flat Earth Science' summed up the enterprise; he insisted that 'Flat Earth is not a religious claim it is a demonstrated fact.'[51] In many ways, while Johnson looked to the Bible, and the Ten Commandments in particular, as an ethical guide and the basis of all that was right in the world, he viewed those who engaged in the rigorous exegesis of scriptural passages with a substantial degree of disdain. He often told reporters that the flat-earth idea was based on real experience rather than 'mathematical scribblings' and he completely rejected theories and hypotheses on the ground that they were 'imaginary'. In his opinion Newton and Darwin had simply invented laws to support their own theories, thus stifling curiosity and preventing people gathering the facts and realizing their full individual potential alongside the truth about the natural world.

As with many zetetics, the democratic concept of common sense was one of Johnson's catchphrases, and he prided himself on his anti-intellectualism, his reliance on common sense and his refusal to listen to an 'expert' élite.[52] Undoubtedly, his position was founded on no-nonsense realism, rather than the mystifying theories peddled by a select professional few, and highlighted that, in common with other flat-earthers, it was not science as such he opposed. His target was more rightly orthodox 'scientific' knowledge, specifically the 'theory' that the earth is a globe, and in this he echoed the 'evolution is just a theory' line of argument propounded by Darwinian critics. The popularity of the globe idea puzzled Johnson and he was sometimes at a loss to understand why the general public did not see the world as he and Marjory did. 'People treat us as if we're some sort of radicals, as if the flat-earth theory is a radical, eccentric idea,' he complained to journalists 'It's like there's some vast body of proof that the Earth is round.' But for the most part, Johnson was unconcerned, for his zetetic 'fact-finding' religion boosted his position and his sense of purpose. It was this rather than fundamentalist Christianity that underpinned his campaign, as he told Leo Ferrari in 1977:

In the past Flat Earth in USA and England has been a kind of branch [of] 'Christian religion.' In any case seeking to prove Earth flat by the 'Bible', I DO NOT do this! I AM NOT a religious fanatic, in fact Christians are my worst enemy! Tho [sic] being sane, I know a creation had to have a creator, I do believe in Logic![53]

Raised as a Nazarene, when he reached adulthood Johnson found he was unable to form a meaningful connection with some of that Church's doctrines. In response he carved out his own personal belief system, a complex and sometimes contradictory interchange of right-wing fundamentalist teachings, countercultural criticism of the military-industrial-academic complex, zetetic philosophy, apocalyptic visions and millenarian hopes for a New Age. Thus 'science' and Christianity were irrevocably linked in his multi-faceted quest: 'It is my sole purpose in life,' Johnson wrote in April 1979, 'to put over Earth is Flat and the Ten Commandments are both reasonable, logical, rational, demonstrable and True.'[54]

While Johnson's complex campaign was sold to the public as a battle for Facts, for God and for Good, it was also presented as a battle for America itself. Amid the cacophony of crisis rhetoric and the chorus of heated demands that thundered through *Flat Earth News*, there were constant calls for patriotism and the protection of freedom, democracy and the American way. Science was killing the American Dream, according to Johnson: the people had been disenfranchised, the 'land of the free' had been invaded. The argument had been advanced before, albeit in a different place and time: at the pinnacle of British imperial, technological and economic superiority, Hampden and Carpenter had branded modern astronomy a threat to the condition of England and rallied the people to fight for their right to discover the 'facts' and to destroy the 'false' knowledge disseminated by a rapidly emerging professional élite. Patriotism and egalitarianism had continued to serve as a double-sided weapon in the zetetic armoury ever since, and they provided ready ammunition for Johnson as heir to the cause. He subsequently translated nationalist imperatives into the contemporary American context during an era when conspiracy theories and public anxiety about individual sovereignty and technocratic entrapment had permeated mainstream debate and the United States was the predominant international force.

In some ways the argument was self-contradictory, for it posted America as the greatest and worst at once, but such inconsistencies were irrelevant to Johnson. With overtones of the New World Order school of analysis, an increasingly familiar frame of reference, he warned his readers that a bell was tolling, that a secret war was being waged against them, that their minds and their liberty were under attack. The only way that they could avoid intellectual captivity, Johnson declared, was to be zetetic, think for themselves and refuse to take 'expert' knowledge for granted. Anything less was complete surrender to the supposed 'authorities' – to the insidious mind-control conspiracy concocted by the government, the media, academia and science.

On repeated occasions, Johnson cited the terrifying vision of George Orwell's novel 1984 for he was certain that a similar phenomenon was occurring around him, if only his fellow Americans could see it. Desperate to spread the word, he reprinted long extracts from Orwell's book in successive issues of Flat Earth News with notable sections in bold. Under a headline that strung together three dystopian classics, '1984 – Brave New World – Animal Farm', Johnson held forth about indoctrination and 'double-think', how those who believed what they were told had created a 'one-world animal farm', and how innocent children were being brainwashed to accept 'the new state religion of greaseball science', the 'popular religion of the animals on animal farm'. Schools were programming children like beasts, Johnson declared; their minds were being warped by what they were taught.[55]

In keeping with the conspiracy rhetoric of his era, the political aspect of Johnson's campaign was particularly blatant. 'Everything is political,' he told Robert Schadewald. 'Everything to do with [opposition to] the flat earth is entirely religious-political. It has nothing to do with the facts.'[56] For Johnson, flat-earth belief was an issue of democracy. A soft-spoken, self-styled prophet, in his more serene moments he hoped that his society would stand at the forefront of a 'New Age of the People' when everyone had a voice, not merely the 'elite experts and pseudo-intellectuals' whose only intention was 'to scoff and debunk'.[57] This was Johnson's Utopia, no different in essence from a 'Science for All' ideology, albeit incontrovertibly opposed in terms of the actual facts.

Johnson was undoubtedly given to rethinking the facts on a grand scale, and he invested much energy into devising a sweeping revisionist narrative of human history — a patent attempt to rationalize a seemingly random, chaotic and threatening world into a comprehensible, meaningful order.[58] His narrative commenced with the contention that God had created the earth as a plane, as the Bible described, while 'Moses and the Prophets and Jesus and all his Disciples taught and knew [that] Earth [was] flat.' The heretical proposition that the earth was a globe was a product of 'Greek superstition', known to the world as Greek science. Despite the falsity of the myth concocted by 'crazy, fire-worshipping pagans', over centuries the idea had become established until Christopher Columbus discovered the earth to be flat.

Johnson was certain that the myth that Columbus was the first to discover the earth to be round on his voyage to the New World was a delusion, and the reality was the reverse. According to Johnson, Columbus discovered the earth was flat and was 'one of our gang'. He had agreed with the 'global nonsense' merely to win favour 'with the scientists and priests in control', but the American public were ignorant of this fact 'because it's not taught in government-run schools'. For Johnson, the 'fact' that gravity did not exist in 1492 made the likelihood of sliding off a round earth much greater than sailing off a flat plane, while the traditional story about the crew of the *Santa Maria* believing the earth to be flat and being fearful of sailing off the edge was unfounded. Johnson began his account with the assertion that the crew of the

Santa Maria actually believed the earth to be a globe, and because there was no gravity, they were frightened that they would sail over a curve and fall off the spherical world. Terrified, they threatened mutiny – but all was not lost. Columbus defended his alleged flat-earth beliefs, 'put them in irons and beat them until he convinced them they weren't going over any curve and they could return'. The harsh tactic was a success, Johnson concluded, and Columbus 'finally calmed them down'.

Besides its falsity, the story, frequently recounted by newspaper journalists, involved some interesting twists: first, the Columbus myth was so established in the popular mind that Johnson was driven to adopt Columbus, an explorer, as a scientific authority to add weight to his case. Furthermore, newspapers frequently printed mocking interviews with Johnson as part of Columbus Day features that perpetuated the myth that the explorer discovered the world to be round. Thus while Johnson was guilty of misinforming the public about the course of history and the shape of the earth, so too, less dramatically, were some of the journalists who ridiculed the unorthodoxy of his ideas.

As myths, misinformation and conspiracy theories became increasingly endemic in late-twentieth-century popular culture, Johnson claimed the same for the sixteenth-century context, arguing that Columbus's view was widely accepted until two events re-introduced the 'spinning ball dogma' to the world. The first was the establishment of the Church of England by Henry VIII and its determination to disobey the Bible by teaching that the world was a globe, a ruse backed by Protestantism writ large. 'The whole hoax is part of the Christian religion invented by Martin Luther and that gang,' Johnson insisted in *Flat Earth News*.[59] Due to its rejection of the flat-earth idea, Johnson argued that Christianity was behind 'the globe business' and in his world of false dilemmas was therefore in league with science. More threatening still for the flat-earth 'fact', he claimed, was the so-called 'Scientific Revolution,' when 'Co-pernicious [*sic*] and co.' revived the Greek globular-earth myth and presented it to the world as truth. Johnson

presented his version of events as a catalogue of disaster: Copernicus invented the idea that the planets revolve round the sun and
that our globular world rotates, Galileo used a telescope to fabricate
experimental findings and to perpetuate the deception, and Sir
Isaac Newton, the 'Priest in England', concocted the laws of gravity
and motion to defend the Church and to conceal Columbus's
rediscovery of the 'fact' that the earth was flat.

In the face of this European conspiracy, Johnson contended
that America was a sole beacon of hope and bastion of true
knowledge in a corrupt world.[60] The guiding light above all others
was the nation's first president, George Washington, a flat-earth
believer who deliberately 'broke with England to get away from
those superstitions'. Safe in the knowledge that the American
Revolution was underpinned by these beliefs, thousands of European settlers fled to the New Land to escape persecution in the
countries of their birth. Although America has long been renowned
for its religious freedom and diversity, spiritual havens and utopian
experiments, this was undoubtedly an original interpretation of the
facts.[61] Johnson himself admitted that it was not widely known but
the United States of America was 'the first flat earth society', and
throughout the twentieth century, the nation had continued to
promote this unorthodox world-view. In the wake of the Second
World War, Franklin D. Roosevelt, Joseph Stalin and Winston
Churchill had finalized plans for an international peacekeeping
force, the United Nations, but again Johnson offered an alternative
version of events:

> Uncle Joe [Stalin], Churchill, and Roosevelt laid the master
> plan to bring in the New Age under the United Nations . . .
> The world ruling power was to be right here in this country.
> After the war the world would be declared flat and Roosevelt
> would be elected first president of the world. When the UN
> Charter was drafted in San Francisco, they took the flat-earth
> map [a polar azimuthal equidistant projection] as their
> symbol.[62]

On these grounds, Johnson argued that the UN was a flat-earth organization and, for him, its insignia served as an emblem of hope. He truly believed that the flat-earth society was 'the oldest continuous society' in the world; it had begun with the creation and Johnson was sure it would survive long after the spinning-ball hoax had passed away. 'We're going to shatter this false [globe] doctrine,' he told Schadewald, 'and we're calling on whomsoever will join us.'[63] It was the ultimate battle of all: an apocalyptic struggle of good against evil underpinned by a millenarian dream. This double-sided approach mirrored Johnson's split personality: at times gentle and reasonable, he could also come across as insane. Frequently fixated with imagery of new ages and Armageddon, he asserted that Satan had been released when Copernicus's *De revolutionibus* was published in 1543 and the world had been under his influence ever since:

> From around that date all the present evil and satanic lies on Earth have come forth, the COPERNICIOUS [*sic*] monstrosity, Galileo, Newton, Martin Luther . . . Descartes, Darwin, Fake Space Program . . . these evil degenerate beasts have come forth . . . now is the season of Satan. Truly has been the DESCENT of Man, from then Werner Von Braun with his hidden friends make the final step to try to **DECEIVE THE WHOLE WORLD!**[64]

While concurring with creationists that evolution was satanic and employing Biblical imagery in comparing scientists to the ten-horned, seven-headed, Antichrist Beast of the Book of Revelation (Revelation 20), Johnson extended the argument by claiming that the space programme was the final frontier, the culmination of a timeline of deceit running from the sixteenth century until the present day. His vision was apocalyptic and his approach to the Apollo missions, an ominous sign of the end, was more straightforward than Shenton's, for Johnson simply denied that they had ever taken place: 'It's one big lie,' he told journalists. 'It's nothing more than a piece of clever stage-managed science-fiction trickery.'

NASA and world leaders knew that the earth was flat, but they had launched the $24-billion-dollar space hoax as a 'scientific plot to hoodwink the public'. From Johnson's perspective, there was no alternative explanation: it was impossible to orbit a flat earth, rockets could not penetrate the firmament of heaven, and such feats were unnecessary because information about the universe and its creation was laid out in the Book of Genesis: 'God created the heavens above and the Earth below and there should be no further doubt about the matter,' Johnson asserted. Yet the 'fact' remained: round-earth belief was rampant, the result of a diabolical web of deception designed to boost profits for big contractors.[65] Reasonable, intelligent people had always recognized that the earth was flat, Johnson claimed; they said otherwise only because round-earth theory and the 'space' programme generated economic opportunities and jobs.[66]

Johnson frequently backed up such sweeping allegations with specifics in a bid to sell his agenda, and the Apollo 11 moon landing was the natural starting point. During numerous interviews he informed reporters that the event was entirely fictitious; the footage of Armstrong and Aldrin on the moon had been scripted and directed by science-fiction writer Arthur C. Clarke and was a Hollywood-movie version of the way things could be if the spherical theory were accurate.[67] Shot at Meteor Crater, near Flagstaff, Arizona, the footage was a 'big, giant hoax', carried on by the government for the entertainment of the masses – the 'animals here in the Animal Farm'.[68] Moreover, the lucrative rights to the production had been allotted to America at a Kennedy–Khrushchev meeting in return for Cuba.[69] Under those circumstances, NASA was a Florida-based entertainment organization and, in Johnson's opinion, a second-rate one. He frequently lambasted what he called the 'grade-Z movie' of the Apollo 11 moon landing, claiming that the Disney corporation, along with various unidentified 'English humorists', had had some involvement.

Also allegedly entangled to varying degrees were the major American television networks, 'Nazi German Scientists' and the

Mormons, who, he claimed, had captured missing millionaire Howard Hughes and assumed control of his companies in order to fund the NASA 'Space Hoax Racket'.[70] Johnson believed Florida to be riddled with drugs and organized crime; he even asserted that the NASA mob were 'dope fiends ... half crazed with smulch'.[71] Yet while accusing the 'enemy' of moral depravity was a well-worn demagogic technique, Johnson truly believed that NASA was capable of anything in its campaign to safeguard the hoax. With a hint of persecution mania, he claimed on one occasion that a 'Seventh Day Adventist and Nazi-NASA stooge' had planned to murder him and Marjory at their isolated desert home.[72]

Such demonization naturally extended to astronauts and Johnson asserted that they had either been force-fed 'drugs that space you out' or brainwashed into believing the round 'Earth and planets nonsense'. 'Some astronauts actually believe they've been in space,' Johnson marvelled, to a reporter from the *San Francisco Examiner*.[73] All he could do was advise Neil Armstrong to 'confess and beg for mercy' for his leading role in the worldwide scam.[74] Johnson was unconcerned when pundits claimed it had cost $24 billion to get Armstrong to the moon and that every day of the mission had been shown on live television; he retorted that there had never been any real proof of space travel. Rocket launches were not evidence in themselves of exploration beyond our planet and all that existed in fact was a sophisticated slice of science-fiction trickery designed to 'finally unhinge the gears of everybody in America by claiming there's proof a man landed on the moon'. '"I saw it on TV,"' Johnson mocked. 'How silly can you be?'[75] Although he remained uncertain about how the fake had been produced, Johnson contended that the details were not important. 'I've seen magicians on the stage saw people in half. I even saw one who made an elephant disappear,' Johnson told a reporter. '*Star Wars* was made out in a fellow's garage in his backyard,' he continued, 'and that was about the whole universe.'[76] How the hoax was perpetrated was not the issue: all that mattered was that people questioned the 'facts' that they had previously taken for granted.[77]

Deception and threat were themes that Johnson continued to repeat throughout the next two decades. When the space shuttle Challenger exploded seventy-three seconds into its tenth mission in January 1986, Johnson speculated that it had either been exploded by NASA or had fallen victim to a divine curse. In a special issue of *Flat Earth News*, headlined 'Challenger Blown up by God', more revelations were to follow. Ronald Reagan was a secret flat-earther, Johnson claimed, for during television interviews about the Challenger disaster he had mentioned God and looked up into the sky. Johnson was jubilant. According to him, Reagan's gesture illustrated his belief that 'heaven is above the Earth, which implies earth is flat'.[78] 'It is impossible for Mr Reagan to believe in God and the Space Shuttle,' Johnson continued, and he left it to his readers to decide where the president's true loyalties lay. Meanwhile, somewhat paradoxically, in subsequent issues of the paper Reagan was lauded as America's greatest president, while his wife Nancy was praised as a 'genius' for her 'Just Say No' anti-drugs campaign.[79]

Johnson continued to exercise his democratic right to freedom of speech through the late 1980s and by the mid-1990s the IFERSA had approximately four thousand members worldwide, although it had been a long haul for Johnson and the work had taken its toll. In 1992, he admitted to filmmaker Robert Abel that defending flat-earth views 'makes you kind of a loner, you know . . . everyone likes to be liked, but you can't be liked. You have to make up your mind to do without that . . . because people stay away. They don't want anything to do with a controversial idea.'[80]

Yet Johnson was not as alone as he feared. In July 1994, the month of the twenty-fifth anniversary of the first moon landing, the *Washington Post* asked a random sample of 1001 Americans whether they thought that astronauts had ever made it to the moon. In response, nine per cent said it was possible the landings never happened and another five per cent were not sure. The findings suggested that up to 20 million Americans might have suspected that the 'great leap for mankind' was a fake.[81] Although believing the moon-landing hoax does not imply belief in a flat earth, in a popular culture beset by conspiracies and subversive discourses from ufology to *The X-Files*, the survey hinted at a level of sympathy for some of Johnson's unorthodox views.[82]

For Johnson, July 1994 was a demanding month. While he gave interviews to accompany stories about the anniversary of Apollo 11, the arrest and preliminary hearing for the trial of ex-NFL player and film actor O.J. Simpson was causing a sensation across the States. At his desert home, Johnson was gripped by television coverage about the upcoming case, for he believed there was an intimate connection between the two apparently unrelated events. Simpson stood accused of stabbing his ex-wife, Nicole, and her male friend, Ron Goldman, to death, but Johnson was certain that he had been framed. For him the reason was obvious: in 1978, Simpson had starred in *Capricorn One*, a movie about a faked NASA mission to Mars, in which he played an astronaut forced to enact a landing on our closest neighbouring planet for the benefit of millions watching on live television. Meanwhile, in real life it was generally accepted that *Capricorn One* was not a classic film: it was 'an expensive, stylistically bankrupt suspense melodrama' in the words of the *New York Times*, although critics could not deny that it was a product of its time. Like many films of its era, it was inspired by burgeoning conspiracy thinking and post-Watergate paranoia, and it drew in good audiences and box-office takings as a consequence.

But for Johnson and fellow conspiracy theorists, *Capricorn One* had a deeper appeal: they claimed it was a documentary about the true practices of science and the moon-landing hoax.[83] O. J. Simpson's leading on-screen role in helping to uncover the *Capricorn One* fake was therefore the true motivation for the murder charges. 'They're finally going after O.J.,' Johnson told the *Washington Post*, 'because he helped to unmask the space hoax.'[84] There was no public comment from Johnson when Simpson was acquitted in October 1995.

Yet as the third millennium advanced, accompanied by escalating public discussion of conspiracy, catastrophe and risk, Johnson became preoccupied with matters closer to home. Besides the flat-earth issue, his most pressing concern during the 1990s had been the deterioration in Marjory's health: she had emphysema, needed a wheelchair and was dependent on supplementary oxygen to breathe. As her condition worsened, so did the couple's situation. Johnson was watching television one Wednesday afternoon in September 1995 when he noticed that the front porch of their house was in flames. Too fierce to control, the fire swept through their home, and although he managed to carry Marjory and their pets to safety, everything they owned – their household goods and personal possessions, the society's library, archives and membership lists – was destroyed. The house and its contents were not covered by insurance, and with no money to rebuild, nowhere to live and no record of their twenty-three-year mission, they moved into a dilapidated trailer next to the site of their former home. The episode was a substantial blow, especially to an elderly couple who had worked for years, despite financial constraints, to abide by their beliefs and, moreover, were given to seeing signs and meanings, divine blessings and curses in external events. It seems that Marjory, Johnson's wife of thirty-five years, never truly recovered from the shock and she died in May 1996 at the age of seventy-five. 'When the house burned with all her treasures,' Johnson subsequently told the *Boston Globe*, 'she gave up the will to live.'[85]

Local authorities subsequently evicted Johnson from his trailer because it lacked the required wooden foundations or sidings. Devastated by the loss of his wife and homeless again, he moved in with his brother in nearby Lancaster. 'It's very lonely without [Marjory],' he told the *Boston Globe*, 'but [she] wanted to keep the work going so I'm forcing myself to go on.' With this aim in mind, he began to rebuild the society with the help of a new secretary.

'As spring comes on,' he confided to Robert Schadewald in March 1997, '[I] think things will be better and better.'[86] Indeed, in July 1999, a Gallup poll found that six per cent of Americans, approximately 12 million people, still believed that NASA had misled the public about the moon landing, but by this time the IFERSA was a shadow of its former self.[87] Old members re-joined, new associates entered the fold and Johnson even gave the odd interview, but the days of public appearances and controversial publications were at an end. On Monday, 19 March 2001, Johnson died at the age of seventy-six. He had fully intended to keep up the 'war against the Grease Ball' for as long as he lived and had remained true to his word. He was buried in the Joshua Memorial Park on the outskirts of Lancaster, while obituaries appeared in papers from Dublin to New York to Canberra to Calgary – from all areas of the globe that he had refused to believe existed.

Epilogue

MYTHS AND MEANINGS

When such a critic says, for instance, that faith kept the world in darkness until doubt led to enlightenment, he is himself taking things on faith, things that he has never been sufficiently enlightened to doubt. That exceedingly crude simplification of human history is what he has been taught, and he believes it because he has been taught. I do not blame him for that; I merely remark that he is an unconscious example of everything that he reviles.

G. K. CHESTERTON

'TRUTH,' Lord Byron wrote, 'is strange, stranger than fiction', and a similar sentiment may be justly applied to the history of the flat-earth idea. It is a complex chronology beset by myths and metaphors, counter-meanings and misinformation circulating in popular culture and the public mind as well as in the minds of advocates who campaigned to establish the authority of this, the most renowned of erroneous concepts. The error is double-sided: that the earth itself is flat, and that educated people who lived after the fifth century BC believed that it was so. In many ways, therefore, just as the history of the flat-earth idea is an account of a misconception, so, too, is it an account of misconceptions about a misconception, one that remains common currency in popular folklore, school textbooks and newspaper reports worldwide. Meanwhile the truth lies in a series of negatives, for medieval people who published an opinion did not generally believe the earth to be flat, Columbus was not the first to discover it was a globe, the belief is not based solely on the Bible, and numerous societies promoting the idea were not established as a joke.

In keeping with familiar misconceptions, flat-earth belief is often held up as the epitome of stupidity, an accusation levelled at deviant contemporaries besides whole generations at a different time. The phrase 'the time when people thought the Earth was flat' is sufficient to evoke images of Ptolemaic maps or unwashed peasants, such is the power of the flat-earth myth, an appeal that Jeffrey Burton Russell has explained with reference to 'presentism' or each culture's tendency to congratulate itself on its knowledge, modernity and progress from the alleged ignorance of the past.[1]

While the so-called 'Dark Ages' were more enlightened than is commonly supposed, conventional wisdom about modern-day believers is similarly limited; they are a minority so much maligned and little understood that their very existence has been the subject of some dispute. Yet contrary to popular perception, a tiny minority of people still believe the earth to be flat and simple stereotypes or invocations of insanity will not do. With Christianity playing such a critical role alongside the complexities of human psychology, flat-earth belief is deserving of notice beyond shallow ridicule and jokes. The zetetic campaigners certainly possessed a rationale, however misguided, for deviating from commonly held opinion in the face of seemingly incontrovertible proof. Neither grasping charlatans nor curious throwbacks, the principal public flat-earth believers were serious-minded individuals, widely read and irrevocably committed to their perception of truth. What they shared was an eccentric standard for the assessment of evidence and a willingness to launch a public campaign in support of a highly unorthodox world-view.

Their reasons for launching a radical challenge to one of the most fundamental tenets of human knowledge were diverse, ranging from a desire to safeguard a literal interpretation of the Bible, the word of God, against the inroads being made by science, to a wish to undercut the increasing professionalization and cultural authority of scientific experts, or a perceived need to defend freedom and democracy, and the rights of the general public to make their own knowledge about the natural world. Driven mainly by religious imperatives, prominent zetetic campaigners co-opted the concept to spearhead a broad social-improvement programme, naming the flat-earth 'fact' as the final solution, the panacea for social ills, the missing link in explanations of social malaise, the ultimate causal explanation for perceived conflict and disharmony.[2] Above all, the zetetics were using the flat-earth idea as an intellectual weapon to restate the place of Christianity in social and spiritual life.[3] From Hampden to Johnson, each had their own ideological axe to grind, as did demagogues and social reformers who employed scientific concepts such as natural selection to further their own far-reaching schemes.

In this way, from the 1840s zetetic astronomy was and remained a political issue: it was framed as a populist cause – as a defence of freedom of thought and of individual rights – the right to discover, the right to own and the right to control knowledge. Through the avenue of 'common sense', flat-earthers claimed intellectual equality with the scientific élite; anyone could practise zeteticism, a genuine high-status science that would democratize knowledge for all.[4] The zetetics sought to make a better world, literal and figurative, devout and democratic, and they cast orthodox science as the chief obstacle blocking their goal. As a central point of conflict between the Biblical interpretation and the findings of science, the flat-earth idea was a critical truth claim and as such was raised to the forefront of their march against conventional knowledge. Yet despite flag-waving, populist rhetoric and social-reformist zeal, leading zetetics were not entirely altruistic: while contesting the cultural authority of science in the name of Christianity or of human rights, they often displayed a desire for self-aggrandizement, a wish to be known as the one who instigated, as Hampden termed it, a 'world-wide revolution in human science'.[5]

Such statements seem vastly inflated, even humorously so, in light of the fact that the earth has been widely believed to be a globe since the fifth century BC, yet, given the hegemony of science and the recent rise of creationism and conspiracy thinking, the existence of a contemporary flat-earth campaign seems less peculiar. Undoubtedly developments in nineteenth-century Britain were the reason that Parallax saw opportunities inherent in a campaign to promote the idea that the earth was not a globe: as one who moved in radical circles he was aware of controversies created by new geological findings and the low scientific milieu in which knowledge was being contested and made. From the early nineteenth century, popular science was a fast-growing genre, and science and religion were hot topics of the day, the subject of discussion in Mechanics' Institutes as well as upper-class drawing rooms. By the latter part of the century, leading men of science were exploiting this new reformation, fighting for professional status and ideological, social and political power on the strength of their credible knowledge. For one key group, it seems in retrospect that the push was influenced by a tendency now known as 'scientific naturalism', a deliberate effort to cast out religious causation from scientific explanations of the natural world. The revival of the flat-earth idea was therefore part of a backlash: sold to the public as a grand defence of Biblical truth at a time when such a campaign was bound to whip up public feeling, draw in audiences and raise a healthy income besides. Beyond Parallax, the campaign continued in the same vein, as a radical response to orthodox scientific authority, materialist 'infidels' and their supposedly irreligious world-view.

While at first glance the history of the flat-earth idea may seem simplistic, quaint or even arcane, in reality it is more complex and wide-ranging than generally supposed. Most notably, the belief is not merely a matter of Biblical literalism, for modern flat-earth campaigners co-opted scientific as well as scriptural proofs in their bids to promote their unorthodox cause. Thus while zetetics emphasized Biblical authority, they also turned to scientific experiments, empirical proof and the mythology of discovery as vital components of their protests. At base level, even zetetic philosophy was promoted as a Baconian fact-finding mission, an objective quest to seek the truth, while rhetoric alluding to common sense, empiricism, reason and logic was as much a feature of zetetic publications as Biblical quotations and statements of faith. The latter point is critical because, from the campaign's inception, flat-earth advocates sought to assume the authority and apparatus of science, the very thing they purported to loathe, and employ it to their own ends. In many respects it was a powerful and timely move. From the early nineteenth century science displaced religion as the primary source of cultural authority and offered a number of well-attested practices that flat-earth believers proceeded to put to novel use. Although, in terms of ideological and organizational sophistication, the movement peaked through the era of Parallax and Lady Blount, Shenton and Johnson went on to reflect such developments in science and society as a whole; in the face of space exploration and shifting world-views they emphasized 'scientific' and political arguments pertaining to democracy and freedom of thought.

The zetetic campaign therefore belies the traditional metaphor of all-out warfare between science and religion, for despite their militaristic, all-guns-blazing, 'anti-science' rhetoric, advocates utilized the language and authority of science throughout the modern public revival of the flat-earth idea. What they actually opposed was orthodox scientific knowledge, specifically the 'theory' of the earth's rotundity, rather than science itself. While 'warfare' was a reality to them, it was not a reality *per se*; just as evolutionary doctrine is not synonymous with atheism, except when viewed from an extreme, so neither is the idea of a spherical earth.[6] As millions know, it is entirely possible to be a devout Christian and believe the earth to be a globe. Hence, the war against science was a construction, a propaganda tactic and an illusion, for the zetetics staked a claim to 'true science' pinned on a challenge to what is generally considered the most essential of natural facts. In terms of an assault on scientific authority, the flat-earth idea was undoubtedly a powerful tool: fundamental, attention-grabbing, with an inherent appeal to human cognition, it has been employed by partisan interests, both religious and secular, for more than a century. In this sense, much like organic evolution, the concept has become heavily politicized; the idea that the earth is flat is a modern heresy, co-opted to support very different ends, even extending beyond its meaning, in the case of the Canadian society, to underscore a deeper philosophical point.

While overarching declarations of warfare are unfounded, the province of polemicists and propaganda campaigns, the history of the flat-earth idea provides another example of where the relationship between science and Christianity is more complex and context-bound than a model of straightforward harmony or conflict between two clearly defined extremes.[7] Reason, facts, logic and common sense were the key points leading to the zetetic conclusion that the earth was flat, and such an approach was not without historical parallels.[8] The comparisons with modern creationism are particularly striking. In the early twentieth century, few advocates insisted on a young earth or a fossil-producing Flood, ideas that had been undermined by geological findings and evolutionary biology in the nineteenth century. Even hardline fundamentalists were able to incorporate the findings of modern geology, including the antiquity of life on earth, into their views of the natural world, commonly applying day-age (interpreting the days of Genesis 1 as vast geological ages) and gap theories (a creation 'in the beginning' as in Genesis 1:1, followed by a later Edenic creation in six twenty-four-hour days) to their understanding of the Biblical account of creation.[9]

This situation only altered with the birth of 'scientific creationism'. This term was adopted in the early 1970s, although the movement itself is generally traced to the work of Seventh Day Adventist geologist George McCready Price (1870–1963) and the subsequent publication of *The Genesis Flood* by John C. Whitcomb

Jr and Henry M. Morris in 1961. The book revived Christian belief in a literal reading of the six twenty-four-hour days of special creation as laid out in Genesis (1:1–2:3), emphasized a young earth (of no more than 10,000 years in age) and explained the palaeontological record with reference to a fossil-producing Flood. Following the publication of *The Genesis Flood*, two major creationist organizations were formed, the Bible-Science Association and the Creation Research Society, while a third well-known group, the Institute for Creation Research, was established under the directorship of Henry Morris in 1972.

Although such creationist organizations recoil from association with flat-earth believers, some general similarities between the two campaigns cannot be overlooked. As Robert Schadewald contended, they concur on a number of issues, including the authority of the scriptures as a scientific guide to the natural world, the limitations of a theory-led approach, the duplicity of conventional scientists, and the impossibility of reconciling orthodox science with the Bible. And just as they have similar foundations and histories, so, too, have they employed similar strategies to promulgate their world-views.[10] Both have challenged the authority of conventional scientific knowledge through lectures, debates and investigations, proposing methodologies and arguments alike in structure, content and tone. More specifically, comparable creationist practices include monetary offers for proof of organic evolution, an emphasis on public debates involving persuasive speakers, and a stress on undertaking original research and field studies where possible, including several expeditions to Turkey's Mount Ararat to locate the remains of Noah's Ark.[11]

The arguments employed by creationists and zetetics also display a marked resemblance. At base level, unlike the majority of Christians, creationists assert that if conventional science is true the Bible must be false; that one cannot believe the Bible and the theory of evolution/the theory that the earth is a globe; that the Genesis account of creation/the earth is flat is what the Bible teaches; that evolution/a globular earth is only a theory;

that evolution/a globular earth is anti-God; that evolution/a globular earth is a satanic device; that children are being corrupted by
the theory of evolution/the theory of a globular earth, and that the
theory of evolution/the theory of a globular earth is a source of
social evil, the cause of imminent chaos and collapse. Moreover,
both creationists and zetetics claim to use the scientific method,
which they define narrowly, and stress a paucity of proof for the
opposing viewpoint; creationists, for example, highlight an alleged
lack of evidence for evolution, especially a lack of transitional forms
from invertebrate to vertebrate, from fish to amphibian, reptile to
bird or ape to man, the so-called 'missing link', an argument hotly
disputed by the evolutionist camp.[12]

Turning to broader tactics, there are similarities in how leading
creationists and zetetics have defined their campaigns. Specifically,
the zetetics claimed to practise the only true science: utilizing
popularist imagery of scientific endeavour, they cast themselves as
the legitimate seekers of truth, guided first and foremost by nature,
seeking to defend an empirically proven scientific fact, backed
by common sense, objective investigation and Biblical authority.
Similarly, in the early 1970s, strict six-day creationists began to
utilize the labels 'creation science' and 'scientific creationism'
to defend their cause.[13] Yet although zetetics and creationists have
packaged their respective issues as science, critical differences
remain. Most notably, the Creation Research Institute and the
zetetics diverge on the issue of Biblical interpretation, for the
former reject the contention that the Bible teaches a flat earth.
Just as important, the zetetic movement, if it can even be classified
as such, never enjoyed the same level of institutionalization or
ideological sophistication as the creation-science movement, or a
comparable level of public support. Despite the growing strength
of Biblical literalism in the United States, there are probably no
more than a few thousand flat-earth believers alive in the world
today and fewer still who would be willing publicly to declare their
conviction. Yet although flat-earth belief is a rare, last-stand
defence against the teachings of science, its grounding in scriptural

authority lends it some resonance at a time when a Gallup poll has revealed that 53 per cent of Americans believe that 'God created human beings in their present form exactly the way the Bible describes it', and a further 31 per cent of respondents hold that, while human beings have evolved over millions of years from other forms of life, this process has been guided by God.[14]

For this reason, the flat-earth idea remains on the peripheries of the debate. Evolutionists have claimed that if the Bible is accepted as an inerrant, infallible, scientifically accurate authority, as creationists argue, then to avoid contradiction they must also believe the earth to be flat.[15] In response, creationists have insisted that the Bible does not teach a flat earth and such claims on the part of evolutionists are evidence of their moral bankruptcy – they are utilizing the flat-earth idea as a slur to disparage the creationist campaign. Further to this, when one critic likened the Creation Research Institute to the Flat Earth Society, a representative of the institute countered that the two organizations did not share any members in common and the linking of them was nothing but a smear.[16] In general, creationists consider flat-earth believers misguided and too extreme, while flat-earth believers consider creationists hypocritical for emphasizing a literal interpretation of the Bible while rejecting the idea that the earth is flat. Biblical interpretation aside, the Creation Research Institute's hostility towards the flat-earth concept set against zetetic disdain for Christians who believe the earth to be a globe illustrates the dangers of lumping together complex belief systems and their adherents, when in reality they may perceive themselves as radically opposed.

For all its apparent absurdity and marginal status, however, the flat-earth idea is significant in two distinct ways: it is arguably one of the most radical and easily refutable Bible-based truth claims about the natural world, while the rotundity of the earth is the most essential of scientific facts, a cornerstone of received knowledge so established it is generally accepted without question. As such, the history of the idea has an important and unique, if somewhat unusual, place in the relationship between science and

Christianity, and provides a valuable comparison to methodologies and arguments utilized in more customary controversies in the field. Meanwhile, the public revival of an antiquated concept in early Victorian England illustrates that the history of ideas is not a simple flow of progress from blind ignorance to far-seeing truth; a procession of heroes setting milestones of discovery on a march towards ultimate knowledge. If anything, the limited success of the zetetic campaign, especially in its earliest incarnation, highlights how knowledge is made in a broad social context and the extent to which people choose their own truth, seemingly influenced by theological and socio-psychological factors – Biblical interpretation, gentlemanly conduct, persuasive arguments, the appearance of authority, personality disorders and so forth – in some cases to a greater extent, it seems, than the reality of the natural world around them.[17]

In this sense, orthodox beliefs, practices and norms can be viewed more clearly in the light of the history of alternative ideas: by turning convention upside-down, deep-seated assumptions, approaches and processes are stripped bare. Although flat-earth belief is dramatic, atypical and, without doubt, restricted to a tiny minority, the public revival of the idea in the mid-nineteenth century and Parallax's ability to gain a level of support tells us much about the nature of the intellectual marketplace, about what wares were being bought and sold. While the majority were not tempted to buy what Parallax *et alii* had to offer, their campaigns, with the responses evoked, say much about how knowledge is created, tested, disseminated and accepted through society.

More generally, the history of the flat-earth idea highlights the precise ways that information can be used and arguments constructed to defend a diversity of opinions and goals. As evidence was filtered, shaped and even denied by zetetics seeking to prove that the earth was flat, their campaigns provide a particularly striking example of the methodologies adopted by individuals seeking to obtain credibility for their views in a modern society loaded with competing and, at times, highly questionable claims to legitimacy, expertise and truth. Science and religion have a central place in such processes as two predominant forces of cultural and epistemological authority, constantly utilized by activists and experts, spokesmen and spin doctors, to publicize and justify a plethora of 'facts'. From evolution to the Atkins diet, in a pluralistic world where one person's crank is another's expert and one person's heresy is another's faith, flat-earth believers illustrate that there will never be a consensus on any given fact, however essential and obvious it may appear. Indeed, efforts to establish credibility, assert authority and build trust are a general feature of our day-to-day lives, besides being integral to how we learn about science and accept technical information as a reliable, trustworthy guide.

From personal interaction, political rhetoric and media coverage, to talk-show interviews and school textbooks, can you really believe what you see or read without investigating the subject at first hand? Do we possess adequate knowledge to trust without doubt what an apparent expert is telling us, or do we have little alternative in the majority of cases but to take them at their word? Besides the photographs from space or a schoolroom globe, how do you really know that the earth is a sphere? Could you refute complex counter-arguments effectively enough to conclusively prove your case? Has a teacher, author or journalist ever told you that medieval people believed the world to be flat, that Columbus discovered it round, that Galileo was persecuted for challenging flat-earth belief or that Darwin's *On the Origin of Species* caused all-out warfare between science and religion from its publication in 1859? Writ large, the issue is how we receive and reject our knowledge and how we accept what is truth.[18]

While the flat-earth idea raises critical questions about human cognition, this issue is in itself a paradox. That the earth is a globe is an exceptionally accessible scientific fact, for one does not require a laboratory, a Ph.D., a time-lapse camera or a telescope to discover the true shape of the earth for oneself. Meanwhile, plentiful evidence of the earth's globularity, incontrovertible to most people's minds, renders the concept highly falsifiable, unlike a range of similarly alternative claims that are neither as easily nor as conclusively refuted either by scientists or the public at large. However, running parallel to this, the earth does indeed seem flat and immobile in terms of everyday sense-experience and the evidence of one's own eyes; it is no accident that a study by geographers at the University of Sussex, undertaken in 1996, revealed that a fifth of children in British primary schools believed the earth to be flat.[19]

The phenomenon is not peculiar to Great Britain; however, studies completed in the United States and Israel indicate that, as late as the age of ten, almost half of the children surveyed likewise believed the earth to be a plane. Of those who accepted that the earth was a globe, the majority believed that humankind inhabited a flat platform on the interior of a ball-shaped earth. Furthermore, the survey highlighted that teachers overestimated pupils' knowledge regarding the earth's shape: teachers of eight-year-olds believed that 95 per cent of their pupils would know the earth was a globe, while in reality only five per cent believed this to be the case. As a result, the investigators concluded that children struggle with an implicit contradiction between reason and instinct, between what they are told about the natural world and the direct experience of their own senses. The idea of a flat earth is therefore a 'common-sense' assumption, known as a naïve theory, comparable to the belief that a cannonball would drop faster than a feather in a vacuum, when in an airless environment all objects fall at the same rate.[20] The longevity of the idea as belief and myth is doubtless partly due to this fundamental appeal in terms of human cognition. It is one of the world's most antiquated concepts, last widely believed long before the birth of Christ, yet still supported by a minority, known to the majority and used by advertising executives, journalists and authors in search of an evocative and eye-catching idea.[21] Sceptic or believer, idealist or cynic, when contemplating science, this concept and its history say much about the sociology of knowledge, about human psychology and cognition, about how we know what we know.

While flat-earth belief often evokes a hostile reaction, ranging from sniggers to outright contempt, in an era when the impact of

Christian fundamentalism and Biblical authority on public policy is the subject of litigation and debate in the United States, and creation-science and scientific illiteracy cause ongoing anxiety on various grounds, the subject takes on a new resonance.[22] Students in high schools across America are taught the Columbus story as historical fact; journalists repeat the myth on Columbus Day each year, and in all probability someone somewhere has been taught that the earth is flat 'because the Bible says so'. While such myths circulate and errors abound, the public understanding of science continues to be a cause for concern. A nationwide telephone survey on heliocentricity on behalf of America's National Science Foundation recently discovered that up to 55 per cent of adult Americans, some 94 million people, potentially do not know that the earth revolves round the sun once a year.[23] Thus, if this is a history that began with myths, it is one that ends in rich irony.

At first glance the modern public revival of the flat-earth idea may seem to epitomize the terrain of the tabloids, the very entrails of popular culture, in fact. Yet in real terms, its history, albeit eccentric, sheds light on a gamut of themes central to the construction of our physical and metaphysical world-views and to the making of human knowledge.

Appendix

SCRIPTURAL 'PROOFS'

Taken from Parallax, *Zetetic Astronomy: Earth not a Globe!* 3rd edn (1881). Passages with original emphasis; Biblical translation/edition unknown.

EARTH NOT A GLOBE

Let us now inquire earnestly, and in all respects fairly, whether the philosophical teachings of the Scriptures are consistent with those of Zetetic Astronomy; or, in other words, are descriptive of that which is, both in nature and in principle, demonstrably true. In the Newtonian astronomy, continents, oceans, seas, and islands are considered as together forming one vast globe of 25,000 English statute miles in circumference. This assertion has been shown to be entirely fallacious, and that it is contrary to the plain literal teaching of Scripture will be clearly seen from the following quotations.

"And God said, Let the waters under the heaven be gathered together unto one place, and let the *dry land* appear. And God called the dry land *earth*; and the gathering together of the waters called The seas." Genesis i., 9–10.

Instead of the word 'earth' meaning both land and water, only the dry land is called earth, and the seas the gathering or collection of the waters in vast bodies. Earth and seas – earth and the great body of waters, are described as two distinct and independent regions, and not as together forming one great globe which modern astronomers call 'the earth'. This description is confirmed by several other passages of Scripture: –

"The earth is the Lord's and the fullness thereof; the world and they that dwell therein; for He hath *founded it upon the seas, and established it upon the floods.*" Psalm xxiv., 1–2.

"O give thanks to the Lord of lords, that by wisdom made the heavens, and that *stretched out the earth above the waters.*" Psalm cxxxvi., 6.

"By the word of God the heavens were of old, and the *earth standing out of the water and in the water.*" 2 Peter iii., 5.

EARTH RESTS ON WATERS

If the earth is a globe, it is evident that everywhere the water of its surface – the seas, lakes, oceans, and rivers – must be sustained or upheld by the land, which must be underneath the water; but being a plane 'founded upon the seas', and the land and waters distinct and independent of each other, then the waters of the 'great deep' must sustain the land as it does a ship, an ice-island, or any other flowing mass, and there must, of necessity, be waters below the earth. In this particular, as in all others, the Scriptures are beautifully sequential and consistent.

"The Almighty shall bless thee with the blessing of Heaven above, and blessings of *the deep that lieth under.*" Genesis xlix., 25.

"Thou shalt not make unto thee any likeness of anything in heaven above, or in the earth beneath, or in the *waters under the earth.*" Exodus xx., 4.

"Take ye, therefore, good heed unto yourselves, and make no similitude of anything on the earth, or the likeness of anything that is in *the waters beneath the earth.*" Deuteronomy iv., 18.

"Blessed be his land, for the precious things of heaven, for the dew, and for *the deep which croucheth beneath*." Deuteronomy xxxiii., 13.

EARTH IMMOVABLE, SUN IN MOTION

The direct evidence of our senses, actual and special observations, as well as the most practical scientific experiments, all combine to make the motion of the sun over the non-moving earth unquestionable. All the expressions of Scripture are consistent with the fact of the sun's motion. They never declare anything to the contrary, but whenever the subject is required to be named, it is expressly in the affirmative: –

"In the heavens hath He set a tabernacle for the sun, which is as a bridegroom coming out of his chamber, and rejoiceth as a strong man *to run a race*. His *going forth* is from the end of the heaven, and his *circuit* unto the end of it." Psalms xix., 4–6.

"The sun also ariseth, and the sun goeth down, and *hasteth* to *his place* where he arose." Ecclesiastes i., 5.

"Let them that love the Lord be as the sun when he *goeth forth* in his might." Judges v., 31.

"The sun *stood still* in the midst of heaven, and *hasted not to go down* about a whole day." Joshua x., 13.

"Great is the earth, high is the heaven, swift is the *sun in his course*." 1 Esdras iv., 34 (Apocrypha)

EARTH THE ONLY MATERIAL WORLD, EARTH NOT A PLANET

One earth only was created; and, in the numerous references to this world contained in the entire Scriptures, no other physical world is ever mentioned. It is never even stated that the earth has companions like itself, or that it is one of an infinite number of worlds which coexist, and were brought into being at the beginning of creation.

The expressions in Hebrews (i.,2) 'By whom also He made the worlds'; 'and (xi., 3) 'Through faith we understand that the worlds were framed', are known to be a comparatively recent rendering from Greek documents. In the later translations either the plural expression 'worlds' was used in order to accord with the astronomical theory then recently introduced, or it was meant to include the earth – the material world and the spiritual world.

"There is no man that hath left house, or parents, or brethren, or wife, or children, for the Kingdom of God's sake, who shall not receive manifold more in this present time, and in the *world to come* life everlasting." Luke xviii., 29–30.

"Whosoever speaketh against the Holy Ghost it shall not be forgiven him, neither in *this world*, neither in *the world to come*." Matthew xii., 32.

Sun, Moon and Stars as Lights, Moon Self Illuminating

The sun, moon, and stars are never referred to as worlds but simply as lights, to rule alternately the day and the night, and to be for signs and for seasons, and for days and years.

"And God said let there be *lights* in the firmament of the heaven to divide the day from the night ... And God made two great lights; the greater light to rule the day, and the lesser light to rule the night." Genesis i., 14–16.

"O give thanks to Him that made *great lights*: ... the sun to rule by day, ... the moon and stars to rule by night." Psalm cxxxvi., 7–9.

"Praise Him, sun and moon; praise Him all *ye stars of light*." Psalm cxlviii., 3.

"Behold even to the moon, and *it shineth not*." Job xxv., 5.

"The light of the moon shall be as the light of the sun, and the light of the sun shall be sevenfold." Isaiah xxx., 26.

"The sun shall be darkened in his going forth, and the moon shall not cause *her light* to shine." Isaiah xiii., 10.

STARS GIVE LIGHT TO EARTH

The same theoretical astronomy teaches that, as the stars are so far away, hundreds of millions of statute miles, they cannot possibly give light upon the earth; that the fixed stars are burning spheres, or suns each to its own system only of planets and satellites; and that millions of miles from the earth their light terminates, or no longer produces an active and visible luminosity. This is an essentially false conclusion, because the proposition is false upon which it depends – that the stars are vast suns and worlds at almost infinite distances. The contrary has been demonstrated by trigonometrical observation; and is again confirmed by the Scriptures.

"He made the stars also, and set them in the firmament to *give light* upon the earth." Genesis i., 16–17.

"For the stars of heaven and the constellations thereof shall not *give their light*." Isaiah xiii., 10.

"The sun and the moon shall be dark, and the *stars* shall withdraw *their shining*." Joel ii., 10.

"They that turn many to righteousness shall *shine as the star* for ever and ever." Daniel xii., 3.

HEAVEN ABOVE, HELL BELOW

If the earth is a globe, revolving at the rate of a thousand miles an hour, all this language of Scripture is necessarily fallacious. The terms 'up' and 'down' and 'above' and 'below' are words without meaning – at best, are merely relative, indicative of no absolute direction.

"Look *down* from Thy holy habitation, from *Heaven*, and bless Thy people Israel." Deuteronomy xxvi., 15.

"For He hath looked *down* from the height of His sanctuary; from Heaven did the Lord behold the earth." Psalm cii., 19.

"Look *down* from Heaven, and behold from the habitation of Thy holiness and of Thy glory." Isaiah lxiii., 15.

"For as the Heaven is high *above* the earth." Psalm ciii., 2.

"And Elijah went *up* by a whirlwind into Heaven." Kings ii., 11.

"So then after the Lord had spoken unto them He was received *up* into Heaven." Mark xvi., 10.

"And it came to pass while He blessed them He was parted from them, and carried *up* into Heaven." Luke xxiv., 51.

'Thou shalt be brought *down* to hell, to the sides of the pit." Isaiah xiv., 15.

"The way of life is *above* to the wise, that he may depart from *hell beneath*." Proverbs xv., 24.

"I cast him *down* to hell. . . . They also went *down* into hell with him." Ezekiel xxxi., 16–17.

"God spared not the angels that sinned, but cast them *down* to hell." 2 Peter ii., 4.

"And the devil that deceived them was cast into the lake of fire and brimstone . . . The sea gave up the dead which were in it; and death and hell delivered up the dead which were in them . . . And death and hell were cast into the lake of fire." Revelation xx., 10, 13–14.

But if there is an endless plurality of worlds, millions upon millions in never-ending succession; if the universe is filled with innumerable systems of burning suns and rapidly revolving planets, intermingled with rushing comets and whirling satellites, all dashing and sweeping through space in directions and with velocities surpassing all human comprehension, and terrible even to contemplate, where is the place of rest and safety? Where is the true and unchangeable 'Palace of God'? In what direction is Heaven to be found?

EARTH WILL BE DESTROYED BY FIRE

The literal teaching of the Old and New Testaments on the subject of the earth's destruction is plain and unmistakable.

"For behold the Lord will come with fire, and with His chariots like a whirlwind, to render His anger with fury, and His rebuke with flames of fire ... The new heavens and the new earth, which I will make, shall remain before Me." Isaiah lxvi., 15–22.

"A fire is kindled in Mine anger, and shall burn unto the lowest hell, and shall consume the earth with her increase, and the foundations of the mountains." Deuteronomy xxxii., 22.

"The heavens and the earth which are now kept in store, reserved unto fire against the Day of Judgment and perdition of ungodly men ... The day of the Lord will come as a thief in the night, in the which the heavens shall pass away with a great noise, and the elements shall melt with fervent heat; the earth also and the works therein shall be burnt up ... All these things shall be dissolved ... the heavens being on fire shall be dissolved. Nevertheless we look for new heavens and a new earth, wherein dwelleth righteousness." 2 Peter iii., 10–13.

Abbreviations

UNB   University of New Brunswick, Harriet Irving Library,
Fredericton, Canada

UWM   University of Wisconsin-Madison, Memorial Library, Madison,
USA

Personal Names

AN	Alden Nowlan
ARW	Alfred Russel Wallace
CEK	Charles Edward Kettle
CJ	Charles Johnson
GBA	George Biddell Airy
JH	John Hampden
LF	Leo Ferrari
RF	Ray Fraser
RS	Robert Schadewald
SS	Samuel Shenton
TP	Sir Thomas Phillipps
WC	William Carpenter

Newspapers, Magazines and Journals

Annals	*Annals of Science*
BJHS	*British Journal for the History of Science*
EM	*English Mechanic*
FEN	*Flat Earth News*
HJ	*Historical Journal*
HOR	*History of Religions*
HS	*History of Science*
ISR	*Interdisciplinary Science Review*
JBAA	*Journal of the British Astronomical Association*
JBS	*Journal of British Studies*
JHB	*Journal for the History of Biology*
NMW	*New Moral World*
OS	*Osiris*
QJRAS	*Quarterly Journal of the Royal Astronomical Society*
SI	*Skeptical Inquirer*
SIC	*Science in Context*
SSS	*Social Studies of Science*
VS	*Victorian Studies*

Notes

Full publishing details are provided in the bibliography.

Prologue – The Columbus Blunder

1 The book was the fifth in a series.
2 A. de Lamartine, *Life of Columbus*.
3 E. M. Bolenius, *The Boys' and Girls' Readers: Fifth Reader*, p. 113.
4 J. B. Russell, *Inventing the Flat Earth*.
5 R. Simek, *Heaven and Earth in the Middle Ages*, p. 24.
6 Russell, *Inventing*, p. 29.
7 A. Koestler, *The Sleepwalkers*, p. 102.
8 D. C. Lindberg, 'The Medieval Church Encounters the Classical Tradition', in D. C. Lindberg and R. L. Numbers (eds), *When Science and Christianity Meet*, pp. 7–32 (pp. 7–8).
9 Russell, *Inventing*, p. 66.
10 See S. Shapin, *Scientific Revolution*.
11 Russell, *Inventing*. Earlier references to Columbus's heroic discovery have been discovered in a book published in 1709, suggesting that the concept was in existence prior to Irving's embellishment. See E. G. R. Taylor, *Ideas on the Shape, Size and Movements of the Earth*, p. 9.
12 W. Irving, *Christopher Columbus*, pp. 61–2.
13 *Ibid*.
14 Russell, *Inventing*, pp. 51–7.
15 Irving, *Christopher Columbus*, p. 64.
16 Russell, *Inventing*, pp. 58–61.
17 *Ibid.*, p. 32.
18 *Ibid.*; Simek, *Heaven and Earth*.
19 Simek, *Heaven and Earth*, p. 37.
20 Russell, *Inventing*, pp. 32–5; Simek, *Heaven and Earth*.
21 J. W. Draper, *History of the Conflict between Religion and Science*, p. 159.
22 D. C. Lindberg, 'Science and the Early Church', in D. C. Lindberg and

R. L. Numbers, *God and Nature*, pp. 19–48 (p. 20); Draper, *History of the Conflict*, p. 157.

23 J. R. Moore, *The Post Darwinian Controversies*, p. 35.

24 *Ibid.*; Lindberg and Numbers, *God and Nature*; Lindberg and Numbers, *When Science and Christianity Meet*.

25 A. D. White, *A History of the Warfare of Science and Theology in Christendom*, vol. I, pp. 89–98.

26 *Ibid.*, p. 97.

27 White, *History of the Warfare*.

28 Moore, *Post Darwinian Controversies*.

29 For erroneous chronologies of flat-earth thinking see D. Boorstin, *The Discoverers*; Koestler, *The Sleepwalkers*.

One – Surveying the Earth

1 A. Chapman, *Gods in the Sky*, pp. 55–7.

2 R. J. Schadewald, 'The Flat Earth Bible', *Bulletin of the Tychonian Society*.

3 *Ibid.*; L. Jacobs, 'Jewish Cosmology', in C. Blacker and M. Loewe (eds), *Ancient Cosmologies*, pp. 66–86.

4 Before this time, *The Iliad* and *The Odyssey*, attributed to Homer, present an image of a flat disc-shaped earth surrounded by a circular river, Okeasnos, where what we would call 'natural forces' were intrinsically connected to the will or the person of various all-powerful Olympian gods.

5 It has been claimed elsewhere that he knew that the earth was a globe; see H. P. Nebelsick, *Circles of God*, p. 9.

6 D. C. Lindberg, *The Beginnings of Western Science*, p. 25; G. E. R. Lloyd, 'Greek Cosmologies', in Blacker and Loewe (eds), *Ancient Cosmologies*, pp. 198–224 (pp. 199–200).

7 Lindberg, *Beginnings*, p. 31.

8 G. E. R. Lloyd, *Early Greek Science*, p. 10.

9 Chapman, *Gods*.

10 J. B. Russell, *Inventing the Flat Earth*, pp. 24–5. Chapman, *Gods*, p. 104.

11 Nebelsick, *Circles*, p. 51.

12 *Ibid.*, see pp. 54–7.

13 Ptolemy's *Geographica* likewise promoted a spherical earth, and includes the grid system of latitude and longitude that is still used today, but the book was not 'recovered' in the West until the fifteenth century. An example of these proofs would be the fact that the sun, moon and stars do not rise and set at the same time for every observer on earth. Furthermore, since the differences in hours seem to be proportional to the distances between the places, one would reasonably suppose the surface of the earth to be spherical.

14 D. C. Lindberg (ed.), *Science in the Middle Ages*, pp. 33–4.

15 R. Simek, *Heaven and Earth in the Middle Ages*, p. 24.

16 D. C. Lindberg, 'The Medieval Church Encounters the Classical Tradition', in D. C. Lindberg and R. L. Numbers (eds), *When Science and Christianity Meet*, pp. 7–32 (pp. 18–19), Russell; *Inventing*, pp. 22–3.

17 Lindberg (ed.), *Science in the Middle Ages*, p. xiii; Russell, *Inventing*, p. 15.

18 For the influence of Aristotelian cosmology see E. Grant, *Planets, Stars and Orbs*.

19 D. C. Lindberg, 'Medieval Science and Religion', in G. B. Ferngren (ed.), *Science and Religion*, pp. 47–56; D. C. Lindberg, 'Early Christian Attitudes Towards Nature', in *ibid.*, pp. 57–72; Lindberg, *Beginnings*.

20 Russell, *Inventing*, pp. 17–18.

21 See S. Shapin, *The Scientific Revolution*.

22 Quoted in R. Yeo, 'Genius, Method and Morality', SIC, p. 258.

23 The term 'scientist' itself was not coined until the 1830s and did not enter common usage until the late-nineteenth century. Before that time scientists were generally dubbed 'philosophers' or later 'men of science'. Science's lack of status as a distinct profession precluded the development and use of special terms; as modern science was carved out from the diffuse practices of natural philosophy and the culture of clerical dominance, so modern terms and definitions emerged.

24 Shapin, *Scientific Revolution*, p. 123.

25 L. Stewart, *The Rise of Public Science*; Stewart, 'The Selling of Newton'.

26 M. Mazzotti, 'Newton for Ladies', *BJHS*. Yeo, 'Genius, Method and Morality', *SIC*.

27 P. Fara, 'Isaac Newton Lived Here', *BJHS*; Yeo, *ibid.*

28 Lamarck's theory of transmutation appeared in Owenite co-operative magazines from the 1820s. See G. J. Holyoake, *A History of Co-operation*, p. 79.

29 For an overview see P. J. Bowler and I. R. Morus, *Making Modern Science*.

30 A. Desmond, *The Politics of Evolution*, p. 117.

31 See *Encyclopaedia Britannica*, vol. IV (1910), p. 558.

32 J. R. Topham, 'Beyond the "Common Context"', *Isis*.

33 A lack of written records makes it impossible to assess the views of uneducated people, although with a topic that involves individual sense-experience one can only guess that common opinion was more diverse. Indeed, Jeffrey Burton Russell has noted there is evidence to suggest that, before 1300, some French people still envisaged the earth as a circular plane, *Inventing*, p. 16. Due to a paucity of sources about rank-and-file believers, this study focuses predominantly on the leading campaigners.

34 R. L. Numbers, *The Creationists*.

Two – A Public Sensation

1 Parallax, *Zetetic Astronomy: Earth not a Globe!*, 3rd edn, p. 91.
2 *Ibid.*, pp. 377–83.
3 Parallax, *Zetetic Astronomy*, p. 130.
4 Zetetics is also a branch of algebra: it relates to the direct search for unknown quantities.
5 E. Royle, *Victorian Infidels*; I. D. McCalman, 'Popular Irreligion in early Victorian England,' in R. W. Davis and R. J. Helmstadter, *Religion and Society in Victorian England*, pp. 51–67.
6 Parallax, *Zetetic Astronomy: Earth not a Globe!*, 2nd edn, p. 4.
7 Samuel Smiles's classic *Self Help* was not published until 1859, albeit written in the 1840s.
8 S. Shapin and S. Schaffer, *Leviathan and the Air Pump*.
9 See R. Yeo, 'Genius, Method and Morality', *SIC*; Shapin and Schaffer, *Leviathan*; S. Shapin, *A Social History of Truth*; S. Sheets-Pyenson, *Scientific Culture in London and Paris, 1820–75*; S. Shapin and A. Thackray, 'Prosopography as a Research Tool', *HS*.
10 Mechanics' Institutes were educational establishments set up to provide adult education for the working classes, although they were also patronized by the middle class. Halls of Science were Owenite educational institutions.
11 I. Inkster, 'Advocates and Audience', *JBAA*.
12 See J. Secord, *A Victorian Sensation*.
13 He had also previously served in Robert Owen's Association of All Classes of All Nations as branch secretary in Stockport, a cotton centre in the Owenite heartland around Manchester, dominated by large-scale textile mills and wealthy cotton lords. In an area beset by disputes between workers and masters, political agitation and calls for parliamentary, social and industrial reform, Parallax's position in the thick of ferment was not for the faint-hearted, while his socialist commitment is further suggested by a later application to serve as an Owenite social missionary.
14 In Owenite lecture theatres, metropolitan anatomy schools and the gutter press, radical artisans were using Lamarckian transmutation to challenge the social and political status quo. If humans evolved from a common ancestry, influenced by environmental factors, and the Bible was mistaken in this respect, then clerical power and the 'God-given' social hierarchy were in error and a democratic society was justified. See A. Desmond, *The Politics of Evolution*; A. Desmond, 'Artisan Resistance and Evolution in Britain, 1819–1848', *OS*.
15 Parallax's action-packed account of his formative years, published in 1872, appears to have been created for propaganda purposes. In it, he maintains

that he converted to flat-earth belief as a seven-year-old, joined the socialist commune merely to enable him to collect experimental proof of the flatness of the earth, and left when free-thinking commune members threatened him with violence as a consequence of his teachings.

16 R. A. Proctor, *Myths and Marvels of Astronomy*, pp. 279–80.

17 Parallax, *The Inconsistency of Modern Astronomy*.

18 *Athlone Sentinel*, 21 May 1851, quoted in Parallax, *Zetetic Astronomy*, p. 403. All media reports taken from this source may have been edited from the original to cast Parallax in a more positive light. However, research indicates that reviews were not invented in their entirety.

19 Proctor, *Myths and Marvels*, pp. 279–80.

20 *Ibid.* Following a lecture in Ely a month later, the *Cambridge Chronicle* was even more complimentary: ' "Parallax" has lectured to respectable and critical audiences in [Ely] Corn Exchange and although it was not thought he gained any disciples to his theory that the earth is a plane and the Newtonian system an error, no one could fail to admire his power as a disputant. After the lectures he met the questions put to him by the most enlightened and scientific citizens with a readiness of reply which astonished his hearers; and he challenged to meet any of them on the points raised . . . but no one accepted his challenge. Report states that he will visit Ely again, when no doubt there will be a full room. Lecturers on the Newtonian system with their orreries, &c., completely fail to interest the people here. "Parallax" has the ability to do this; he met even the "sledge-hammer" of Mr. Burns with only a gentlemanly retort.' *Cambridge Chronicle*, 27 December 1856, p. 7.

21 Cambridge University Library (CUL), Greenwich Royal Observatory Collection (GROC), RGO6/251/440, Aylesbury poster.

22 CUL, GROC, RGO6/251/439, W. H. Smyth to George Biddell Airy (GBA), 7 November 1857.

23 CUL, GROC, RGO6/251/441–2, GBA to W. H. Smyth, 11 November 1857.

24 *South Midland Free Press*, 14 August 1858, quoted in Parallax, *Zetetic Astronomy*, p. 405.

25 I am extremely grateful to Rowbotham's great-great-granddaughter, Ann Coltman, for information about Rowbotham's family life, some of which is taken from the private memoirs of his granddaughter, Gudrun Hope Kassner.

26 *Mechanics' Magazine*, 29 March 1861, p. 219.

27 For discussion of Jean Bernard Léon Foucault's famous pendulum demonstration of the earth's rotation in the Panthéon in Paris in 1851, and subsequent 'pendulum mania' in Europe and the States, see M. F. Conlin, 'The Popular and Scientific Reception of the Foucault Experiment in the United States', *Isis*; W. Tobin and B. Pippard, 'Foucault, his Pendulum and the Rotation of the Earth', *ISR*.

28 *Greenwich Free Press*, 11 May 1861, quoted in Parallax, *Zetetic Astronomy*, p. 406.

29 See W. E. Hatcher, *John Jasper*, pp. 44–6.

30 *Greenwich Free Press*, 19 May 1862, quoted in Parallax, *Zetetic Astronomy*, pp. 406–7.

31 *Gosport Free Press*, 14 May 1864, quoted in Parallax, *Zetetic Astronomy*, p. 408.

32 *Western Daily Mercury*, 28 September 1864, p. 4.

33 D. Gooding, T. Pinch and S. Schaffer, 'Introduction', *The Uses of Experiment*, p. xiii; Shapin and Schaffer, *Leviathan*; Shapin, *Social History of Truth*.

34 *Western Daily Mercury*, 13 October 1864, p. 2.

35 *Devonport Independent and Plymouth and Stonehouse Gazette*, 15 October 1864, p. 5.

36 *Western Daily Mercury*, 14 October 1864, p. 4.

37 *Ibid.*, 21 October 1864, p. 5.

38 *Western Daily Mercury*, 17 October 1864, p. 4.

39 Further evidence of Parallax's quick thinking is provided in the memoirs of his granddaughter, Gudrun Hope Kassner. According to a well-known family story, Parallax was running to catch a train after delivering a lecture when he was accosted by two men in masks with blunderbusses, who demanded, 'Your money or your life.' Apparently Parallax's presence of mind led him to respond, 'Run for your lives. I've just burgled that house and the police are after me.' He escaped as a consequence with health and finances intact. Many thanks to Ann and Felicity Coltman for this information.

40 Proctor, *Myths and Marvels*, pp. 281–3.

41 Gooding, Pinch and Schaffer, *Uses of Experiment*, p. 18; J. Golinski, *Making Natural Knowledge*; Shapin, *Social History of Truth*.

42 *Western Daily Mercury*, 3 November 1864, p. 4.

43 *Ibid.*, 4 November 1864, p. 1.

44 Common Sense, *Theoretical Astronomy Examined and Exposed*, 2nd edn. (1869).

45 *News of the World*, 8 May 1864, quoted in Common Sense, *Theoretical Astronomy*, 2nd edn (1869).

46 *Era*, 4 November 1866, quoted in W. Carpenter, *'Bosh' and 'Bunkum'*.

47 *Observer*, 15 January 1865, quoted in Common Sense, *Theoretical Astronomy*.

48 *Army and Navy Gazette*, 29 December 1865, quoted in Carpenter, *'Bosh' and 'Bunkum'*.

49 *Weekly Times*, 16 December 1866, quoted in W. Carpenter, *Wallace's Wonderful Water*.

50 See Appendix, p. 363.

51 Shapin and Schaffer, *Leviathan*; Shapin, *Social History of Truth*; Golinski, *Making Natural Knowledge*.

52 *Stroud Journal*, 28 October 1865.

53 CUL, GROC, RGO6/256/490–1, William Pumfrey to GBA, 14 October 1865.

54 CUL, GROC, RGO6/256/501, B. Gott to GBA, 6 May 1867.

55 *Leeds Times*, 11 May 1867, p. 3.

56 CUL, GROC, RGO6/256/499, C. J. Whitmell to GBA, 3 May 1867.

57 After the event, the *Leeds Mercury* reflected that, all in all, he had proved himself 'easily capable' of maintaining 'a mastery of his subject which . . . induces his audience to manifest symptoms of scientific unbelief'. CUL, GROC, RGO6/256/512, *Leeds Mercury*, 8 May 1867.

58 See Shapin's *Social History of Truth* for discussion of the role of gentlemanly conduct in scientific practice.

59 *Leeds Times*, 11 May 1867, p. 8.

60 R. J. Cooter and S. Pumfrey, 'Separate Spheres and Public Places', *HS*; Golinski, *Making Natural Knowledge*.

61 A confirmed globularist who attended the Leeds Stock Exchange lecture later reminisced, 'The language and sentiment of [Parallax's flyer] address [to the 'Men of Leeds'], together with what I heard at the lectures, made such an impression on my mind that it was not many weeks before I felt myself unable to longer believe earth to be a globe. At this time I am a firm convert to zetetic astronomy.' B. of Hull, to *Zetetic*, vol. 1, no.1, July 1872, p. 79.

62 See observations in J. Endersby, 'Escaping Darwin's Shadow'.

63 One such interpreter was Major Rider Bresher, vicar at St Martin's in Coney Street, York, who published *The Newtonian System of Astronomy with a Reply to the Various Objections made against it by Parallax* in 1868. Having observed Parallax's performance in his hometown, Bresher noted that many who listened to his fervid denunciations of the Newtonian system 'seem to have been carried away and have become converts to his system'. According to Bresher, the subject had caused 'considerable talk and great controversy' in York for weeks afterwards and as a result he had felt compelled to write a book 'to help my fellow citizens to form a correct estimate of the merits of modern astronomy'.

64 CUL, GROC, RGO6/256/506, W. George Scott to GBA, 3 June 1867.

65 CUL, GROC, RGO6/256/509–10, GBA to J. E. Clarke, 8 July 1867.

66 For a broader context of etiquette in cross-class scientific correspondence see A. Secord, 'Corresponding Interests', *BJHS*.

67 See E. Clodd, *Memories*, pp. 57–8. Sir Robert Stawell Ball (1840–1913), popularizer and director of Cambridge University Observatory, also had a paradox box but, unlike Airy, refused to correspond with flat-earthers, who he said 'may believe the earth to be concave if they like'. For Ball,

'controversy' was only to be conducted with members of his profession, for only then would there be 'some chance of a clear issue, and a certainty that each will at least understand and respect the other'. See W. V. Ball, *Reminiscences and Letters of Sir Robert Ball*, p. 230.

68 J. R. Moore, *The Post Darwinian Controversies*; D. C. Lindberg and R. L. Numbers, *God and Nature*; D. C. Lindberg and R. L. Numbers, *When Science and Christianity Meet*; P. J. Bowler, *Reconciling Science and Religion*; J. H. Brooke, *Science and Religion*; G. B. Ferngren (ed.), *Science and Religion*.

69 Parallax, *Zetetic Philosophy*.

70 CUL, GROC, RGO6/257/614–5, William Carpenter to GBA, 26 November 1868.

71 John Hampden's family was indirectly descended from that of the great parliamentarian and leading figure of the English Revolution, John Hampden 'the patriot' (1594–1643), and shared the same coat of arms. I am very grateful to Roy Bailey and members of the John Hampden Society for information regarding Hampden genealogy.

72 Renn Dickson Hampden was embroiled in an ongoing public controversy through the 1830s and 1840s as a result of his liberal views. Despite vociferous opposition, he was consecrated Bishop of Hereford in 1848.

73 See G. Cantor and S. Shuttleworth (eds), *Science Serialized*.

74 I. Inkster, 'Advocates and Audience', *JBAA*.

75 Parallax, *Experimental Proofs that the Surface of Standing Water is not Convex*, pp. 1–3.

Three – The Infamous Flat-earth Wager

1 D. Knight, 'Scientists and their Publics: Popularization of Science in the Nineteenth Century', in M. J. Nye (ed.), *The Cambridge History of Science: The Modern Physical and Mathematical Sciences*, p. 73.

2 *Ibid.*, p. 87; A. Desmond, 'Re-defining the X-Axis', *JHB*.

3 F. J. Turner, *Between Science and Religion*; F. J. Turner, 'The Victorian Conflict Between Science and Religion', *Isis*; A. Desmond, *Huxley*; R. Barton, 'Huxley, Lubbock, and Half a Dozen Others', *Isis*.

4 B. Lightman, 'Victorian Sciences and Religions: Discordant Harmonies', in J. H. Brooke, M. J. Osler and J. van der Meer, *Science in Theistic Contexts*, pp. 343–66.

5 S. Shapin, *A Social History of Truth*; S. Shapin and S. Schaffer, *Leviathan and the Air Pump*.

6 Parallax, *Experimental Proofs that the Surface of Standing Water is not Convex*, pp. 1–3.

7 Reprinted in 'Curious Notes', *English Mechanic*, 21 January 1870, p. 466.

8 'A Fellow of the Royal Astronomical Society' to *English Mechanic*, 13 May 1870, p. 165.

9 For discussion see J. R. Moore, 'Wallace's Malthusian Moment', in R. Lightman (ed.), *Victorian Science in Context*, pp. 290–311.

10 J. Marchant, *Alfred Russel Wallace*, Alfred Russel Wallace (ARW) to Henry Walter Bates, 24 December 1860, p. 59.

11 For Wallace's earlier interest in spiritualism see E. Clodd, *Memories*, pp. 65–6.

12 J. R. Durant, 'Scientific Naturalism and Social Reform in the Thought of Alfred Russel Wallace', *BJHS*, p. 46; M. J. Kottler, 'Alfred Russel Wallace, the Origin of Man and Spiritualism', *Isis*.

13 Turner, *Between Science and Religion*; J. R. Durant, 'Scientific Naturalism and Social Reform in the Thought of Alfred Russel Wallace'; C. H. Smith, 'Alfred Russel Wallace: Philosophy of Nature and Man', *BJHS*; Lightman, 'Victorian Sciences and Religions: Discordant Harmonies', in Brooke, Osler and van der Meer, *Science*, pp. 343–66.

14 Moore, 'Deconstructing Darwinism', *JHB*.

15 Quoted in A. Desmond and J. R. Moore, *Darwin*, pp. 569–70.

16 Turner, *Between Science and Religion*.

17 A. R. Wallace, *My Life*, vol. II, p. 365.

18 A. R. Wallace, *Reply to Mr. Hampden's Charges against Mr. Wallace*, pp. 1–2.

19 D. Gooding, T. Pinch and S. Schaffer, *Uses of Experiment*.

20 Wallace to Newton, 18 January 1870, in A. F. R. Wollaston, *Life of Alfred Newton*, pp. 269–70.

21 R. A. Proctor, 'Our Earth – Its Figures and Motions', *English Mechanic*, 22 October 1869, p. 118.

22 Cambridge University Library (CUL), Greenwich Royal Observatory Collection (GROC), RGO6/258/336, John Hampden (JH) to George Biddell Airy (GBA), 4 February 1870.

23 *Ibid.*, Hampden, 'The Bible versus Falsehood', CUL, GROC, RGO/258/348, *Weston-super-Mare Gazette*, 5 February 1870.

24 CUL, GROC, RGO6/258/337–8, Charles Edward Kettle (CEK) to JH, 7 February 1870.

25 *Ibid.*, RGO6/258/339–40, JH to CEK, 8 February 1870.

26 CUL, GROC, RGO6/258/344, JH to CEK, 11 February 1870.

27 *Ibid.*, RGO6/258/345–7, CEK to JH, 12 February 1870.

28 CUL, GROC, RGO6/258/343, GBA to CEK, 18 February 1870.

29 *Ibid.*, RGO6/258/350, JH to GBA, 25 February 1870.

30 P. Duhem, *The Aim and Structure of Physical Theory*; Shapin, *Social History*; Shapin and Schaffer, *Leviathan*.

31 Quoted in Hampden, *Is Water Level or Convex After All?*, p. 15.

32 Quoted in W. Carpenter, *Water not Convex, The Earth not a Globe!*, p. 5.

33 *Ibid.*, p. 8.

34 Carpenter, *Water not Convex*, p. 7.

35 *Ibid.*, pp. 5–6.

36 J. Dyer, *The Spherical Form of the Earth*, pp. iii–iv, pp. 7–9.

37 'Reviews', *English Mechanic*, 4 March 1870, p. 600.

38 R. A. Proctor to *English Mechanic*, 6 May 1870, p. 159.

39 J. Dyer to *English Mechanic*, 20 May 1870, p. 204.

40 Wallace, *My Life*, vol. II, pp. 365–70.

41 Carpenter, *Water not Convex*, p. 15.

42 R. J. Cooter and S. Pumfrey, 'Separate Spheres and Public Places', *HS*; A. Ophir and S. Shapin, 'The Place of Knowledge', *SIC*.

43 Wallace, *My Life*, vol. II, p. 368.

44 Carpenter to *Earth not a Globe Review*, 26 March 1894, p. 166.

45 Carpenter, *Water not Convex*, p. 28.

46 Bodleian Library, University of Oxford (BOD), Papers of Sir Thomas Phillipps, MS Phillipps-Robinson (MSPR), e.48 fols 74v-5, Thomas Phillipps (TP) to JH, 24 March 1870.

47 *Ibid.*, c.616 fols 40–41, JH to TP, 24 March 1870.

48 BOD, MSPR, d.100 fols 7–9, TP to JH, n.d.

49 *Ibid.*, b.179 fols 80–81, TP to JH, 26 March 1870.

50 Hampden, *Is Water Level?*, p. 8.

51 'The Convexity of Water', *Field*, 26 March 1870, p. 285.

52 'The Experimental Proof of the Convexity of Water', *Field*, 26 March 1870, p. 266.

53 J H to *Field*, 2 April 1870, p. 305.

54 'P.S.W.' to *Field*, 9 April 1870, p. 312.

55 'The Convexity of Water Painfully Demonstrated', *Scientific Opinion*, 27 April 1870, p. 377.

56 R. A. Proctor to *English Mechanic*, 10 November 1871, p. 193.

57 Quoted in R. Colp, '"I will Gladly do my Best": How Charles Darwin Obtained a Civil List Pension for Alfred Russel Wallace', *Isis*, p. 11.

58 R. Bellon, 'Joseph Dalton Hooker's Ideals for a Professional Man of Science', *JHB*.

59 Endersby, 'Escaping Darwin's Shadow'; Colp, '"I will Gladly do my Best"'.

60 'WG' to *English Mechanic*, 20 October 1871, p. 117.

61 W. Carpenter, 'The Flying Philosophers', in *West Londoner*, 25 November 1871.

62 *Scientific Opinion*, 27 April 1870, p. 377.

63 Hampden, *Is Water Level?*, p. 6.

64 *Ibid.*, p. 10.

65 CUL, GROC, RGO6/258/360, JH, *Public Notice to All Whom it May Concern*, May 1870.
66 S. Shapin, 'Discipline and Bounding', *HS*.

Four – Trials and Tribulations

1 Parallax, *Experimental Proofs that the Surface of Standing Water is not Convex but Horizontal*, p. 23.
2 University of Liverpool (ULIV), F.2.21(15), Report of a Survey made for Mr. Hampden.
3 *Nature*, 14 July 1870, p. 214.
4 *Ibid.*, 21 July 1870, p. 236. Conscious of maintaining an authoritative tone, it was the last time that *Nature* courted controversy with flat-earthers directly, although it continued to publish the occasional sideswipe from Hampden.
5 *Nature*, 2 February 1871, p. 267.
6 A. R. Wallace, 'On the Attitude of Men of Science towards Investigators of Spiritualism', in *The Year-book of Spiritualism for 1871*.
7 C. H. Smith, 'Alfred Russel Wallace on Spiritualism, Man and Evolution: An Analytical Essay', at www.wku.edu/~smithch/essays/ARWPAMPH.htm/, M. J. Kottler, 'Alfred Russel Wallace, the Origin of Man, and Spiritualism', *Isis*.
8 Wallace, 'Review of *The Descent of Man and Selection in Relation to Sex*', *Academy*.
9 J. Marchant, *Alfred Russel Wallace: Letters and Reminiscences*, Wallace (ARW) to Darwin (CD), 14 May 1871, pp. 216–17.
10 ULIV, F.2.21(6), Hampden, *British Science Outlawed*.
11 Cambridge University Library (CUL), Greenwich Royal Observatory Collection (GROC), RGO6/258/362–3, John Hampden (JH) to George Biddell Airy (GBA), 2 June 1871.
12 A. R. Wallace, *A Reply to Mr. Hampden's Charges*, p. 7; A. R. Wallace, *My Life*, vol. II, p. 371.
13 Hope Library, Oxford (HLO), ARW to MacLachlan, 17 May 1871.
14 ULIV, F.2.21(14), Hampden, *After all the Commotion, John Hampden Triumphant*.
15 *Daily News*, 28 July 1871, in G. Peacock, *Is the World Flat or Round?*, pp. 21–3.
16 J. Marchant, *Alfred Russel Wallace*, CD to ARW, 12 July 1871, p. 220.
17 ULIV, F.2.21(14), Hampden, *After all the Commotion*.
18 Peacock, *Is the World Flat or Round?*, p. 2.
19 *Ibid.*, p. 32.

20 *English Mechanic*, 13 October 1871, p. 94.

21 In his autobiography, science popularizer Edward Clodd remembered Proctor fondly: 'To the outer world a polemic, ever delighting in controversy and fulminating against abuse in high places, [yet] a more tender hearted mortal never lived.' See E. Clodd, *Memories*, p. 58.

22 Parallax claimed that the apparent rising and setting of the sun were caused by a daily increase and decrease of its 2000-mile distance above the plane earth. *English Mechanic*, 20 October 1873, p. 117.

23 Royal Geographical Society (RGS), CB6, Hampden, J., JH to RGS, 23 October 1871.

24 HLO, ARW to MacLachlan, 25 October 1871.

25 *English Mechanic*, 27 October 1871, p. 145.

26 *Ibid.*, 3 November 1871, p. 170.

27 *English Mechanic*, 10 November 1871, p. 193.

28 *Ibid.*, 17 November 1871, p. 220.

29 *The Times*, 22 November 1871, p. 12 col. b.

30 ULIV, F.2.21(13), Hampden, *An Anticipated Increase in the Waste Paper Department*.

31 ULIV, F.2.21(7), Hampden, *Facts or Fiction, Science or Truth, Working Men, Which is it to be?*

32 RGS, CB6, Hampden, J., JH to RGS, 12 May 1872.

33 Marchant, *Alfred Russel Wallace*, ARW to CD, 31 August 1872, pp. 226–7.

34 *The Times*, 10 October 1872, p. 11 col. e.

35 *Ibid.*; ULIV, F.2.21(16), JH, *To A. R. Wallace, Esq., Professor, F.R.G.S. & c.*, p. 2.

36 *The Times*, 20 November 1872, p. 11 col. d.

37 From memoirs of Rowbotham's granddaughter, Gudrun Hope Kassner, held in private hands.

38 Wallace, *My Life*, vol. II, p. 364.

39 Wallis, *Autobiography of Thomas Wilkinson Wallis*, ARW to Wallis, 20 April 1874, p. 181.

40 Two days after the aborted test, a more reliable report of the proceedings appeared in the *Daily Telegraph*, stating that 'atmospheric indistinctness' had been too great to allow completion of the experiment and the referees and umpire had simply agreed to postpone it until a more suitable time. *Daily Telegraph*, 28 August 1873, p. 3.

41 *The Times*, 17 December 1873, p. 10 col. d.

42 Marchant, *Alfred Russel Wallace*, ARW to CD, 6 December 1874, p. 233.

43 *English Mechanic*, 6 September 1874, p. 211.

44 It read: 'During a conversation in a railway carriage, on the 27th of August [1874], a gentleman speaking of [you], said "Yes, indeed that man is a self

convicted thief, who is content to wait 8 months for a legal trial of his villainy, when 8 hours would suffice if he knew he was innocent."' *The Times*, 10 February 1875, p. 5 col. f.

45 *Ibid.*

46 *Chelmsford Chronicle*, 5 March 1875, p. 5.

47 *Ibid.*, 12 March 1875, p. 7. -

48 British Library (BL), Macmillan Archive (MA), Add 55221 ff. 9–10, ARW to Macmillan, 28 March 1875.

49 *Chelmsford Chronicle*, 12 March 1875, p. 5.

50 *Ibid.*

51 *Nature*, 11 November 1875, p. 29.

52 *The Times*, 18 January 1876, p. 11 col. b.

53 Wallace, *My Life*, vol. II, p. 370.

54 *The Times*, 26 January 1876, p. 12 col. a.

55 Kottler, 'Alfred Russel Wallace, the Origin of Man, and Spiritualism', *Isis*.

56 See also *Wallace's Wonderful Water*, Carpenter's libellous wager account; *Proctor's Planet Earth*, his energetic assault on Proctor's primer for schoolchildren; *Lessons in Elementary Astronomy*, and *Mr. Lockyer's Logic*, a critique of eminent astronomer and *Nature* editor, J. Norman Lockyer, and his *Science Primer for Elementary Schools*.

57 BL, MA, Add 55221 ff. 15–16, ARW to Macmillan, 14 July 1878.

58 J. R. Durant, 'Scientific Naturalism and Social Reform in the Thought of Alfred Russel Wallace', *BJHS*; F. J. Turner, *Between Science and Religion*, p. 69; G. Jones, 'Alfred Russel Wallace, Robert Owen and the Theory of Natural Selection', *BJHS*.

59 R. Colp, '"I will Gladly do my Best"', *Isis*.

60 Lockyer and Proctor were old enemies: on numerous occasions Proctor had publicly criticized the accuracy of Lockyer's astronomical work, even accusing him of plagiarism; see J. Meadows, *Science and Controversy*. For Proctor's disgust at the poor quality of contemporary popular science writing, see Proctor to Clodd, 31 August 1887, in E. Clodd, *Memories*, p. 61.

61 See Lightman, '*Knowledge* confronts *Nature*': Richard Proctor and Popular Science Periodicals', in L. Henson, G. Cantor, G. Dawson, R. Noakes, S. Shuttleworth and J. R. Topham (eds), *Culture and Science in the Nineteenth-Century Media*, pp. 199–210.

62 S. Sheets-Pyenson, *Low Scientific Culture in London and Paris, 1820–1875*; W. H. Brock, 'Science, Technology and Education in the *English Mechanic*', in W. H. Brock, *Science for All*, pp. 1–13.

63 In fact, while Wallace had publicly sponsored spiritualism, Proctor had defended the existence of extraterrestrial life as a legitimate area for scientific research; see M. J. Crowe, *The Extraterrestrial Life Debate, 1750–1900*, pp. 368–77; B. Lightman, 'Astronomy for the People: R. A.

Proctor and the Popularization of the Victorian Universe', in van der Meer (ed.), *Facets of Faith and Science*, pp. 31–45.

64 *Knowledge*, 30 March 1883, p. 198.

65 Proctor made copious use of Parallax's experimental demonstrations to expose the falsity of flat-earth theory; for example, see his series of articles in *Knowledge*, 'Pretty Proofs of the Earth's Rotundity', and his book *Myths and Marvels of Astronomy*.

66 *Knowledge*, 30 November 1883, p. 336.

67 *Ibid.*, 14 December 1883, p. 362.

68 *Knowledge*, 4 April 1884, p. 233.

69 *Ibid.*, 2 May 1884, pp. 313–14.

70 Lightman, '*Knowledge* confronts *Nature*'.

71 *Knowledge*, 4 September 1885, p. 204, quoted in *ibid.*

72 Rowbotham is remembered in his granddaughter's memoirs as a flat-earth believer, but critics continued to claim that he was a fraud who sought either to test the boundaries of popular belief or simply to make money.

73 *Bookseller*, 7 January 1885, p. 12.

74 *Earth not a Globe Review*, no. 4, October 1893, p. 14.

75 These included *Genesis or Geology; Moses or Mathematics; Inspiration or Isaac Newton: A Dialogue on the Elementary Principles of Physical Cosmology* and *The Infidel Globe or Scientific Witchcraft, The Emblem of Paganism and the Refuge of the Atheist.*

76 J. Hampden, *John Hampden's Letter to Professor Huxley*, p. 3.

77 Wallace, *My Life*, vol. II, p. 364.

78 *Ibid.*, p. 376.

79 See Crowe, *Extraterrestrial Life Debate*.

80 *Illustrated London News*, 31 January 1891, p. 134.

81 *Daily Graphic*, 27 January 1891, p. 6.

82 *English Mechanic*, 6 February 1891, p. 506.

Five – Lady Blount and the New Zetetics

1 General Register Office (GRO), marriage certificate, 16 May 1874.

2 W. Elliott, 'The Blount Family', *Cleobury Chronicles*, pp. 29–40.

3 Zetetes, '*Cranks,*' or False Theories of '*Science*' versus the Truth of Nature and the Bible, p. 8.

4 See R. L. Numbers, *The Creationists*, for the vast diversity of creationist thought.

5 *Atlanta Constitution*, 21 May 1905, p. 4.

6 P. J. Bowler, *Reconciling Science and Religion*. See also J. R. Moore, *Post Darwinian Controversies*; J. H. Brooke, *Science and Religion*; D. C.

Lindberg and R. L. Numbers, *God and Nature*; D. C. Lindberg and R. L. Numbers, *When Science and Christianity Meet*; G. B. Ferngren (ed.), *Science and Religion*; F. J. Turner, *Between Science and Religion*; A. Desmond, *Huxley*.

7 J. R. Moore, *The Post Darwinian Controversies*; W. H. Brock and R. M. Macleod, 'The Scientists' Declaration: Reflexions on Science and Belief in the Wake of Essays and Reviews', *BJHS*; Numbers, *The Creationists*.

8 See D. J. Wertheimer, The Victoria Institute, 1865–1919.

9 *Earth not a Globe Review* (*ENAGR*), vol. 1, no. 1, January 1893, p. 24.

10 *Ibid.*, vol. 1, no. 2, April 1893, p. 15.

11 W. Carpenter, *One Hundred Proofs that the Earth is not a Globe*.

12 *Ibid.*, p. 36.

13 'William Carpenter Tried to Prove World is Flat and Had Many Followers', *Washington Post*, 17 May 1908, p. 36.

14 M. Janvier, *Baltimore in the Eighties and Nineties*, p. 175. I am grateful to Allender Sybert, Maryland Genealogical Society, for this reference.

15 *Baltimore Sun*, 2 September 1896, p. 1.

16 'Obituary of William Carpenter', *ENAGR*, vol. 3, no. 6, September–December 1896, p. 58.

17 Baltimore City Register of Wills, SRM 85, pp. 387–8, William Carpenter, 11 December 1900.

18 E. A. Randolph, *The Life of the Reverend John Jasper*.

19 *Portland Maine Press Herald*, 24 August 1960. Many thanks to Jean Hankins for information relating to Joe Holden.

20 Maine Historical Society, *Bridgton Scrapbook*, MTBM64.1, vol. 2, p. 22.

21 *Ibid.*

22 *Semi-Weekly Landmark*, 27 April 1900, p. 2. Although members of the East Otisfield Free Baptist Church do not share these views about the shape of the earth, they still hold an annual picnic in Joe Holden's memory, with traditional treats of popcorn, peanuts and strawberry ice-cream paid for by a bequest from Holden's estate. *Region*, 31 August 1987, p. 11.

23 *Earth: A Magazine of Sense and Science*, vol. 6, nos 72 and 73, July–August 1906, p. 5.

24 Report of the Seventy-First Meeting of the British Association for the Advancement of Science, pp. 725–6.

25 *Earth*, vol. 3, nos 25 and 26, August–September 1902, pp. 13–14.

26 *Ibid.*, vol. 5, nos 49 and 50, August–September 1904, pp. 16–17.

27 *Earth*.

28 *Ibid.*, vol. 3, nos 27 and 28, October–November 1902, pp. 69–70.

29 *Earth*, vol. 5, nos 49 and 50, August–September 1904, pp. 16–17.

30 D. W. Scott, *Terra Firma*, p. iv.

31 UWM (University of Wisconsin-Madison), RSPC, Breach file, Carton 10 (3:1), *Hampshire Telegraph*, 28 March 1896.

32 Zetetes, *Is the Earth a Whirling Globe as Assumed and Taught by Modern Astronomical 'Science'?*, p. 34.

33 *Ibid.*, p. 51.

34 Zetetes, *Is the Earth a Whirling Globe?*, p. 67.

35 Report from *Essex Weekly News*, reprinted in *Earth*, vol. 5, nos 49–50, August–September 1904, pp. 150–51.

36 E. A. M. Blount, 'Is Water Level? The Bedford Level Experiments by Lady Blount and Party', *Earth*, vol. 4, nos 47 and 48, June–July 1904, pp. 389–91.

37 *Atlanta Constitution*, 21 May 1905, p. 4.

38 Blount, 'Is Water Level? The Bedford Level Experiments by Lady Blount and Party', *Earth*, vol. 4, nos 47 and 48, June–July 1904, pp. 389–91.

39 *Earth*, vol. 5, nos 49 and 50, August–September 1904, pp. 1–2.

40 *Ibid.*, p. 53.

41 'Kappa' to *English Mechanic*, 19 August 1904, p. 40.

42 E. A. M. Blount to *English Mechanic*, 25 November 1904, p. 366.

43 J. E. Gore to *English Mechanic*, 13 January 1905, pp. 524–5.

44 C. Stretton to *English Mechanic*, 20 January 1905, p. 544.

45 E. A. M. Blount to *English Mechanic*, 13 January 1905, p. 525.

46 *Earth*, vol. 3, nos 27 and 28, October and November 1902, p. 17.

47 *Atlanta Constitution*, 21 May 1905, p. 4.

48 Blount, *Magnetism as a Curative Agency*.

49 *Portsmouth Daily Herald*, 13 April 1906, p. 6.

50 British Library (BL), Alfred Russel Wallace Papers (ARW), Add 46437 ff. 242–7, E. A. M. Blount to ARW, 17 October 1907.

51 *Washington Post*, 23 October 1909, p. 5.

52 Public Record Office (PRO), Board of Trade (BT) file, BT 31/18754/10198.

53 'A Blackguard Under Royal Patronage: The National Dental Aid Society', *John Bull*, 17 December 1910, p. 965.

54 PRO, BT, BT 58/34/COS 1011, Minute from Companies Department, 19 March 1910.

55 *Ibid.*, Kennedy to Companies Registration Office, 18 March 1910.

56 PRO, BT, BT 58/34/COS 1011, Report of Sergeant Prothero, 22 February 1910.

57 PRO, BT, BT 31/18754/10198.

58 GRO, Grant of Probate, 21 May 1916.

59 E. A. M. Blount, *The Origin and Nature of Sex*, p. 55.

60 GRO, marriage certificate, 28 August 1923.

61 Many thanks to Sir Walter Blount for family reminiscences about his grandmother.

Six – Flat-earth Utopia

1 T. Forby, 'A Brief Authentic Biography of the Rev. Wilbur Glenn Voliva', *Leaves of Healing*, 1 September 1928, p. 1.
2 See G. Wacker, *Heaven Below*.
3 G. Lindsay, *John Alexander Dowie*, p. 104.
4 *Leaves of Healing*, 7 April 1906, p. 446.
5 P. L. Cook, *Zion City, Illinois: Twentieth-Century Utopia*, p. 192.
6 Lindsay, *John Alexander Dowie*, pp. 106–7.
7 Pentecostals later adopted similar arguments, see Wacker, *Heaven Below*, pp. 191–2.
8 *Ibid.*, p. 165. For Dowie's arrests for practising medicine without a licence, see Wacker, *Heaven Below*, p. 189.
9 R. M. Kanter, *Commitment and Community*; R. S. Fogarty, *All Things New*; Cook, *Zion City*, pp. 2–3.
10 Dowie, *The City of Zion: A Résumé* (n.d.), p. 1.
11 Forby, 'A Brief Authentic Biography of the Rev. Wilbur Glenn Voliva', *Leaves of Healing*, 1 September 1928, p. 5.
12 *Ibid.*, p. 1.
13 Cook, *Zion City*, p. 23. In several respects, Dowie's church was a forerunner of Pentecostalism, a radical branch of charismatic Christianity characterized by the practice of speaking in tongues.
14 *Ibid.*, pp. 57–8. Such proclamations were not without parallels. During this period, Frank Weston Sandford, a similarly charismatic healer and leader of a Maine commune, likewise pronounced himself the reincarnation of the prophet Elijah. Although Dowie had markedly more success than Sandford, due largely to his superior business sense, both were ridiculed for their megalomania by the religious and the secular press. Sandford was later prosecuted for the manslaughter of several followers when he failed to provide sufficient provisions on a round-the-world missionary cruise. Wacker, *Heaven Below*, pp. 155–6; Fogarty, *All Things New*, pp. 91–3.
15 Zion Historical Society (ZHS), *Wilbur Glenn Voliva, Continuing History of Zion 1901–61*, p. 7.
16 Cook, *Zion City*, p. 198.
17 *Leaves of Healing*, 7 April 1906, p. 453.
18 *Ibid.*, p. 438.
19 *Ibid.*, pp. 149–50.
20 *Washington Post*, 6 April 1906, p. 6.
21 *Elyria Chronicle*, 6 April 1906, p. 4.
22 The lace factory was sold by the receiver to Marshall Field in 1907.
23 Wacker, *Heaven Below*, p. 6.
24 See G. Wacker 'Travails of a Broken Family: Radical Evangelical

Responses to the Emergence of Pentecostalism in America, 1906–16', in E. L. Blumhofer, R. P. Spittler and G. Wacker, *Pentecostal Currents in American Protestantism*, pp. 23–49.

25 *Newark Advocate*, 6 March 1911, p. 1.

26 A. Prowitt, 'Croesus at the Altar', *American Mercury*, April 1930, p. 404.

27 *Wisconsin Rapids Daily Tribune*, 26 February 1921, p. 2.

28 See Wacker, *Heaven Below*.

29 *Appleton Post-Crescent*, 18 January 1921, p. 4.

30 *Port Arthur Daily News*, 10 April 1921, p. 5.

31 T. O'Hara, 'Wilbur Glenn Voliva', *Chicago Magazine*, 2 August 1911, p. 429.

32 *Mansfield News*, 25 August 1922, p. 1.

33 For discussion, see Wilbur Glenn Voliva, 'Which will You Accept? The Bible the Inspired Word of God or the Infidel Theories of Modern Astronomy', *Leaves of Healing*, 10 May 1930, pp. 131–59.

34 Day-age theory involves interpreting the days of creation in Genesis 1 as vast geological ages, while gap theory posits a creation 'in the beginning' as in Genesis 1:1, followed by a later Edenic creation in six twenty-four-hour days. See R. L. Numbers, *The Creationists*, for comparisons.

35 Numbers, *Creationists*, pp. 17–18.

36 Quoted in Schadewald, 'The Earth was Flat in Zion', p. 72.

37 *Ibid.*, p. 72. For example, see Isaiah 11:12 and Job 28:24.

38 R. J. Schadewald, 'The Earth was Flat in Zion', *Fate*. Carpenter's pamphlet was reissued in 1929. See special issue of *Leaves of Healing*, 10 May 1930.

39 *Ibid.*; see M. Gardner, *Fads and Fallacies*, p. 18.

40 *Decatur Review*, 12 March 1922, p. 17.

41 *Daily Northwestern*, 11 June 1928, p. 8.

42 W. Davenport, 'They Call me a Flathead', *Colliers*, 14 May 1927, p. 31.

43 See S. Shapin and S. Schaffer, *Leviathan and the Air Pump*; D. Gooding, T. Pinch and S. Schaffer, *The Uses of Experiment*.

44 *Wichita Daily Times*, 29 September 1921, p. 2.

45 *Bismarck Tribune*, 10 May 1928, p. 11.

46 *Iowa Recorder*, 13 June 1928, p. 6; *Kingsport Times*, 16 September 1921, p. 1.

47 *Ibid.*

48 *Leaves of Healing*, 10 May 1930, p. 130.

49 *Decatur Review*, 19 June 1929, p. 7.

50 G. M. Marsden, *Fundamentalism and American Culture*; G. M. Marsden, *Understanding Fundamentalism and Evangelicalism*; R. L. Numbers, *Darwinism Comes to America*.

51 E. Larson, 'The Scopes Trial in History and Legend', in D. C. Lindberg and R. L. Numbers, *When Science and Christianity Meet*, pp. 245–64;

J. T. Scopes and J. Presley, *Center of the Storm: Memoirs of John T. Scopes*, p. 60.

52 M. Noll, 'Evangelicalism and Fundamentalism', in G. P. Ferngren, *Science and Religion*, pp. 261–76, (p. 275); Larson, 'The Scopes Trial', *ibid*, pp. 289–98; *ibid.*, pp. 245–64.

53 L. S. de Camp, *The Great Monkey Trial*, p. 106.

54 *Haywood Review*, 11 July 1925, p. 2.

55 de Camp, *Great Monkey Trial*, p. 106.

56 *Sheboygan Press Telegram*, 19 October 1921, p. 2.

57 *Ibid.*

58 Davenport, 'They Call me a Flathead', *Colliers*, 14 May 1927, p. 30.

59 See I. Wallace, *The Square Pegs*.

60 *Ibid.*, p. 10.

61 *National Cyclopedia of American Biography*, 31, p. 439.

62 Wacker, *Heaven Below*, p. 4; S. Hunt (ed.), *Christian Millenarianism*.

63 S. O'Leary, *Arguing the Apocalypse*.

64 *Helena Daily Independent*, 12 March 1931, p. 3.

65 *Reno Evening Gazette*, 19 April 1934, p. 1.

66 O'Leary, *Arguing the Apocalypse*.

67 *Reno Evening Gazette*, 20 April 1934, p. 9.

68 *Ironwood Daily Globe*, 10 September 1935, p. 1.

69 *Newark Advocate*, 3 January 1935, p. 1.

70 ZHS, *Wilbur Glenn Voliva*, p. 34.

71 *Oshkosh Northwestern*, 23 April 1938, p. 18.

72 Schadewald, 'The Earth was Flat in Zion', p. 79.

Seven – Man on the Moon?

1 University of Liverpool (ULIV), Science Fiction Foundation Collection (SFF), Flat Earth Society Papers (FESP), Samuel Shenton (SS), 'Blunder or Crime?', *Channel* (1962).

2 *Ibid.*

3 P. Moore, *Can You Speak Venusian?*, p. 20.

4 *Ibid.*, p. 21.

5 *Coshocton Tribune*, 10 May 1961, p. 13.

6 ULIV, SFF, FESP, Shenton notebook.

7 *Ibid.*, SS to K. Bennett, n.d.

8 ULIV, SFF, FES, SS, draft article for *Parsec*, n.d.

9 J. Glenn, 'Friendship 7: Twenty Years Later, Random Remembrances', http://www.lib.ohio-state.edu/arvweb/glenn/legacy/20year.htm.

10 *Birmingham Evening Echo*, 17 April 1969, p. 8.

11 University of Wisconsin-Madison (UWM), Schadewald Pseudo-Science

Collection (SPSC), 11 (3:1), Ellis Hillman file, SS to E. Hillman, 6 October 1963.

12 ULIV, SFF, FES, 'Enoch Sends Flat-earth Men Round the Bend', *Evening News and Star* (London), 25 June 1964, p. 9. I am grateful to Elisabeth Novitski, British Library, Colindale, for this reference.

13 *Ibid.*, J. Lawson Short to SS, 16 April 1965.

14 ULIV, SFF, FES, J. Mitchell and colleagues to SS, 9 June 1965.

15 *Ibid.*, D. M. Lewis to SS, 14 June 1965.

16 ULIV, SFF, FES, *Ibid.*

17 *Observer*, 28 August 1966, p. 2.

18 ULIV, SFF, FES, anonymous correspondent to SS, 20 September 1966.

19 *Ibid.*, T. Killian to SS, *c.* 4 October 1966.

20 ULIV, SFF, FES, L. Myers to SS, n.d. (1966).

21 *Ibid.*, W. Cook to SS, *c.* 10 October 1966.

22 ULIV, SFF, FES, SS to W. Cook, 20 June 1967.

23 *Ibid.*, J. Rodgers to SS, 9 September 1966.

24 ULIV, SFF, FES, J. H. Green to SS, 16 September 1966.

25 *Ibid.*, SS to *Dover Express*, 24 December 1966.

26 ULIV, SFF, FES, SS to *Daily Express*, 25 April 1967.

27 *Ibid.*, Mr Dalquist and members of the 1966–7 Physics Class to SS, 16 February 1967.

28 ULIV, SFF, FESP *Buffalo Courier Express*, 2 April 1967.

29 J. R. Moore, *The Post Darwinian Controversies*; J. H. Brooke, *Science and Religion*; D. C. Lindberg and R. L. Numbers, *God and Nature*; D. C. Lindberg and R. L. Numbers, *When Science and Christianity Meet*; P. J. Bowler, *Reconciling Science and Religion*. For the warfare myth in action, see J. W. Draper, *History of the Conflict*; A. D. White, *A History of the Warfare of Science with Theology in Christendom*.

30 The argument echoed the insistence of fundamentalists that true faith rested on one's evolutionary views, when it has been argued that there need be no contradiction between evolutionary doctrine and Christian faith; see G. M. Marsden, *Understanding Fundamentalism*.

31 Shenton refused to attend church because he could not locate one that taught that the earth was flat, in keeping with his interpretation of the Bible.

32 ULIV, SFF, FES, SS to R. Morley, 30 May 1967.

33 *Ibid.*, SS to *Points of View*, 27 June 1967.

34 *The Times*, 17 August 1967, p. 1.

35 NASA, B. Normyle to O. W. Nicks, NASA Headquarters Library, Washington, 19 September 1967.

36 *Sun*, 17 August 1967, p. 3.

37 ULIV, SFF, FES, SS to Brethren and Sisters, Lavington, 6 March 1968.

38 *Ibid.*, SS to *Daily Express*, 25 March 1968.

39 ULIV, SFF, FES, G. Svenson and J. Skaar to SS, 20 May 1968.

40 *Ibid.*, G. St. John-Culdwart to SS, 4 July 1968.

41 ULIV, SFF, FES, G. St. John-Culdwart to SS, 18 July 1968.

42 *Ibid.*, G. St. John-Culdwart to SS, 10 September 1968.

43 ULIV, SFF, FES, H. S. Wentz and friends to SS, 28 October 1968.

44 *Ibid.*, R. M. Lynch to SS, 13 October 1968; T. T. Gardner to SS, 13 October 1968.

45 ULIV, SFF, FES, F. I. Kelberger and B. D. St. Cyr to SS, 12 October 1968.

46 *Ibid.*, SS to M. A. Sagarji, 12 November 1968.

47 ULIV, SFF, FES, SS to Mr Gentieu, 14 November 1968.

48 *Ibid.*, SS to *Manchester Guardian*, 17 December 1968.

49 ULIV, SS to *Manchester Guardian*, 27 December 1968.

50 A rhetorical technique common in conspiracy theory and apocalyptic literature; see S. O'Leary, *Arguing the Apocalypse*.

51 For fundamentalist tendency towards polarized thinking, see Marsden, *Understanding Fundamentalism*; G. M. Marsden, *Fundamentalism and American Culture*; R. L. Numbers, *Darwinism Comes to America*.

52 Opinions on origins actually cover a wide spectrum of views through naturalistic evolutionists, theistic evolutionists, intelligent-design theorists, scientific creationists and more; see Numbers, *Darwinism Comes to America*, pp. 11–12; R. L. Numbers, *The Creationists*.

53 P. J. Bowler and I. R. Morus, *Making Modern Science*; Numbers, *The Creationists*.

54 *Appleton Post-Crescent*, 6 January 1969, p. 9.

55 'Where are they now? The Flat Earthers', *Newsweek*, 13 January 1969, p. 8.

56 See O'Leary, *Arguing the Apocalypse*, for further discussion in relation to apocalyptic eschatologies.

57 R. J. Schadewald, 'Scientific Creationism', *SI*. For varieties of creationist thought, see Numbers, *The Creationists*.

58 S. W. Leslie, *The Cold War and American Science*; E. Mendelsohn, 'Science, Scientists and the Military', in J. Krige and D. Pestre (eds), *Science in the Twentieth Century*, pp. 175–202.

59 The seminal work in this respect was Thomas Kuhn's *Structure of Scientific Revolutions*, published in 1962.

60 Leslie, *Cold War and American Science*, p. 233.

61 For conspiracy thinking and apocalypse rhetoric see M. F. Lee (ed.), *Millennial Visions*; O' Leary, *Arguing the Apocalypse*; J. Parish and M. Parker, *The Age of Anxiety*.

62 ULIV, SFF, FES, D. Ruhnke to SS, 20 January 1969.

63 *Ibid.*, G. St. John-Culdwart to SS, 10 January 1969.

64 ULIV, SFF, FES, *ibid.*
65 Moore, *Can You Speak Venusian?*, p. 15. The BBC programme *One Pair of Eyes* was aired on Saturday, 10 May 1969.
66 ULIV, SFF, FES, SS to B. Roberts, 7 February 1969.
67 *Ibid.*, G. St. John-Culdwart to SS, 3 March 1969.
68 ULIV, SFF, FES, L. S. Jackman to SS, 1969.
69 *Ibid.*, Class 9a, Västerhaninge, Stockholm, to SS, 17 April 1969.
70 ULIV, SFF, FES, SS to J. Kelsey, 24 April 1969.
71 *Ibid.*
72 ULIV, SFF, FES, M. J. Moriarty to SS, 25 May 1969.
73 *Ibid.*, J. Huider to SS, 19 May 1969.
74 ULIV, SFF, FES, R. J. Rennison to SS, 10 April 1969.
75 *San Jose Mercury*, 17 July 1969, p. 2.
76 ULIV, SFF, FES, J. A. Terhune and friends to SS, 14 July 1969.
77 *Ibid.*, J. Puckett Jnr to SS, 3 August 1969.
78 ULIV, SFF, FES, A. Powell to SS, 13 November 1969.
79 *Evening Standard*, 26 April 1996, p. 21.
80 UWM, SPSC, 11 (3:3), Ellis Hillman file, E. Hillman to R. Schadewald, 25 February 1990.
81 ULIV, SFF, FES, A. Jay to SS, 3 December 1969.
82 *Ibid.*
83 ULIV, SFF, FES, SS to friends, 18 February 1970.
84 *Ibid.*, SS to V. Safar, 23 May 1970; 3 July 1970.
85 The concept of time is common in apocalyptic and conspiracy rhetoric; see O' Leary, *Arguing the Apocalypse*.
86 *Ibid.* Other ideas deemed 'alternative' or 'pseudo-scientific' are less readily refuted. For discussion of broad issues connected with science and 'anomalistics' see H. Bauer, *Science or Pseudoscience*.
87 ULIV, SFF, FES, E. G. Patten to SS, 5 November 1970.
88 *Ibid.*, C. Beverley to SS, 12 March 1971.

Eight – The View from the Edge

1 An example of the flat-earth myth in action; see Prologue.
2 University of Liverpool (ULIV), Science Fiction Foundation Collection (SFF), Flat Earth Society (FES), Leo Ferrari (LF) to Samuel Shenton (SS), 26 November 1970.
3 Appropriately, 676 Windsor Street is now the University of New Brunswick's postgraduate student bar, decorated with Nowlan memorabilia.
4 P. Toner, *If I Could Turn and Meet Myself*, p. 133.
5 *Ibid.*, p. 204.

6 University of New Brunswick (UNB), Leo Ferrari papers (LFP),
    MGH168, (Series) 2, (Subseries) 2, (Box) 6, (File) 23–1, *The Earth is Flat*
    MS, p. 210.

7 University of Calgary Library (UC), Special Collections, Alden Nowlan
    fonds (ANF), #40.12.23.150, Alden Nowlan (AN) to Ray Fraser (RF), 19
    December 1970.

8 UC, ANF, #40.12.6.37a, J. Stewart to FES, 27 November 1970.

9 *Ibid.*, #40.12.6.250, A. Pittman to AN, n.d.

10 UC, ANF, #40.12.6.34, AN to L. Chudy, 26 November 1970.

11 *Ibid.*, #40.12.23.150, AN to RF, 19 December 1970.

12 UC, ANF, #40.12.6.300, AN to RF, 18 November 1970.

13 *Ibid.*

14 UC, ANF, #40.12.6.330, RF to AN, 25 November 1970.

15 *Ibid.*, #40.12.24.100, AN to RF, 11 January 1971.

16 UC, ANF, #40.12.24.2, RF to AN, 27 January 1971.

17 *Ibid.*, #40.12.24.5a, RF to AN, 25 February 1971.

18 UNB, LFP, MGH168, 2, 2, 6, 23–1, *The Earth is Flat* MS, p. 220.

19 Flyer, 'Introducing the Flat Earth Society', in author's possession.

20 UC, ANF, #40.12.24.66, AN to RF, 1 March 1971.

21 *Ibid.*

22 UC, ANF, #40.12.24.12, RF to AN, 9 June 1971.

23 *Ibid.*, #40.12.24.13a, AN to RF, 14 June 1971.

24 UNB, LFP, MGH168, 2, 2, 6, 23–1, *The Earth is Flat* MS, p. 214.

25 UC, ANF, #40.12.24.14a, AN to RF, 18 June 1971.

26 UNB, LFP, MGH168, 2, 2, 6, 23–1, *The Earth is Flat* MS, pp. 215–16.

27 UC, ANF, #40.12.24.14a, AN to RF, 18 June 1971.

28 *Ibid.*, #40.12.24.17, RF to AN, 25 June 1971.

29 UC, ANF, #40.12.24.25, AN to RF, 3 September 1971.

30 *Ibid.*, #40.12.24.29, RF to AN, 15 October 1971.

31 UC, ANF, #40.12.24.37, AN to RF, 2 February 1972.

32 *Ibid.*, #40.12.38, RF to AN, 17 February 1972.

33 UC, ANF, #40.12.56a, J. Stewart to AN, 5 March 1972.

34 *Ibid.*, #40.12.6.59, AN to G. MacEwen, 12 March 1972.

35 UC, ANF, #40.12.24.4, AN to RF, 20 April 1972.

36 Toner, *If I Could*, p. 228.

37 UC, ANF, #40.12.24.54, RF to AN, 28 December 1972.

38 UNB, LFP, MGH168, 2, 1, 6, 6 Leo Ferrari (LF) to P. Soles, 5 February
    1973.

39 UC, ANF, #40.12.6.83, G. Hartnup to LF, 27 December 1972.

40 *Ibid.*, #40.12.6.24, G. F. Phanent to FES, n.d.

41 UC, ANF, #40.12.6.92, B. Andrew to LF, 3 January 1973.

42 *Ibid.*, #40.12.6.96, J. Lemke to LF, 17 January 1973.

43 UC, ANF, #40.12.6.9, H. Castner to LF, 30 January 1973.

44 UNB, LFP, MGH168, 2, 1, 7, *Daily Gleaner*, 27 January 1973, p. 1.

45 UC, ANF, #40.12.6.103, S. Noonan to LF, 11 March 1973.

46 *Ibid.*, #40.12.6.102a, L. Dillon to LF, 23 February 1973.

47 UNB, LFP, MGH168, 2, 2, 6, 23–1, *The Earth is Flat* MS, p. 218.

48 J. Johnson, 'Flat Earth Society', *Saturday Review of the Sciences*, May 1973, p. 5.

49 UNB, LFP, MGH168, 2, 1, 6, 5–1, M. Baranelli to LF, 7 May 1973.

50 *Ibid.*, MGH168, 2, 1, 7, 3–1, B. Andrew, 'In Defence of the Flat Earth Society', *A.R.O. Observer*, July 1973.

51 *Ibid.*

52 P. J. Bowler and I. R. Morus, *Making Modern Science*; S. Shapin and S. Schaffer, *Leviathan and the Air Pump*; S. Shapin, *A Social History of Truth*.

53 UC, ANF, #40.11.52.17, LF to AN, 11 June 1973.

54 *Ibid.*, #40.11.50.240, LF to AN, 28 July 1973.

55 UNB, LFP, MGH168, 2, 1, 7, 30, *Santa Cruz Sentinel*, 26 July 1973.

56 UC, ANF, #40.12.6.116, R. Miller to AN, 22 March 1974.

57 UNB, LFP, MGH168, 2, 2, 6, 23–2, LF to J. Monroe, 30 January 1974.

58 *Ibid.*, MGH168, 2, 2, 6, 23–1, *The Earth is Flat* MS, p. 217.

59 UNB, LFP, MGH168, 2, 1, 7, 30, *Weekend*, 14 December 1974, pp. 16–18.

60 *Ibid.*, 19 December 1974.

61 UC, ANF, #40.11.52.34, AN to LF, 16 March 1975.

62 *Ibid.*, #40.11.52.33, LF to AN, 7 March 1975.

63 UNB, LFP, MGH168, 2, 1, 6, 5–2, LF to D. Gontar, 25 February 1977.

64 UC, ANF, #40.12.24.86, AN to RF, 3 September 1975.

65 UNB, LFP, MGH168, 2, 1, 6, 5–2, LF to B. Andrew, 7 March 1975.

66 *Ibid.*, MGH168, 2, 2, 6, 23–1, *The Earth is Flat* MS, p. 228.

67 Toner, *If I Could*, p. 221.

68 *Ibid.*

69 UNB, LFP, MGH168, 2, 1, 7, 30, R. Fraser, 'Royally Wronged: Why James III is just plain Jim', *Weekend*, 17 January 1976, pp. 8–9.

70 Information from Ray Fraser, 25 July 2004.

71 Toner, *If I Could*, p. 262.

72 *Ibid.*, p. 263.

73 UC, ANF, #40.12.6.129, LF to J. Stewart, RF and AN, 10 March 1977.

74 Toner, *If I Could*, p. 284.

75 UC, ANF, #40.12.6.1, T. Dolan to AN, 16 February 1981.

76 *Ibid.*, #40.12.6.138, AN to T. Dolan, 24 February 1981.

Nine – The Californian Connection

1 University of Wisconsin-Madison (UWM), Schadewald Pseudo-Science Collection (SPSC), 12 (3:2), International Flat Earth Research Society of America (IFERSA) file, Charles Johnson (CJ) to Robert Schadewald (RS), interview transcript (IT) (n.d.).

2 *Ibid.*

3 UWM, SPSC, 12 (3:2), IFERSA file, CJ to RS, IT.

4 *Chicago Tribune*, 26 August 1977, p. 1.

5 It has been argued that creationism possesses an appeal on 'common-sense' grounds, for besides its grounding in Biblical authority, the doctrine of divine origins may seem more readily acceptable to a non-expert audience than the heady complexities of evolutionary theory. See G. M. Marsden, *Understanding Fundamentalism*.

6 UWM, SPSC, 12 (3:2), IFERSA file, CJ to RS, IT.

7 *Ibid.*, 12 (3:2), CJ to RS, 25 August 1982.

8 UWM, SPSC, 12 (3:2), IFERSA file, CJ to RS, IT.

9 *Ibid.*

10 UWM, SPSC, 12 (3:2), IFERSA file, CJ to RS, IT.

11 *Evening Standard*, 26 April 1996, p. 21.

12 The Science Fiction Foundation Collection, including Shenton's papers, was transferred to the University of Liverpool in 1993.

13 *Evening Standard*, 26 April 1996, p. 21.

14 Ironically, the largest set of Johnson's own publications (in the form of his paper, *Flat Earth News*) outside private hands is held as part of the J. Lloyd Eaton Collection of Science Fiction, Fantasy, Horror and Utopian Literature at the University of California, Riverside – the most extensive collection of its type worldwide.

15 UWM, SPSC, 12 (3:2), CJ file, CJ to RS, 10 July 1987.

16 *Ibid.*, CJ to RS, 2 August 1985.

17 *Boston Globe*, 14 October 1996.

18 Quoted in Schadewald, 'The Plane Truth', *TWA Ambassador*.

19 University of Liverpool (ULIV), Science Fiction Foundation Collection (SFF), Flat Earth Society Papers (FESP), CJ to C. Barren, SFF, 3 May 1973.

20 'Keeping the Flat-earth Faith', *Newsweek*, 2 July 1984, p. 12.

21 *Montréal Gazette*, 27 January 1978.

22 Surprisingly, Johnson never published his experimental data, a key source if he was seeking legitimacy and support for his views.

23 UWM, SPSC, 12 (3:2), CJ file, CJ to RS, 28 October 1977.

24 *Ibid.*, 12 (3:2), IFERSA file, D. Blankenship to RS, 1 April 1978.

25 UWM, SPSC, 10 (3:1), general file, CJ to RS, n.d.

26 UWM, SPSC, 12 (3:2), CJ file, CJ to RS, 24 October 1978.
27 Johnson saw himself and Marjory as simply two witnesses, rather than *the* two witnesses referred to in Revelation 11.
28 R. J. Schadewald, 'The Flat-out Truth', *Science Digest*.
29 UWM, SPSC, 12 (3:2), CJ file, CJ to RS, 18 October 1979.
30 University of New Brunswick (UNB), Leo Ferrari Papers (LFP), MGH168, (Series) 2, (Subseries) 1, (Box) 6, (File) 16, CJ to LF, 14 March 1977.
31 UWM, SPSC, 12 (3:2), CJ file, M. Johnson to R. Schadewald, 27 June 1986.
32 UNB, LFP, MGH168, 2, 1, 6, 16, CJ to LF, 2 September 1976.
33 *Ibid.*, CJ to LF, 14 March 1977.
34 UWM, SPSC, 12 (3:2), Flat Earth Society of Canada (FESC) file, LF to RS, 27 January 1980.
35 UNB, LFP, MGH168, 2, 1, 6, 1, CJ to LF, 28 June 1977.
36 UWM, SPSC, 12 (3:2), IFERSA file, CJ to RS, 3 February 1980.
37 *Ibid.*, 12 (3:2), FESC file, LF to RS, 27 January 1980.
38 University of California, Riverside (UCR), Eaton Collection (EC), *Flat Earth News*, 52, 1984, p. 2.
39 UWM, SPSC, 12 (3:2), CJ file, CJ to RS, 25 August 1982.
40 *Ibid.*
41 UWM, SPSC, 13 (3.2), Gerardus Bouw file, RS to G. Bouw, 6 April 1987.
42 *San Jose Mercury News*, 17 July 1994, p. 1.
43 UWM, SPSC, 12 (3:2), IFERSA file, CJ to RS, 15 August 1980.
44 The flyer is reproduced verbatim by Robert P. J. Day at 'The International Flat Earth Society', http://www.talkorgins.org/faqs/flatEarth.html.
45 UCR, EC, *Flat Earth News*, 54 (1985), p. 3.
46 See Parallax's criticism of the 'pseudo-philosophers' in *Zetetic Astronomy*.
47 UCR, EC, *Flat Earth News*, 54 (1985), p. 3. For fundamentalist views of the concept of 'survival of the fittest', see O. Lindermayer, 'Europe as Anti-Christ', in S. Hunt, *Christian Millenarianism*, pp. 39–49 (p. 46).
48 UWM, SPSC, 12 (3:2), CJ file, CJ to RS, 16 March 1981.
49 For the complex and context-specific relationship between science and religion see P. J. Bowler, *Reconciling Science and Religion*; P. J. Bowler and I. R. Morus, *Making Modern Science*; J. H. Brooke and G. Cantor, *Reconstructing Nature*; J. H. Brooke, *Science and Religion*; G. B. Ferngren (ed.), *Science and Religion*; F. J. Turner, *Between Science and Religion*; D. C. Lindberg and R. L. Numbers, *God and Nature*; D. C. Lindberg and R. L. Numbers, *When Science and Christianity Meet*; J. R. Moore, *The Post Darwinian Controversies*.
50 UWM, SPSC, 12 (3:2), CJ file, CJ to RS, 23 March 1990.
51 *Ibid.*, CJ to RS, 3 March 1983. For Parallax's insistence that the issue

should be removed from the realms of belief and that experimental investigations were the only authoritative source of knowledge in matters of 'matter', see *Zetetic Astronomy* (1881), p. 399.

52 For 'common sense' and inductive aspects of fundamentalist thinking, see Marsden, *Understanding Fundamentalism*, pp. 118–19.

53 UNB, LFP, MGH168, 2, 1, 6, 16, CJ to LF, 14 March 1977.

54 UWM, SPSC, 12 (3:2), CJ file, CJ to RS, 27 April 1979.

55 *Daily Ledger-Gazette* (Antelope Valley), 14 October 1980, p. 1.

56 UWM, SPSC, 12 (3:2), CJ file, CJ to RS, IT.

57 *Ibid.*, 7 February 1977.

58 In the realm of conspiracy thinking such fantastic narratives are far from rare; see S. O'Leary, *Arguing the Apocalypse*; J. Parish and M. Parker (eds), *The Age of Anxiety*; M. F. Lee (ed.), *Millennial Visions*. Johnson's world history narrative was somewhat piecemeal and frequently subject to alteration.

59 UCR, EC, *Flat Earth News*, 52 (1984), p. 2.

60 For millenarian, fundamentalist belief that the European Union is the state of the Antichrist and the New World Order, see O. Lindermayer, 'Europe as Anti-Christ', in C. Hunt, *Christian Millenarianism*.

61 Between the Revolutionary and Civil Wars, *c.*1780–1860, approximately a hundred known Utopian communities were established in the US. See R. M. Kanter, *Commitment and Community*, p. 61.

62 R. J. Schadewald, 'The Flat-out Truth', *Science Digest*.

63 Schadewald, 'Plane Truth'.

64 UCR, EC, *Flat Earth News*, 55 (1985), p. 4. See Darwin's *Descent of Man* (1871), in which natural selection is applied to human evolution, and Revelation (12:9) for the 'ancient serpent called the Devil or Satan, that deceived the whole world'.

65 Schadewald, 'Flat-out Truth'.

66 Schadewald, 'Plane Truth'.

67 *Ibid.*

68 'Are 20 Million Americans out to Launch?', *Washington Post*, 20 July 1994, p. 1.

69 UCR, EC, 'John F. Kennedy and Crime Partner Nikita Kruschev: Satanic Beasts are Burying America', *Flat Earth News*, 50 (1984), p. 1.

70 *Delaware Gazette*, 9 October 1978, p. 12; UCR, EC, *Flat Earth News*, 57 (1985), p. 2.

71 UCR, EC, *Flat Earth News*, 57 (1985), covering letter 3 October 1986.

72 *Ibid.*, 53 (1985), p. 4.

73 *San Francisco Examiner*, 7 October 1984, p. 1.

74 UCR, EC, *Flat Earth News*, 54 (1985), p. 4.

75 *San Jose Mercury News*, 17 July 1994, p. 1.

76 *San Francisco Examiner*, 7 October 1984, p. 1.

77 *Washington Post*, 20 July 1994, p. 1.

78 For Christ lifting his eyes to heaven as proof of the zetetic world-view, see Parallax, *Zetetic Astronomy* (1881), p. 389.

79 UCR, EC, *Flat Earth News*, 65 (1988), p. 3.

80 *Independent*, 30 March 2001, p. 6.

81 *Washington Post*, 20 July 1994, p. 1.

82 Parish and Parker (eds), *Age of Anxiety*.

83 *New York Times*, 2 June 1978, p. 8.

84 *Washington Post*, 20 July 1994, p. 1.

85 *Boston Globe*, 14 October 1996, p. 1.

86 UWM, SPSC, 10 (3.1), general file, CJ to RS, 4 March 1997.

87 P. Plait, *Bad Astronomy*, p. 156.

Epilogue – Myths and Meanings

1 J. B. Russell, *Inventing the Flat Earth*.

2 J. Parish and M. Parker (eds), *The Age of Anxiety*, p. 7.

3 While ordinary flat-earth believers did not have any wide-ranging reformist goal in mind, it is evident that prominent zetetic campaigners utilized the concept as an arresting headline for their broader social improvement programmes. See G. Cantor, 'The Edinburgh Phrenology Debate', *Annals*, G. Cantor, 'A Critique of Shapin's Social Interpretation of the Edinburgh Phrenology Debate', *Annals*; R. J. Cooter, *The Cultural Meaning of Popular Science*; S. Shapin, 'Phrenological Knowledge and the Social Structure of Early Nineteenth-Century Edinburgh', *Annals*; S. Shapin, 'History of Science and its Sociological Reconstructions', *HS*; S. Shapin, *A Social History of Truth*; J. van Whye, *Phrenology and the Origins of Victorian Scientific Naturalism*; van Whye, 'Was Phrenology a Reform Science?', *HS*.

4 For status claims of other alternative sciences, see van Whye, *Phrenology*; A. Winter, *Mesmerized*; H. Bauer, *Science or Pseudoscience*.

5 See van Whye, *Phrenology*; van Whye, 'Reform Science'.

6 G. M. Marsden, *Fundamentalism and American Culture*; P. J. Bowler, *Reconciling Science and Religion*.

7 J. H. Brooke, *Science and Religion*; see also Bowler, *Reconciling Science*; G. B. Ferngren, *Science and Religion*; D. C. Lindberg and R. L. Numbers, *God and Nature*; D. C. Lindberg and R. L. Numbers, *When Science and Christianity Meet*; J. R. Moore, *The Post Darwinian Controversies*.

8 For wider discussion see M. Noll, 'Evangelicalism and Fundamentalism', in Ferngren, *Science and Religion*; Marsden, *Understanding Fundamentalism and Evangelicalism*; Marsden, *Fundamentalism and American Culture*; R. L. Numbers, *The Creationists*.

9 Numbers, *Creationists*, pp. 335–6; R. L. Numbers, *Darwinism Comes to America*.

10 R. J. Schadewald, 'Scientific Creationism, Geocentricity and the Flat Earth', *SI*; R. J. Schadewald, 'The Evolution of Bible Science', in L. Godfrey, *Scientists Confront Creationism*.

11 R. L. Ecker, *Dictionary of Science and Creationism*, p. 30; Numbers, *Creationists*.

12 *Ibid.*

13 See Numbers, *Creationists*, pp. 242–3.

14 Only 12 per cent of the sample of 1005 adults nationwide believed that human beings had evolved without any involvement from God: Gallup poll, September 2005. For previous statistics see Numbers, *Darwinism*, pp. 9–10; Numbers, *Creationists*, p. ix.

15 For example, see Ecker, *Science and Creationism*, p. 70.

16 *Acts and Facts*, May 1979, p. 3; see Schadewald, 'Evolution of Bible Science', p. 293. Unfortunately a lack of sources means that the interaction between creationists, geocentrists and flat-earth believers cannot be traced in any depth.

17 R. J. Cooter and S. Pumfrey, 'Separate Spheres and Public Places', *HS*; A. Ophir and S. Shapin, 'The Place of Knowledge', *SIC*; S. Shapin, 'Discipline and Bounding', *HS*.

18 Shapin, *Social History of Truth*; J. Golinski, *Making Natural Knowledge*.

19 *Evening Standard*, 26 April 1996, p. 21.

20 The investigators experimented with various teaching demonstrations showing children that something can look flat even when it is not, thereby illustrating that appearances can be deceptive. A. Lightman and P. Sadler, 'The Earth is *Round*? Who are you Kidding?', *Science and Children*, 25, 5 (1988), pp. 25–6.

21 The idea has recently been used as the key motif for a connected earth in Thomas L. Friedman's globalization bestseller, *The World is Flat: A Brief History of the 21st Century*.

22 For a survey of debates over the teaching of evolution see Numbers, *Darwinism*; Numbers, *Creationists*.

23 Twenty-one per cent of the sample thought that the sun revolves round the earth and seven per cent said they did not know either way. Meanwhile, of the 72 per cent who answered correctly, 45 per cent said it takes one year for the earth to orbit the sun, 17 per cent said one day, two per cent said one month and nine per cent did not know.

Bibliography

Archives and Unpublished Sources

Where works were published under a pseudonym, the author's true name is added in brackets. Place of publication is London, unless otherwise stated. PU has been used where place is unknown.

Abbott, E. A., *Flatland* (Seeley, 1884)

Airy, W. (ed.), *Autobiography of Sir George Biddell Airy* (Cambridge, Cambridge University Press, 1896)

Allen, D. E., *The Naturalist in Britain: A Social History*, 2nd edn (Princeton, Princeton University Press, 1994)

Anderson, R. M., *Visions of the Disinherited: The Making of American Pentecostalism* (Oxford, Oxford University Press, 1979)

Armytage, W. H. G., *Heavens Below: Utopian Experiments in England, 1560–1960* (Routledge & Kegan Paul, 1961)

Ashworth, W. J., 'The Calculating Eye: Baily, Herschel, Babbage and the Business of Astronomy', *BJHS*, 27 (1994), pp. 409–41

Ball, W. V. (ed.), *Reminiscences and Letters of Sir Robert Ball* (Cassell, 1915)

Barnes, B., Bloor, D., and Henry, J., *Scientific Knowledge: A Sociological Analysis* (Chicago, Chicago University Press, 1996)

Barnes, B., and Edge D., *Science in Context* (Milton Keynes, Open University Press, 1982)

Barnes, B., and Shapin, S., *Natural Order: Historical Studies of Scientific Culture* (Sage, 1979)

Barrow, L., *Independent Spirits: Spiritualism and English Plebians, 1850–1910* (Routledge & Kegan Paul, 1986)

Barton, R., 'Just before *Nature*: The Purposes of Science and the Purposes of Popularization in some English Popular Science Journals of the 1860s', *Annals*, 55 (1998), pp. 1–33

———, '"Huxley, Lubbock, and Half a Dozen Others": Professionals and Gentlemen in the Formation of the X Club, 1851–1864', *Isis*, 89 (1998), pp. 410–44

———, '"Men of Science": Language, Identity and Professionalization in the Mid-Victorian Scientific Community', *HS*, 41 (2003), pp. 73–119

Bartrip, P. W. J., *Mirror of Medicine: A History of the British Medical Journal* (Oxford, Clarendon Press and BMJ, 1990)

Basbanes, N. A., *A Gentle Madness: Bibliophiles, Bibliomanes and the Eternal Passion for Books* (New York, St Martin's Press, 1999)

Bathgate, W., *The Shape of the Earth* (Liverpool, Parker, 1872)

Bauer, H., *Science or Pseudoscience: Magnetic Healing, Psychic Phenomena and Other Heterodoxies* (Urbana and Chicago, Illinois University Press, 2001)

Baumgartner, F. J., *Longing for the End: A History of Millennialism in Western Civilization* (New York, Palgrave, 1999)

Beardsley, J., *The Earth Proved not to be a Sphere* (F. Pitman, 1872)

Bedford, J., *The Bedfordian System of Astronomy*, 3rd edn (H. Vickers, 1881)

Belchem, J., *Popular Radicalism in Nineteenth-Century Britain* (Macmillan, 1996)

Bellon, R., 'Joseph Dalton Hooker's Ideals for a Professional Man of Science', *JHB*, 34 (2001), pp. 51–82

Berman, D., *A History of Atheism in Britain from Hobbes to Russell* (Croom Helm, 1988)

Berry, A. (ed.), *Infinite Tropics: An Alfred Russel Wallace Anthology* (Verso, 2002)

Birley, S. [Samuel Birley Rowbotham], *Phosphorus, as Discovered and Prepared by Dr. Birley* (Gordon Murray, 1881)

———, *A Scientific Discourse for Thoughtful Minds* (Gordon Murray, 1902)

Blacker, C., and Loewe, M., *Ancient Cosmologies* (George Allen & Unwin, 1975)

Blount, E. A. M., *Questions and Answers* [on religion] (Malvern, C. E. Brooks, 1897)

———, *Adrian Galilio, or a Song Writer's Story* (Malvern, C. E. Brooks, 1898)

———, *Magnetism as a Curative Agency*, 3rd edn (Appareil Magnétique, 1905)

———, *The Secrets of Nature Exhumed* (Worthing, the author, 1913)

———, *Our Enclosed World – Extracts from Lectures* (Worthing, the author, 1914)

———, *The Origin and Nature of Spiritism* (Long, 1919)

———, *The Dreamer* (Brighton, the author, 1921)

———, *The Origin and Nature of Sex*, 6th edn (Health Promotion, 1921)

———, *The Origin of the Soul* (Worthing, the author, 1922)

———, *The Origin and Basis of All Revealed Knowledge* (Brighton, Southern Publishing Company, 1923)

Blount, E. A. M. and Zetetes [Albert Smith], *Zetetic Astronomy: Or the Sun's Motions North and South; With the Moon's Motions; Fancied and Real: Showing the Uselessness of the Gravitational Theory etc.* (Kingston Hill, the author, 1906)

Blumhofer, E. L., Spittler, R. P., and Wacker, G. A., *Pentecostal Currents in American Protestantism* (Urbana and Chicago, Illinois University Press, 1999)

Bolenius, E. M., *The Boys' and Girls' Reader: Fifth Reader* (New York, Houghton, Mifflin, 1919)

Boorstin, D., *The Discoverers* (New York, Random House, 1983)

Bowler, P. J., *The Non-Darwinian Revolution: Reinterpreting a Historical Myth* (Baltimore, Johns Hopkins University Press, 1992)

———, *The Eclipse of Darwinism: Anti-Darwinian Evolution Theories in the Decades Around 1900* (Baltimore, Johns Hopkins University Press, 1992)

———, *Reconciling Science and Religion* (Chicago, Chicago University Press, 2001)

———, *Evolution: The History of an Idea*, 2nd edn (Berkeley, California University Press, 1989)

Bowler, P. J., and Morus, I. R., *Making Modern Science: A Historical Survey* (Chicago, Chicago University Press, 2005)

Bramhall, W., *The Great American Misfit* (New York, C. N. Potter, 1982)

Breach, E., *Twenty Reasons against Newtonianism* (Southsea, S. Phillips, n.d.)

Bresher, M. R., *The Newtonian System of Astronomy* (Whittaker, 1868)

Brock, W. H., *Science for All* (Aldershot, Variorum, 1996)

Brock, W. H., and Macleod, R. M., 'The Scientists' Declaration: Reflexions on Science and Belief in the Wake of Essays and Reviews', *BJHS*, 9 (1976), pp. 39–66

Brooke, J. H., *Science and Religion: Some Historical Perspectives* (Cambridge, Cambridge University Press, 1991)

Brooke, J. H., and Cantor, G., *Reconstructing Nature: The Engagement of Science and Religion*, 2nd edn (Oxford, Oxford University Press, 2000)

Brooks, J. L., *Just before the Origin: Alfred Russel Wallace's Theory of Evolution* (New York, Columbia University Press, 1984)

Brown, P. S., 'Social Context and Medical Theory in the Demarcation of Nineteenth-Century Boundaries', in Bynam, W. F., and Porter, R. (eds), *Medical Fringe and Medical Orthodoxy, 1750–1850* (Croom Helm, 1987), pp. 216–33

Browne, J., *Charles Darwin: Voyaging* (Cape, 1995)

———, *Charles Darwin: The Power of Place* (Pimlico, 2003)

Burkhardt, F. (ed.), *Charles Darwin's Letters: A Selection 1825–1859* (Cambridge, Cambridge University Press, 1996)

Bynam, W. F., and Porter, R. (eds), *Medical Fringe and Medical Orthodoxy, 1750–1850* (Croom Helm, 1987)

———, *Health for Sale: Quackery in England 1660–1850* (Manchester, Manchester University Press, 1989)

Camerini, J. R., *The Alfred Russel Wallace Reader: A Selection of Writings from the Field* (Baltimore, Johns Hopkins University Press, 2002)

Cannon, S. F., *Science in Culture: The Early Victorian Period* (New York, Science History Publications, 1978)

Cantor, G., 'The Edinburgh Phrenology Debate: 1803–1828', *Annals*, 32 (1975), pp. 195–218

————, 'A Critique of Shapin's Social Interpretation of the Edinburgh Phrenology Debate', *Annals*, 33 (1975), pp. 245–56

Cantor, G., and Shuttleworth, S. (eds), *Science Serialized: Representations of Science in the Nineteenth Century Media* (Cambridge, Massachusetts Institute of Technology, 2004)

Carey, J., *E. W. Bullinger: A Biography* (Grand Rapids, Kregel, 2000)

Carpenter, W., *'Bosh' and 'Bunkum'!* (Heywood & Co., 1868)

————, *The Flying Philosophers* (British & Colonial Publishing Co., 1871)

————, *Water not Convex, The Earth not a Globe! Demonstrated by Alfred Russel Wallace on the 5th March 1870* (W. Carpenter, 1871)

————, *The 'Bedford Level' Experiment* (W. Carpenter, 1872)

————, *Sense versus Science* (W. Carpenter, 1873)

————, *Proctor's Planet Earth* (W. Carpenter, 1875)

————, *Wallace's Wonderful Water* (Abel Heywood, 1875)

————, *Mr. Lockyer's Logic: An Exposition of Mr. J. Norman Lockyer's Astronomy* (W. Carpenter, 1876)

————, *The Delusion of the Day or Dyer's Reply to Parallax* (Abel Heywood, 1877)

————, *One Hundred Proofs that the Earth is not a Globe* (Baltimore, W. Carpenter, 1885)

Chadwick, O., *The Victorian Church*, 2 vols (New York, Oxford University Press, 1966–70)

————, *The Secularization of the European Mind in the 19th Century* (Cambridge, Cambridge University Press, 1975)

Chaikin, A., *A Man on the Moon: The Voyages of the Apollo Astronauts* (Harmondsworth, Penguin, 1994)

Chapman, A., *The Victorian Amateur Astronomer: Independent Astronomical Research in Britain, 1820–1920* (Chichester, John Wiley, 1998)

————, *Astronomical Instruments and their Users* (Aldershot, Variorum, 1996)

————, *Gods in the Sky: Astronomy from the Ancients to the Renaissance* (Macmillan, 2002)

Claeys, G. (ed.), *The Selected Works of Robert Owen*, 4 vols (Pickering, 1993)

Clerke, A. M., *A Popular History of Astronomy during the Nineteenth Century* (Edinburgh, A. & C. Black, 1885)

Clodd, E., *Memories*, 2nd edn (Watts, 1926)

Collins, H. M., *Changing Order: Replication and Induction in Scientific Practice*, 2nd edn (Chicago, Chicago University Press, 1992)

Colp, R., '"I will Gladly do my Best"': How Charles Darwin Obtained a Civil List Pension for Alfred Russel Wallace', *Isis*, 83 (1992), pp. 3–26

Common Sense [William Carpenter], *Communion with 'Ministering Spirits'* (William Horsall, 1858)

————, , *Theoretical Astronomy Examined and Exposed* (W. Carpenter, 1869)

————, *The Earth not a Globe* (Job Caudwell, 1864)

————, *Something about Spiritualism* (Job Caudwell, 1865)

Conlin, M. F., 'The Popular and Scientific Reception of the Foucault Pendulum in the United States', *Isis*, 90 (1999), pp. 181–204

Cook, F. H., *The Terrestrial Plane: The True Figure of the Earth* (1908)

Cook, P. L., *Zion City, Illinois: Twentieth-Century Utopia* (New York, Syracuse University Press, 1996)

Cooter, R. J., 'Phrenology: The Provocation of Progress', *HS*, 14 (1976), pp. 211–34

———, *The Cultural Meaning of Popular Science: Phrenology and the Organization of Consent in Nineteenth-Century Britain* (Cambridge, Cambridge University Press, 1984)

Cooter, R. J., and Pumfrey, S., 'Separate Spheres and Public Places: Reflections on the History of Science Popularisation and Science in Popular Culture', *HS*, 32 (1994), pp. 237–67

Crowe, M. J., *The Extraterrestrial Life Debate, 1750–1900* (Cambridge, Cambridge University Press, 1986)

Curd, M., and Cover, J. A., *Philosophy of Science: The Central Issues* (New York, W. W. Norton, 1998)

Curry, P., *A Confusion of Prophets: Victorian and Edwardian Astrology* (Collins & Brown, 1992)

Danielson, D. R., *The Book of the Cosmos: Imagining the Universe from Heraclitus to Hawking* (Cambridge, Perseus, 2000)

Darwin, C., *On the Origin of Species by Means of Natural Selection, or the Preservation of favoured races in the struggle for life*, 6th edn (John Murray, 1888)

———, *The Descent of Man, and Selection in relation to sex*, 2nd edn (John Murray, 1874)

Davies, C. M., *Heterodox London*, 2 vols (Tinsley Brothers, 1874)

Davis, R. W., and Helmstadter, R. J., *Religion and Irreligion in Victorian Society* (Routledge, 1992)

———, *Religion and Society in Victorian England* (HarperCollins, 1992)

Dear, P., *Revolutionizing the Sciences: European Knowledge and its Ambitions 1500–1700* (Basingstoke, Palgrave, 2001)

De Camp, L. S., *The Great Monkey Trial* (New York, Doubleday, 1968)

De Grazia, A., *The Velikovsky Affair* (Sidgwick and Jackson, 1966)

de Morgan, A., *A Budget of Paradoxes* (Longmans, Green, 1872)

de Morgan, S., *Memoir of Augustus de Morgan* (Longmans, Green, 1882)

Desmond, A., 'Artisan Resistance and Evolution in Britain 1819–1848', *OS*, 3 (1987), pp. 77–110

———, *The Politics of Evolution: Morphology, Medicine, and Reform in Radical London* (Chicago, Chicago University Press, 1989)

———, *Huxley* (Harmondsworth, Penguin, 1998)

———, 'Redefining the X-Axis: "Professionals", "Amateurs" and the Making of Mid-Victorian Biology – A Progress Report', *JHB*, 34 (2001), pp. 3–50

Desmond, A., and Moore, J. R., *Darwin* (Michael Joseph, 1991)

Downham Market and District Amenity Society, *A History of Downham Market* (Downham Market, Downham Market and District Amenity Society, 1999)

Draper, J. W., [1874], *History of the Conflict between Religion and Science*, new edn (Pioneer Press, 1923)

Dreyer, J. L. E., *A History of Astronomy from Thales to Kepler*, 2nd edn (New York, Dover, 1953)

Duhem, P., [1906], *The Aim and Structure of Physical Theory* (New York, Athenaeum, 1974)

Durant, J. R., 'Scientific Naturalism and Social Reform in the Thought of Alfred Russel Wallace', *BJHS*, 12 (1979), pp. 31–58

Dyer, J., *The Spherical Form of the Earth. A Reply to 'Parallax'* (Trübner & Co., 1870)

————, *Thoughts on the Laws of Health and Suggestions for their Promotion* (W. Tweedie, 1866)

Ecker, R. L., *Dictionary of Science and Creationism* (New York, Prometheus, 1990)

Edgell, W., *Does the Earth Rotate?* (Radstock, the author, 1914)

Elliott, W., 'The Blount Family', *Cleobury Chronicles*, vol. 1 (Cleobury Mortimer, 1991), pp. 29–40

Encyclopaedia Britannica, IV (Cambridge, Cambridge University Press, 1910)

Endersby, J., 'Escaping Darwin's Shadow', *JHB*, 36 (2003), pp. 385–403

Evans, C. W. de Lacy, *Can we Prolong Life?* (Baillière, Tindall and Cox, 1879)

————, *Consumption* (Baillière, 1881)

Fara, P., 'Isaac Newton lived here: sites of memory and scientific heritage', *BJHS*, 33 (2000), pp. 407–26

Ferngren, G. B. (ed.), *Science and Religion: A Historical Introduction* (Baltimore, Johns Hopkins University Press, 2002)

Fichman, M., *An Elusive Victorian: The Evolution of Alfred Russel Wallace* (Chicago, Chicago University Press, 2003)

Fogarty, R. S., *All Things New: American Communes and Utopian Movements, 1860–1914* (Chicago, Chicago University Press, 1990)

Fort, C. H., *New Lands*, 2nd edn (John Brown, 1996)

French, R. D., *Antivivisection and Medical Science in Victorian Society* (Princeton, Princeton University Press, 1975)

Friedman, T. L., *The World is Flat: A Brief History of the 21st Century* (New York, Farrar, Straus and Giroux, 2005)

Fyfe, A., *Science and Salvation: Evangelical Popular Science Publishing in Victorian Britain* (Chicago, Chicago University Press, 2004)

Gardiner, F. J., *History of Wisbech and Neighbourhood 1848–98* (Wisbech, Gardiner, 1898)

Gardner, M., *Fads and Fallacies in the Name of Science* (New York, Dover, 1952)

————, *The Night is Large: Collected Essays 1938–1995* (Harmondsworth, Penguin, 1996)

Garwood, C., 'John Hampden (1819–1891)', in Lightman, B. (ed.), *The Dictionary of Nineteenth-Century British Scientists* (Thoemmes, 2004)

————, 'Samuel Birley Rowbotham (1816–1884)', in Lightman, B. (ed.), *The Dictionary of Nineteenth-Century British Scientists* (Thoemmes, 2004)

Gay, H., and Gay, J., 'Brothers in Science: Science and Fraternal Culture in Nineteenth-Century Britain', *HS*, 35 (1997), pp. 425–53

Gillispie, C. C., *Genesis and Geology: A Study in the Relations of Scientific Thought, Natural Theology, and Social Opinion in Great Britain, 1790–1850* (New York, Harper and Row, 1951)

Gingerich, O., *The Eye of Heaven: Ptolemy, Copernicus, Kepler* (New York, American Institute of Physics, 1992)

Gleason, A., *Is the Bible from Heaven? Is the Earth a Globe? Scientifically and Theologically Demonstrated*, 2nd edn (Buffalo, Buffalo Electrotype and Engraving, 1893)

Golinski, J., *Making Natural Knowledge: Constructivism and the History of Science* (Cambridge, Cambridge University Press, 1998)

Gooding, D., Pinch, T., and Schaffer, S., *The Uses of Experiment* (Cambridge, Cambridge University Press, 1989)

Gore, J. E., *Astronomical Curiosities: Facts and Fallacies* (Chatto & Windus, 1909)

Goudey, H. J., *Earth not a Globe: Scientifically, Geometrically, Philosophically Demonstrated* (Boston, H. J. Goudey, 1930)

Gould, S. J., 'The Persistently Flat Earth', *Natural History*, 103 (1994), pp. 14–19

Grant, E., *Planets, Stars and Orbs: Medieval Cosmos, 1200–1687* (Cambridge, Cambridge University Press, 1994)

Greenberg, J., *The Problem of the Earth's Shape from Newton to Clairaut* (Cambridge, Cambridge University Press, 1995)

Gregory, J., *The Vegetarian Movement in Britain c.1840–1901. A Study of its Development, Personnel and Wider Connections* (unpublished PhD thesis, University of Southampton, 2002)

Hall, T. H., *The Spiritualists* (Duckworth, 1962)

Hampden, H., *Some Memorials of Renn Dickson Hampden* (Longmans, Green, 1871)

Hampden, J., *The Rampart of Steel* (Canterbury, G. Birch, 1852)

————, *Church of England! Reformation or Ruin* (1861)

————, *The Popularity of Error and the Unpopularity of Truth* (Swindon, Alfred Bull, 1869)

————, *Report of a Survey made for Mr. Hampden to Test the Truthfulness and Honesty of the Decision made by Mr. Walsh* (1870)

————, *Is Water Level or Convex after All? The Bedford Canal Swindle Detected and Exposed* (Swindon, Alfred Bull, 1870)

———, *Astronomy as Learnt from the Bible: A Familiar Dialogue upon the Earth, Sun, Moon and Stars, Shewing the Situation the Earth Occupies in Creation* (Swindon, Alfred Bull, 1870)

———, *Scientific Professors and their Principles* (Swindon, 1871)

———, *A Wrinkle or Two* (Swindon, A. C. Dore, 1871)

———, *Touching the Death of Sir John Herschel* (PU, 1871)

———, *After All the Commotion John Hampden Triumphant! Always has been and always means to be* (Chippenham, PU, 1871)

———, *Facts or Fiction, Science or Truth, Working Men, Which is it to be?* (the author, 1871)

———, *British Science Outlawed* (Chippenham, the author, 1871)

———, *An Appeal to our Mathematical Professors* (Chippenham, the author, 1872)

———, *The 'Bedford Level' Experiment* (W. Carpenter, 1872)

———, *An Open Letter to Alfred Russel Wallace* (PU, 1872)

———, *A Few Scriptural Inconsistencies with the Newtonian Theory* (Croydon, Thomas Beck, 1873)

———, *A Panic Among the Globe Makers and Educational Professors* (Swindon, A. R. Dore, 1873)

———, *Geographical Hydrophobia* (Samuel Palmer, 1873)

———, *Science Prizes – Theory is not Science* (Croydon, the author, 1874)

———, *Description and Specification of J.H.'s Improvements in Artillery* (J. Cook, 1876)

———, *The New Manual of Biblical Cosmography* (Beaumont, 1877)

———, *The Sun Dial and its Lessons* (Chelsea, the author, 1879)

———, *John Hampden's Challenge to Richard A. Proctor Esq. and all Globe Makers and Globular Theorists in Europe* (the author, 1879)

———, *The Earth in its Creation, its Chronology, its Physical Features, and the One Alone Portion of the Universe Adapted to Man's Occupation and Service* (W. H. Guest, 1880)

———, *Genesis or Geology; Moses or Mathematics; Inspiration or Isaac Newton: A Dialogue on the Elementary Principles of Physical Cosmology* (Civil Service Printing & Publishing, 1884)

———, *Modern Christianity – What Is It?* (PU, the author, 1884)

———, *The Infidel Globe* (Wade, 1884)

———, *John Hampden's Letters to Alfred Russel Wallace* (Balham, the author, 1885)

———, *For Distribution at the Exhibition of Geographical Appliances* (PU, the author, 1885)

———, *John Hampden's Letter to Professor Huxley* (Wade, 1886)

———, *Modern Education Conducted on Wrong Principles* (Wade, 1886)

———, *The globular theory as adopted and formulated by Sir Isaac Newton* (Croydon, the author, 1887)

———, *The Zetetic Society and Modern Astronomy* (W. Reeves, 1887)

————, *The Astronomer's Globe: The Witchcraft of Science, the Refuge of the Atheist: a Mischievous Counterfeit, Contradicted alike by Scripture, by Reason, and by Fact* (Partridge, 1887)

————, *The Newtonian or Solar System: Is It Scientific?* (Croydon, Zetetic Society, 1890)

————, *Outbreak of Rabies at the Greenwich Observatory: The Professors Frantic – Astronomy Doomed* (Croydon, John Hampden, 1890)

————, *Strictures on Popular Astronomy by Various Authors* (Swindon, the author, n.d.)

————, *A Nut for the British Scientific Association* (PU, n.d.)

————, *To the Principals, Tutors, Professors and Students of any Public, Private, Military or Commercial School or College in Great Britain* (Chippenham, the author, n.d.)

————, *Words of Warning to the Unsound, Ill-taught, Half Reformed Protestant Ministers and People of England* (Croydon, the author, n.d.)

————, *God's Truth or Man's Science? Which Shall Prevail?* (PU, the author, n.d.)

————, *What is a Swindler, and who is the Man that is Compelled to Submit to the Term?* (PU, the author, n.d.)

————, *Fact or Fiction – Science or Truth, Working Men, Which is it to be?* (PU, the author, n.d.)

————, *The Fall of the Apple, or the Tipsy Philosopher* (Bayswater, the author, n.d.)

————, *Britons Never Will be Slaves* (Bayswater, the author, n.d.)

————, *The Proposed Arctic Balloon Expedition* (Bayswater, the author, n.d.)

————, *The Mosaic Cosmogony Proved to be a Fact, Modern Astronomy Proved to be a Fable* (Bayswater, the author, n.d.)

Hardy, D., *Alternative Communities in Nineteenth Century England* (Longman, 1979)

Harrison, J. F. C., *Robert Owen and the Owenites in Britain and America: Quest for the New Moral World* (New York, Charles Scribner's Sons, 1969)

————, *The Second Coming: Popular Millenarianism 1780–1850* (Routledge & Kegan Paul, 1979)

————, 'Early Victorian Radicals and the Medical Fringe', in Bynam, W. F., and Porter, R. (eds), *Medical Fringe and Medical Orthodoxy, 1750–1850* (Croom Helm, 1987), pp. 198–215

————, *Late Victorian Britain 1875–1901* (Fontana, 1990)

Hatcher, W. E., *John Jasper: The Unmatched Negro Philosopher and Preacher* (New York, Felming H. Revell, 1908)

Helmstadter, R. J., and Lightman, B. (eds), *Victorian Faith in Crisis: Essays on Continuity and Change in Nineteenth-Century Religious Belief* (Macmillan, 1990)

Henry, J., *The Scientific Revolution and the Origins of Modern Science* (Basingstoke, Palgrave, 2002)

Henson, L., Cantor, G., Dawson, G., Noakes, R., Shuttleworth, S., and Topham J. (eds), *Culture and Science in the Nineteenth-Century Media* (Aldershot, Ashgate, 2004)

Heppenheimer, T. A., *Countdown: A History of Space Flight* (New York, Wiley, 1997)

Hetherington, N. S., 'Early Greek Cosmology: A Historiographical Review', *Culture and Cosmos*, 1 (1997), pp. 10–33

Hingley, P. (ed.), *A Far off Vision: A Cornishman at Greenwich Observatory* (Truro, Royal Institution of Cornwall, 1999)

Holyoake, G. J., *Sixty Years of an Agitator's Life*, 2 vols (T. Fisher Unwin, 1900)
——, *A History of Co-operation* (T. Fisher Unwin, 1906)

Hoskin, M. (ed.), *The Cambridge Concise History of Astronomy* (Cambridge, Cambridge University Press, 1999)

Hunt, S. (ed.), *Christian Millenarianism: From the Early Church to Waco* (Bloomington, Indiana University Press, 2001)

Inkster, I., 'Advocates and Audience: Aspects of Popular Astronomy in England 1750–1850', *JBAA*, 92 (1982), pp. 117–24

Inkster, I., and Morrell, J. (eds), *Metropolis and Province: Science in British Culture 1750–1850* (Hutchinson, 1983)

Irving, W., *The Life and Voyages of Christopher Columbus*, later edn (New York, John B. Alden, 1877)

Janvier, M., *Baltimore in the Eighties and Nineties* (Baltimore, Roebuck, 1933)

Jastrow, J., *Error and Eccentricity in Human Belief* (New York, Dover, 1962)

Jones, G., 'Alfred Russel Wallace, Robert Owen and the Theory of Natural Selection', *BJHS*, 35 (2002), pp. 73–96

Kanter, R. M., *Commitment and Community: Communes and Utopias in Sociological Perspective* (Cambridge, Harvard University Press, 1972)

Knight, D. M., *The Age of Science: The Scientific World-View in the Nineteenth Century* (Oxford, Blackwell, 1996)
——, 'Scientists and their Publics: Popularization of Science in the Nineteenth Century', in Nye, M. J., *The Cambridge History of Science: The Modern Physical and Mathematical Sciences* (Cambridge, Cambridge University Press, 1993), pp. 72–90

Koestler, A., *The Sleepwalkers: A History of Man's Changing Vision of the Universe*, 3rd edn (Hutchinson, 1961)

Kohn, D. (ed.), *The Darwinian Heritage* (Princeton, Princeton University Press, 1985)

Kottler, M. J., 'Alfred Russel Wallace, the Origin of Man, and Spiritualism', *Isis*, 65 (1974), pp. 145–92

Krige, J., and Pestre, D., *Science in the Twentieth Century* (Reading, Harwood, 1997)

Kuhn, T. S., *Structure of Scientific Revolutions* (Chicago, Chicago University Press, 1962)

Lamartine, A. de, *Life of Columbus*, later edn (New York, Hurd and Houghton, 1877)

Langdon, J. C., *Pocket Editions of the New Jerusalem: Owenite Communitarianism in Britain 1825–1855* (unpublished PhD thesis, University of York, 2000)

Laurent, J., 'Science, Society and Politics in Late Nineteenth-Century England: A Further Look at Mechanics' Institutes', *SSS*, 14 (1984), pp. 585–619

Lee, M. F. (ed.), *Millennial Visions: Essays on Twentieth Century Millenarianism* (Westport, Praeger, 2000)

Leslie, S. W., *The Cold War and American Science: The Military-Industrial-Academic Complex at MIT and Stanford* (New York, Columbia University Press, 1993)

Lightman, A., and Sadler, P., 'The Earth is *Round*? Who are you Kidding?', *Science and Children* (1988), pp. 25–6

Lightman, B., *The Origins of Agnosticism: Victorian Unbelief and the Limits of Knowledge* (Baltimore, Johns Hopkins University Press, 1987)

———, 'Astronomy for the People: R. A. Proctor and the Popularization of the Victorian Universe', in van der Meer, J. (ed.), *Facets of Faith and Science* (Lanham, New York and London: Pascal Centre for Advanced Studies in Faith and Science and University Press of America, 1996), pp. 31–45

———, (ed.), *Victorian Science in Context* (Chicago, Chicago University Press, 1997)

———, 'The Visual Theology of Victorian Popularizers of Science', *Isis*, 91 (2000), pp. 651–80

———, 'Victorian Sciences and Religions: Discordant Harmonies', in J. H. Brooke, Ostler, M. J., and van der Meer, D. (eds), *Science in Theistic Contexts* (Chicago, Chicago University Press, 2001), pp. 343–66

———, '*Knowledge* confronts *Nature*: Richard Proctor and Popular Science Periodicals', in Henson, L., Cantor, G., Dawson, G., Noakes, R., Shuttleworth, S., and Topham, J. R. (eds), *Culture and Science in the Nineteenth-Century Media* (Aldershot, Ashgate, 2004), pp. 199–210

———, (ed.), *Science in the Middle Ages* (Chicago, Chicago University Press, 1978)

Lindberg, D. C., and Westman, R. S. (eds), *Reappraisals of the Scientific Revolution* (Cambridge, Cambridge University Press, 1990)

Lindberg, D. C., and Numbers, R. L. (eds), *God and Nature: Historical Essays on the Encounter Between Science and Christianity* (Berkeley, California University Press, 1986)

———, *When Science and Christianity Meet* (Chicago, Chicago University Press, 2003)

Lindberg, D. C., *The Beginnings of Western Science* (Chicago, Chicago University Press, 1992)

Lindermayer, O., 'Europe as Anti-Christ', in Hunt, C., *Christian Millenarianism* (Bloomington, Indiana University Press, 2001), pp. 39–40

Lindsay, G., *John Alexander Dowie: A Life Story of Trials, Tragedies and Triumphs* (Dallas, Christ for the Nations, 1986)

Lloyd, G. E. R., *Early Greek Science: Thales to Aristotle* (Chatto & Windus, 1970)

———, *Greek Science after Aristotle* (Chatto & Windus, 1973)

———, *Magic, Reason and Experience: Studies in the Origin and Development of Greek Science* (Cambridge, Cambridge University Press, 1979)

———, *The Revolutions of Wisdom: Studies in the Claims and Practices of Ancient Greek Science* (Berkeley, California University Press, 1989)

———, *Adversaries and Authorities: Investigations into Ancient Greek and Chinese Science* (Cambridge, Cambridge University Press, 1996)

———, *The Ambitions of Curiosity: Understanding the World in Ancient Greece and China* (Cambridge, Cambridge University Press, 2002)

Loudon, I., 'The Vile Race of Quacks with which this Country is Infested', in Bynam, W. F., and Porter, R. (eds), *Medical Fringe and Medical Orthodoxy, 1750–1850* (Croom Helm, 1987), pp. 106–28

Mahoney, M. J., 'Psychology of the Scientist', *SSS*, 9 (1979), pp. 349–75

Marchant, J., *Alfred Russel Wallace: Letters and Reminiscences* (New York, 1916)

Marsden, G. M., *Fundamentalism and American Culture: Shaping of Twentieth-Century Evangelicalism 1870–1925* (Oxford, Oxford University Press, 1983)

———, *Understanding Fundamentalism and Evangelicalism* (Grand Rapids, Eerdmans, 1991)

Maunder, E. W., *The Astronomy of the Bible* (T. Sealey, Clark, 1908)

Mazzotti, M., 'Newton for Ladies: Gentility, Gender and Radical Culture', *BJHS*, 37 (2004), pp. 119–46

McCalman, I. D., 'Popular irreligion in early Victorian England: infidel preachers and radical theatricality in 1830s London', in Davis, R. W., and Helmstadter, R. J., *Religion and Irreligion in Victorian Society* (Routledge, 1992), pp. 51–67

McCready, W. D., 'Isidore, the Antipodeans, and the Shape of the Earth', *Isis*, 87 (1996), pp. 108–27

McIver, T., *Anti-Evolution: A Reader's Guide to Writings before and after Darwin* (Baltimore, Johns Hopkins University Press, 1988)

McLaughlin-Jenkins, E., 'Common Knowledge: Science and the Late Victorian Working Class Press', *HS*, 39 (2001), pp. 445–65

———, 'Common Knowledge: The Victorian Working Class and the Low Road to Science 1870–1900 (unpublished PhD thesis, York University, 2001)

Meadows, J., *Science and Controversy: A Biography of Sir Norman Lockyer* (Macmillan, 1972)

Mendelsohn, E., 'Science, Scientists and the Military', in Krige, J., and Pestre, D., *Science in the Twentieth Century* (Reading, Harwood, 1997), pp. 175–98

Michell, J., *Eccentric Lives and Peculiar Notions* (Thames & Hudson, 1984)

Middleton, E. E., *Controversy on the Shape of the Earth, between a Newtonian Astronomer and a Poet* (Weston-super-Mare, Clarke, 1872)

———, *Middleton's Impeachment of Modern Astronomy* (Judd, 1879)

————, *On the Variation of the Needle in Connection with the Shape of the Earth* (Judd, 1876)

————, *The Cruise of 'The Kate'*, 2nd edn (Rupert Hart-Davis, 1953)

Milner, R., 'Charles Darwin and Associates, Ghostbusters', *Scientific American*, 275, 4 (1996), pp. 96–101

Moore, J. R., *The Post Darwinian Controversies: A Study of the Protestant Struggle to come to terms with Darwin in Great Britain and America, 1870–1900* 2nd edn (Cambridge, Cambridge University Press, 2003)

————, (ed.), *History, Humanity and Evolution: Essays for John C. Greene* (Cambridge: Cambridge University Press, 1989)

————, 'Theodicy and Society: The Crisis of the Intelligentsia', in Helmstadter, R. J., and Lightman, B. (eds), *Victorian Faith in Crisis* (Macmillan, 1990), pp. 153–86

————, 'Deconstructing Darwinism: The Politics of Evolution in the 1860s', *JHB*, 24 (1991), pp. 353–408

Moore, P., *Can you speak Venusian? A Guide to Independent Thinkers* (Newton Abbot, David & Charles, 1972)

Morrison, R. J., *The New Principia or True System of Astronomy* (J. G. Berger, 1869)

Morse, C. W., *Unpopular Truth against Popular Error in Reference to the Shape of the Earth* (Boston, C. J. F. Fletcher, 1913)

National Cyclopedia of American Biography, 31 (New York, James T. White, 1944)

Naylor, J., *Bedford Canal not Convex, The Earth not a Globe* (Samuel Palmer, 1873)

Nebelsick, H. P., *Circles of God: Theology and Science from the Greeks to Copernicus* (Edinburgh, Scottish Academic Press, 1985)

Noakes, R. J., '"Cranks and Visionaries": Science, Spiritualism and Transgression in Victorian Britain' (unpublished PhD thesis, Corpus Christi College, Cambridge, 1998)

Nowlan, A., *Double Exposure* (Fredericton, Brunswick Press, 1978)

Numbers, R. L., *The Creationists* (Berkeley, California University Press, 1992)

————, *Darwinism Comes to America* (Cambridge, Harvard University Press, 1998)

Numbers, R. L. and Butler, J. M., *The Disappointed: Millerism and Millenarianism in the Nineteenth Century* (Knoxville, Tennessee University Press, 1993)

Nye, M. J. (ed.) *The Cambridge History of Science: The Modern Physical and Mathematical Sciences* (Cambridge, Cambridge University Press, 2003)

Olby, R. C., Cantor, G. N., Christie, J. R. R., and Hodge, M. J. S., *Companion to the History of Modern Science* (Routledge, 1990)

O'Leary, S., *Arguing the Apocalypse: A Theory of Millennial Rhetoric* (Oxford, Oxford University Press, 1998)

Ophir, A., and Shapin, S., 'The Place of Knowledge: A Methodological Survey', *SIC*, 4 (1991), pp. 3–21

Oppenheim, J., *The Other World: Spiritualism and Psychical Research in England, 1850–1914* (Cambridge, Cambridge University Press, 1985)

Orwell, G., *Nineteen Eighty-Four* (Harmondsworth, Penguin, 1997)

Ostler, M. J. (ed), *Rethinking the Scientific Revolution* (Cambridge, Cambridge University Press, 2000)

Owen, R., *A New View of Society or Essays on the Formation of Human Character* (1813–16) (Macmillan, 1972)

Parallax [Samuel Birley Rowbotham], *Zetetic Astronomy: A Description of Several Experiments which Prove that the Surface of the Sea is a Perfect Plane, and that the Earth is not a Globe* (Birmingham, W. Cornish, 1849)

———, *The Inconsistency of Modern Astronomy and its Opposition to the Scriptures!* (E. Farrington, 1849)

———, *Zetetic Astronomy: Earth not a Globe! An Experimental Inquiry into the True Figure of the Earth* (Bath, S. Haywood, 1865)

———, *Zetetic Philosophy: Patriarchal Longevity, its Reality, Causes, Decline and Possible Reattainment* (Heywood, 1867)

———, *Experimental Proofs that the Surface of Standing Water is not Convex, but Horizontal* (William Mackintosh, 1870)

———, *Zetetic Astronomy: Earth not a Globe! An Experimental Inquiry into the True Figure of the Earth*, 2nd edn (John B. Day, 1873)

———, *Zetetic Astronomy: Earth not a Globe! An Experimental Inquiry into the True Figure of the Earth*, 3rd edn (John B. Day, 1881)

———, *Patriarchal Longevity Reattainable* (J. Snow, 1883)

Parish, J., and Parker, M. (eds), *The Age of Anxiety: Conspiracy Theory and the Human Sciences* (Oxford, Blackwell, 2001)

Park, R., *Voodoo Science: The Road from Foolishness to Fraud* (Oxford, Oxford University Press, 2000)

Peacock, G., *Is the World Flat or Round?* (Gloucester, John Bellows, 1871)

Pearsall, R. P., *The Table-Rappers* (Michael Joseph, 1972)

Phin, J., *The Seven Follies of Science* (Archibold Constable, 1906)

Plait, P., *Bad Astronomy: Misconceptions and Misuses Revealed from Astrology to the Moon Landing 'Hoax'* (New York, John Wiley, 2002)

Porter, R., 'Gentlemen and Geology: The Emergence of a Scientific Career, 1660–1920', *HJ*, 21 (1978), pp. 809–36

Proctor, R. A., *Myths and Marvels of Astronomy* (Chatto & Windus, 1878)

———, *Old and New Astronomy* (Longmans, Green, 1892)

Pumfrey, S., 'Ideas Above his Station: A Social Study of Hooke's Curatorship of Experiments', *HS*, 29 (1991), pp. 1–44

Raby, P., *Alfred Russel Wallace: A Life* (Chatto & Windus, 2001)

Randolph, E. A., *The Life of the Reverend John Jasper* (Richmond, T. Hill, 1884)

Rectangle [Thomas Winship], *Zetetic Cosmogony, or Conclusive Evidence that the World is not a Rotating-Revolving Globe but a Stationary Plane Circle*, 2nd edn, (Durban, T. L. Cullingworth, 1899)

Reddie, J., *The Mechanics of the Heavens, and the New Theories of the Sun's Electro-Magnetic and Repulsive Influence: Being an Essay on Revolving Bodies and Centripetal Forces* (R. Hardwicke, 1862)

——, *On Revolving Bodies and Centripetal Forces: An Essay to Prove that the Theory of Universal Gravitation is not Founded on True Mathematical Principles* (Bradbury & Evans, 1862)

——, *Vis Inertiae Victa, or Fallacies Affecting Science: An Essay Towards Increasing our Knowledge of Some Physical Laws, and a Review of Certain Mathematical Principles of Natural Philosophy* (Bradbury & Evans, 1862)

——, *Scientia Scientiarum. Being Some Account of the Origin and Objects of the Victoria Institute, or Philosophical Society of Great Britain by a Member* (R. Hardwicke, 1865)

——, *Current Physical Astronomy Critically Examined and Confuted* (Charles Griffin, 1865)

Report of the Seventy-first Meeting of the British Association for the Advancement of Science (John Murray, 1901)

Ritvo, H., *The Animal Estate: The English and Other Creatures in the Victorian Age* (Cambridge, Harvard University Press, 1987)

Rose, R. N., *The Field 1853–1953: A Centenary History* (Michael Joseph, 1953)

Rousseau, G. S., and Porter, R., *The Ferment of Knowledge: Studies in the Historiography of Eighteenth-Century Science* (Cambridge, Cambridge University Press, 1980)

Royle, E., 'Mechanics' Institutes and the Working Classes, 1840–1860', *HJ*, 14 (1971), pp. 305–21

——, *Victorian Infidels: The Origins of the British Secularist Movement 1791–1866* (Manchester, Manchester University Press, 1974)

——, *The Infidel Tradition from Paine to Bradlaugh* (Macmillan, 1976)

——, *Robert Owen and the Commencement of the Millennium* (Manchester, Manchester University Press, 1998)

Rudwick, M. J. S., *The Great Devonian Controversy: The Shaping of Scientific Knowledge among Gentlemanly Specialists* (Chicago, Chicago University Press, 1985)

Russell, J. B., *Inventing the Flat Earth: Columbus and Modern Historians* (Westport, Praeger, 1991)

Schadewald, R. J., 'The Plane Truth', *TWA Ambassador*, December 1977, pp. 42–3

——, 'He knew Earth is Round, but his Proof fell Flat', *Smithsonian*, April 1978, pp. 101–13

——, 'The Flat-Out Truth', *Science Digest*, 88 (1980), pp. 58–63

——, 'Is the World in Curious Shape?', *Isaac Asimov's Science Fiction Magazine*, December 1980, pp. 97–106

——, 'Scientific Creationism, Geocentricity and the Flat Earth', *SI* 1981–2, pp. 41–8

———, 'The Evolution of Bible Science', in Godfrey, L. (ed.), *Scientists Confront Creationism* (New York, W. W. Norton, 1983), pp. 283–99

———, 'The Flat Earth Bible', *Bulletin of the Tychonian Society*, 44 (1987), pp. 13–17

———, 'Some Like it Flat', in Schultz, T., *The Fringes of Reason: A Whole Earth Catalogue* (New York, 1989), pp. 86–8

———, 'The Earth was Flat in Zion', *Fate*, May 1989, Harmony Books, pp. 70–79

———, 'Looking for Lighthouses', *Creation/Evolution*, 31 (1992).

Schaffer, S., 'Herschel in Bedlam: Natural History and Stellar Astronomy', *BJHS*, 13 (1980), pp. 211–39

———, 'Natural Philosophy and Public Spectacle in the Eighteenth Century', *HS*, 21 (1983), pp. 1–43

———, 'Accurate Measurement is an English Science', in Wise, M. N. (ed.), *The Values of Precision* (Princeton, Princeton University Press, 1995)

Schiaparelli, G., *Astronomy in the Old Testament* (Oxford, Clarendon, 1905)

Schwartz, J. S., 'Darwin, Wallace and the *Descent of Man*', *JHB*, 17 (1984), pp. 271–89

———, 'Robert Chambers and Thomas Henry Huxley, Science Correspondents: The Popularization and Dissemination of Nineteenth Century Natural Science', *JHB*, 32 (1999), pp. 343–83

———, 'Darwin, Wallace and Huxley, and *Vestiges of the Natural History of Creation*', *JHB*, 23 (1990), pp. 127–53

Scoevola [B. Charles Brough], *What is the Shape of the Earth? A Letter to George Peacock* (Stafford, Joseph Halden, 1871)

Scopes, J. T., and Presley, J., *Center of the Storm: Memoirs of John T. Scopes* (New York, Henry Holt, 1967)

Scott D. W., *Terra Firma: The Earth not a Planet* (Simpkin, Marshall, Hamilton, Kent, 1901)

———, *Hades and Beyond: With some Sidelights Along the Way* (James Clarke, 1892)

Secord, A., 'Corresponding Interests: Artisans and Gentlemen in Nineteenth-Century Natural History', *BJHS*, 27 (1994), pp. 383–408

Secord, J., *A Victorian Sensation: The Extraordinary Publication, Reception, and Secret Authorship of Vestiges of the Natural History of Creation* (Chicago, Chicago University Press, 2000)

Shank, M. (ed.), *The Scientific Enterprise in Antiquity and the Middle Ages* (Chicago, Chicago University Press, 2000)

Shapin, S., 'Phrenological Knowledge and the Social Structure of Early Nineteenth-Century Edinburgh', *Annals*, 32 (1975), pp. 219–43

———, 'History of Science and its Sociological Reconstructions', *HS*, 20 (1982), pp. 157–211

———, 'Discipline and Bounding: The History and Sociology of Science as seen through the Externalism-Internalism Debate', *HS*, 30 (1992), pp. 333–69

———, *A Social History of Truth: Civility and Science in Seventeenth-Century England* (Chicago, Chicago University Press, 1994)

———, *The Scientific Revolution* (Chicago: Chicago University Press, 1996)

Shapin, S., and Barnes, B., 'Science, Nature and Control: Interpreting Mechanics' Institutes', *SSS*, 7 (1977), pp. 31–74

Shapin, S., and Thackray, A., 'Prosopography as a Research Tool in the History of the British Scientific Community', *HS*, 12 (1974), pp. 1–28

Shapin, S., and Schaffer, S., *Leviathan and the Air Pump: Hobbes, Boyle and the Experimental Life* (Princeton, Princeton University Press, 1985)

Shaw, G. B., *Everybody's Political What's What* (Constable, 1944)

Sheets-Pyenson, S., 'Scientific Culture in London and Paris, 1820–1875 (unpublished PhD thesis, University of Pennsylvania, Philadelphia, Pennsylvania, 1976)

———, 'Popular Science Periodicals in Paris and London: The Emergence of a Low Scientific Culture', *Annals*, 42 (1985), pp. 549–72

Shermer, M., *The Borderlands of Science: Where Sense Meets Nonsense* (Oxford, Oxford University Press, 2001)

———, *Why People Believe Weird Things: Pseudoscience, Superstition and Other Confusions of our Time*, 2nd edn (New York, Henry Holt, 2002)

Simek, R., *Heaven and Earth in the Middle Ages: The Physical World before Columbus* (Woodbridge, Boydell, 1996)

Sitwell, E., *English Eccentrics* (Harmondsworth, Penguin, 1958)

Smith, A., *The Sea Earth Globe and its Monstrous Hypothetical Motions* (Northampton, the author, 1918)

Smith, C. H., 'Alfred Russel Wallace on Spiritualism, Man and Evolution: An Analytical Essay', at www.wku.edu/~smithch/essays/ARWPAMPH.html

Smith, K. A., *Is the World a Globe?* (Leicester, Willsons, 1904)

Smith, R., 'Alfred Russel Wallace: Philosophy of Nature and Man', *BJHS*, 6 (1972), pp. 177–99

Solomon, J. R., *Objectivity in the Making: Francis Bacon and the Politics of Inquiry* (Baltimore, Johns Hopkins University Press, 2003)

Spencer, C., *The Heretic's Feast: A History of Vegetarianism*, 2nd edn (Hanover, University Press of New England, 1993)

Spiegel-Rösing, I., and de Solla Price, D., *Science, Technology and Society* (Sage, 1977).

Stephens, W. N., 'The Figure of the Earth in Isidore's "De Natura Rerum"', *Isis*, 71 (1980), pp. 268–77

Stewart, L., 'The Selling of Newton: Science and Technology in Early Eighteenth-Century England', *JBS*, 25 (1986), pp. 179–92

———, *The Rise of Public Science* (Cambridge, Cambridge University Press, 1992)

Taylor, A., *Visions of Harmony: A Study in Nineteenth-Century Millenarianism* (Oxford, Oxford University Press, 1987)

Taylor, E. G. R., *Ideas on the Shape, Size and Movements of the Earth* (Historical Association, 1943)

Thompson, E. P., *The Making of the English Working Class*, 3rd edn. (Gollancz, 1980)

Timbs, J., *English Eccentrics and Eccentricities* (Chatto and Windus, 1875)

Tobin, W., and Pippard, B., 'Foucault, His Pendulum and the Rotation of the Earth', *ISR*, 19 (1994), pp. 326–37

Toner, P., *If I Could Turn and Meet Myself: The Life of Alden Nowlan* (Fredericton, Goose Lane, 2000)

Topham, J. R., 'Beyond the "Common Context": The Production and Reading of the Bridgewater Treatises', *Isis*, 89 (1998), pp. 233–62

Tryon [Samuel Birley Rowbotham], *Biology: An Inquiry into the Cause of Natural Death* (Manchester, Abel Heywood, 1845)

Turner, F. J., *Between Science and Religion: The Reaction to Scientific Naturalism in Late Victorian England* (New Haven, Yale University Press, 1974)

———, 'The Victorian Conflict Between Science and Religion: A Professional Dimension', *Isis*, 69 (1978), pp. 356–76

———, *Contesting Cultural Authority: Essays in Victorian Intellectual Life* (Cambridge, Cambridge University Press, 1993)

Urbach, P., *Francis Bacon's Philosophy of Science: An Account and Reappraisal* (La Salle, Open Court, 1987)

van der Meer, J. (ed.), *Facets of Faith and Science* (Lanham, New York and London: The Pascal Center for Advanced Studies in Faith and Science and University Press of America, 1996), pp. 31–45

van Whye, J., 'Was Phrenology a Reform Science? Towards a New Generalization for Phrenology', *HS*, 43 (2004), pp. 313–31

———, *Phrenology and the Origins of Victorian Scientific Naturalism* (Aldershot, Ashgate, 2004)

Vincent, D., *Bread, Knowledge and Freedom: A Study of Nineteenth Century Working Class Autobiography* (Methuen, 1981)

Voliva, W. G., 'Which Will You Accept?' The Bible and the Inspired Word of God or the Infidel Theories of Modern Astronomy', *Leaves of Healing*, 10 May 1930, pp. 131–59

Vox [Edmund Parsloe], *Astronomy and the Bible Reconciled* (Birmingham, Journal Printing Works, 1896)

Wacker, G., *Heaven Below: Early Pentecostals and American Culture* (Cambridge, Harvard University Press, 2001)

Wallace, A. R., *A Reply to Mr. Hampden's Charges against Mr. Wallace* (J. J. Tiver, 1871)

———, 'Review of *The Descent of Man and Selection in Relation to Sex*', *Academy*, 15 May 1871, pp. 177–83

———, 'On the Attitude of Men of Science towards Investigators of Spiritualism', in *The Year-book of Spiritualism for 1871* (Boston, William White, 1871) pp. 28–31

———, *Miracles and Modern Spiritualism*, rev. edn (G. Redway, 1896)

———, *The Geographical Distribution of Animals*, 2 vols (Macmillan, 1876)

———, *The Wonderful Century: Its Successes and its Failures* (Swan Sonnenschein, 1901)

———, *My Life: A Record of Events and Opinions*, 2 vols (Chapman & Hall, 1905)

Wallace, I., *The Square Pegs* (New York, Knopf, 1957)

Wallis, R. M. (ed.), *On the Margins of Science* (Keele, Keele University Press, 1979)

Wallis, T. W., *Autobiography of Thomas Wilkinson Wallis* (Louth, J. W. Goulding, 1899)

Walters, A. N., 'Conversation Pieces: Science and Politeness in Eighteenth Century England', *HS*, 35 (1997), pp. 121–54

Warner, C. D., *Washington Irving* (Boston, Houghton, Mifflin, 1884)

Wertheimer, D. J., 'The Victoria Institute, 1865–1919: A Study in Collective Biography Meant as an Introduction to the Conflict of Science and Religion after Darwin'(unpublished typescript, 1971)

White, A. D., *A History of the Warfare of Science with Theology in Christendom*, 2 vols (New York, D. Appleton, 1896)

Wiener, J. H., *Radicalism and Freethought in Nineteenth-Century Britain: The Life of Richard Carlile* (Westport, Greenwood, 1983)

Winter, A., 'Mesmerism and Popular Culture in Early Victorian England', *HS*, 32 (1994), pp. 317–43

———, *Mesmerized* (Chicago, Chicago University Press, 1998)

Wollaston, A. F. R., *Life of Alfred Newton* (John Murray, 1921)

Wolpert, L., *The Unnatural Nature of Science* (Faber and Faber, 1992)

Wright, D. G., *Popular Radicalism: The Working Class Experience 1780–1880* (Longman, 1988)

Yeo, R., 'Science and Intellectual Authority in Mid-Nineteenth-Century Britain: Robert Chambers and *Vestiges of the Natural History of Creation*', *VS*, 28 (1984), pp. 5–31

———, 'Genius, Method and Morality: Images of Newton in Britain, 1760–1860', *SIC*, 2 (1988), pp. 257–84

Zetetes [Albert Smith], *Is the Earth a Whirling Globe as Assumed and Taught by Modern Astronomical 'Science'?* (Leicester, the author, 1904)

———, *Cranks or False Theories of 'Science' versus the Truth of Nature and the Bible* (Leicester, the author, n.d.)

Zion Historical Society, *Wilbur Glenn Voliva*, Continuing History of Zion 1901–61, Series 7 (Zion, Zion Historical Society, n.d.)

Internet Sources

Carpenter, W., *One Hundred Proofs that Earth is not a Globe* (Baltimore, William Carpenter, 1885), http://www.geocities.com/lclane2/hundredc.html

Glenn, J., 'Friendship 7: Twenty Years Later, Random Remembrances;, http://www.lib.ohio-state.edu/arvweb/glenn/legacy/20year.htm

IFERSA membership flyer, http://www.talkorigins.org/faqs/flatearth.html

Jasper, J., *De Sun Do Move*, http://www.library.vcu.edu/jbc/speccoll/vbha/6th5.html

Schadewald, R. J., Various articles at 'Bob Schadewald's Corner', http://www.lhup.edu/~dsimanek/schadew.htm

Wallace, A. R., A wealth of material is available at Charles Smith's Alfred Russel Wallace website, www.wku.edu/~smithch/home

Zetetic/Flat Earth Society Journals

The Armourer (1870)

Cosmos: A Geographical, Philosophical and Educational Review, Nautical Guide and General Students' Manual (1883–5)

The Earth: A Monthly Magazine of Sense and Science (1900–1906)

The Earth and Its Evidences: Scripturally, Rationally and Practically Described – A Geographical, Philosophical and Educational Review, Nautical Guide, and General Students' Manual (1886–8)

Earth Life: A Monthly Journal and Record of all such Facts, Principles and Discoveries as Relate to the Improvement and Preservation of Earthly Existence (1873)

The Earth not a Globe Review: A Magazine of Cosmographical Science (1893–7)

The Flat Earth News (1984–8)

John Hampden's Monthly: The Truth Seekers Oracle and Scriptural Science Review (1876)

The Official Organ [of the Flat Earth Society of Canada] (1971–7)

Parallax: A Geographical, Philosophical and Educational Review (1885)

The Zetetic: A Monthly Journal of Cosmographical Science (1872–3)

Newspapers

Appleton Post-Crescent
Atlanta Constitution
Baltimore Sun

Birmingham Evening Echo
Bismarck Tribune
Boston Globe
Bridgton Scrapbook
Buffalo Courier Express
Cambridge Chronicle
Chelmsford Chronicle
Chicago Tribune
Coshocton Tribune
Daily Express
Daily Gleaner (Fredericton, NB)
Daily Graphic
Daily Ledger-Gazette (Antelope Valley, CA)
Daily Northwestern
Daily Telegraph
Decatur Review
Delaware Gazette
Detroit News
Devonport Independent and Plymouth and Stonehouse Gazette
Dover Express
Elyria Chronicle
Essex Weekly News
Evening News and Star (London)
Evening Standard (London)
Hampshire Telegraph
Haywood Review
Helena Daily Independent
Illustrated London News
Independent
Iowa Recorder
Ironwood Daily Globe
John Bull (London)
Kingsport Times
Leeds Mercury
Leeds Times
Le Madawaska (Edmunston, NB)
Manchester Guardian
Mansfield News
Miami Herald
Montréal Gazette
Newark Advocate
New York Times
Observer

Oshkosh Northwestern
Port Arthur Daily News
Portland Maine Press Herald
Portsmouth Daily Herald
Region
Reno Evening Gazette
San Francisco Examiner
San Francisco Sunday Examiner and Chronicle
San Jose Mercury News
Santa Cruz Sentinel
Semi-Weekly Landmark
Sheboygan Press Telegram
Star (Johannesburg, SA)
Star in the East (Wisbech)
Stroud Journal
Sun
The Times
Washington Post
Western Daily Mercury
Wichita Daily Times
Wisconsin Rapids Daily Tribune

Magazines and Periodicals

Academy
Acts and Facts
American Mercury
A.R.O. Observer
Athenaeum
Bookseller
Channel (Dover)
Chicago Magazine
Colliers
English Mechanic
Knowledge
Leaves of Healing (Zion, Illinois)
Mechanics' Magazine
Nature
New Moral World
Newsweek
Punch
Saturday Review of the Sciences

Science and Children
Science Digest
Scientific Opinion
Social Pioneer or Record of the Progress of Socialism (Manchester)
T.W.A. Ambassador
Weekend
Working Bee and Herald of the Hodsonian Community Society (Manea)

MS Collections

Alden Nowlan fonds (ANF), University of Calgary Library, UC
Alfred Russel Wallace Papers (ARW), BL
Flat Earth Society Papers (FESP), SFF, ULIV
Greenwich Royal Observatory Collection (GROC), CUL
J. Lloyd Eaton Collection (EC), UCR
John Couch Adams Papers (JCAP), SJCC
Leo Ferrari Papers (LFP), UNB
Macmillan Archive (MA), BL
Raymond Fraser Papers, UNB
Robert Schadewald Pseudo-Science Collection (SPSC), UWM
Science Fiction Foundation Collection (SFF), ULIV
Sir Thomas Phillipps Papers (MS Phillipps-Robinson) (TPP), BOD
Wallace Collection, HLO
Wallace Papers, Natural History Museum, London
Wallace Papers, RGS
Various, RAS

Index

Index